Time Distortions in Mind

Time Distortions in Mind

Temporal Processing in Clinical Populations

Edited by

A. Vatakis
M.J. Allman

BRILL

LEIDEN | BOSTON

Cover design: Özlem Altin

Library of Congress Cataloging-in-Publication Data

Time distortions in mind : temporal processing in clinical populations / edited by A. Vatakis (Cognitive Systems Research Institute, Athens, Greece), M. Allman (Michigan State University, East Lansing, USA).
pages cm
Includes bibliographical references and index.
ISBN 978-90-04-23064-4 (hardback : alk. paper) -- ISBN 978-90-04-23069-9 (e-book) 1. Time perception. 2. Cognitive psychology. I. Vatakis, Argiro, editor. II. Allman, Melissa J., editor.

BF468.T547 2015
153.7'53--dc23

2015010275

This publication has been typeset in the multilingual "Brill" typeface. With over 5,100 characters covering Latin, IPA, Greek, and Cyrillic, this typeface is especially suitable for use in the humanities. For more information, please see www.brill.com/brill-typeface.

ISBN 978-90-04-23064-4 (hardback)
ISBN 978-90-04-23069-9 (e-book)

This book is printed on acid-free paper.

Contents

Preface

Examinations of aspects of temporal processing in clinical populations may inform not only typical psychological functioning – but may also elucidate the psychological consequences of any pathophysiological differences in temporal processing. This line of scientific enquiry may improve understanding, and potential remediation, of the psychological condition.

Humans are equipped with the remarkable ability to estimate event duration, recall past events, remember to execute future tasks, and perceive a multisensory and temporally unified world. This ability enables us (and other animals) to anticipate, learn, and adapt to temporal regularities and dynamics in the social and non-social environment. Moreover, the perception of time, typically (although not always) reveals hallmarks shared by other forms of perception (e.g., Weber's law). Thus, even though our subjective experience of time might be quite different from the actual, physical duration of an event, temporal experience remains an important and necessary part of our everyday living and ultimately forms the fabric of our thoughts and behaviors.

Distortions in aspects of temporal processing from the range of milliseconds to seconds, to conceptual notions of a timeline for time (past, present, future) have been separately reported for a variety of psychological disorders. Aside from adages such as "a watched pot never boils" and "time flies when you're having fun," the experience of time distortions by the healthy individual may require oneself to imagine a world in which aspects of our temporal experiences are "off." For example, you have some sense of how long you've been reading this. Now imagine that this was suddenly uncertain for you (minutes, hours?). You would most likely feel anxious and aroused and would likely engage in different patterns of behavior and decision-making. Go a step further and imagine that you experience events that are disorganized in terms of order, duration, and sensory integration. How would that feel? There is no mental disorder of timing; and it is fair to say that deficits in timing are not characterizing deficits of any mental disorder. But it is also fair to say that the way in which aspects of temporal perception and sensitivity influence our psychology is profound, and there are many interesting findings implicating aspects of timing to the neurological, behavioral, and cognitive profiles characteristic of a number of mental disorders (many of which are described in this book!).

Perception, cognition, and action are tightly intertwined with events unfolding in time and, thus, many processes such as working memory, attention, decision making etc. are linked to timing, and implicated in successful interval timing models. Disturbance of one or more of these related processes could lead to time distortions or vice versa. Some researchers argue that certain disorders (e.g., dyslexia, aphasia) may potentially have their root solely on time distortions. For instance, Frederick T. Melges devoted his career to defining the role of time distortions in a number of psychiatric disorders. In his book, *Time and the inner future*, he supported that time disturbances distort a person's view of his/her future, thus disrupting goal-directed behavior and rendering the person unable to live a normal life.

It is widely believed that (within the bounds of the psychological present, 3-5-8 s) people with schizophrenia experience an "overload" of sensory input and lack the executive functioning skills to compensate and "sift-through" information. Schizophrenics often produce "word salads" and lose track of a train of thought, failing to link different sentences appropriately. Theories of autism also emphasize differences in "neural signatures," "temporal binding" or the adaptive functioning of long-range (multiple neural systems), and a form of sensory "overload" is posited with certain senses tending to become highly arousing (visual) or aversive (auditory, tactile). It has been shown recently that children with autism spectrum disorder reveal characteristic differences in *supra*-second timing, "bind" sensory information over an extended time window, and the relative functioning of aspects of temporal processing might correspond to characteristic features of the disorder (e.g., language, sensory processing).

The neural basis of timing and time perception is currently a "hot topic" in empirical psychology, but the extent to which the clinical features of certain disorders are related to time distortions per se is currently unknown, although this is beginning to be elucidated. Time perception in the seconds-to-minutes range recruits cortico-striatal and cortico-cerebellar circuits (including various neurotransmitter systems) – brain circuits embedded in the frontal lobes and the basal ganglia, which are modulated by the dopamine transmitter system. Thus, patients with frontal lobe injuries display impaired estimation of time intervals, while patients with Parkinson's disease show decreased dopaminergic function in the basal ganglia accompanied by interval discrimination deficits. In the millisecond range, patients with dyslexia, lesions in the left hemisphere or aphasia, reveal various temporal processing difficulties; e.g., order distortions in aphasia, severe deficits in time perception and motor timing of short intervals in individuals with cerebellar damage, and a processing deficit for rapidly presented speech and non-speech stimuli in dyslexia. Children with attention deficit hyperactivity disorder and adults with aberrant impulsive behaviors, tend to overestimate and under-produce given time intervals, and appear less tolerant of delays in the imagined "future" – as if they consider them "too long." Time perception is also crucial in decision-making and impulse control. For example, a side effect of L-Dopa treatment for Parkinson's disease can be to increase impulsivity (and inappropriate behaviors) and the subjective 'warping' of time. This relationship of impulsivity and time perception has been attributed to an increased cognitive processing speed (and, thus, increased clock speed), given research on time perception showing that increased arousal often leads to interval overestimations. Disturbances in cognitive functioning and emotional processing can also modulate ones temporal percept.

In testament to the multiplicity of temporal processing to the human condition, we also think about the consequences of our actions in hypothetical time; discount events based on imagined delay; and temporal dynamics influence our valuations of future rewards and past experience. As it may relate to human's capacity for 'mental time travel', and the extent to which this forms the scaffold for our lives, there is some historical basis for positing there may be fundamental pathophysiological differences in certain clinical (psychiatric, psychological, neurological) populations. Disorientation in time and inability to coordinate and orientate oneself temporally in the external world is routinely observed in the majority of neurological patients, and those with states of mental confusion (e.g., intellectual disability, dementia), who also present with poorly developed abstract concepts of time.

The problem for modern psychiatry as a form of medicine is that it has a nomenclature of mental illness, but little known etiologies. The mind is considered an emergent property of the brain, not a "product" like urine from a kidney. The presence of mind is based on the interactions of elements in the brain, and the adequate organization of elements is fundamental for all conscious mental features (faculties, drives, responses). One of the first psychiatrists to characterize time in psychopathological minds was Eugene Minkowski. Minkowski proposed that individuals with schizophrenia lack a 'feeling of time': They understand 'before' and 'after', but live very much in the 'present'. Melges described paranoid psychiatric patients as having a form of "temporal disintegration" – memories of the past, perceptions in the present, and expectations in the future may be confused and appear to be happening at the same time, thus they may appear interconnected. Feelings of sadness, Minkowski noted, often accompany thoughts of the past, and people suffering from depression and compulsive neuroticism may tend to "live in the past," while neurotics in general tend to neglect the "present." People who are depressed tend to overestimate intervals, possibly due to increased allocation of attention to the passage of time (or rumination), which in turn leads to a higher processing of time units and, thus, overestimation of an event. It has been reported that children with autism also experience difficulty thinking forwards and backwards in time.

As of today, we might not be able to claim that certain disorders are due to time distortions *per se*, but experimentation, imaging, and modeling allow us to better understand how timing is represented in the normal and disordered brain and mind. Recently, there has been a burgeoning interest in timing and time perception in certain disorders with a known neuropathology in identified interval-timing regions (Parkinson's disease, aphasia) or by virtue of diagnostic psychopathological distortions in the apparent temporal organization of cognition or behavior (e.g., attention deficit hyperactivity disorder, autism, schizophrenia). Deviations from "normal" timing patterns have been obtained with a variety of targeted psychological, psychiatric, and neurological patient populations on a variety of timing and time perception tasks. Timing is critical in everyday functioning and a crucial factor for studying (and maybe rehabilitating) the disturbed brain and mind.

It is pertinent, therefore, to collect the current knowledge on temporal processing in certain clinical populations. This book serves as such a collection, bringing together new and old research and ideas on aspects of temporal processing in clinical populations, in the ultimate hope that it will elucidate the interdependence between perturbations in timing and disturbances in mind and brain. The book as a whole does not provide the answer to the relative function of temporal processing within these populations (although this is reviewed). Instead it offers a window into cutting edge research examining temporal processing in these disorders, and as it relates to how these disorders present (their phenotypes). This collection of accumulated knowledge will be an excellent reference for the student and scientist interested in the topic of temporal processing, and abnormal psychology, but it will also serve as the stepping-stone to share ideas and push forward the advancement in understanding how distorted timing can lead to a disturbed brain and mind or vice versa.

Time Distortions in Mind – Temporal processing in clinical populations was inspired from the TIMELY (COST Action TD0904) Training School on "Temporal Processing in Clinical Populations" that was held in Thessaloniki, Greece on March 26–29, 2012. We would, thus, like to thank the TIMELY network and COST funding for supporting this endeavor and our authors who contributed to this book and patiently waited for its completion and publication. We would also like to *dearly* thank two people: Özlem Altin, who provided us with the cover art for this book (we encountered this work in a gallery in Berlin and immediately wished for it to be the book's cover art; Özlem was gracious to immediately reply to our email request and work with us in terms of the image's quality and presentation), and Mary Kostaki, who worked on the design of the cover for this book (and showed patience to all the changes – big or small – so as to reach the appropriate result for this excellent collection of chapters).

Melissa J. Allman and Argiro Vatakis

List of Figures

List of Tables

CHAPTER 1

Distorted Multisensory Experiences of Order and Simultaneity

Argiro Vatakis and Alexandra Elissavet Bakou***

1 Introduction

> In order to deal appropriately with the events in its environment, an animal must perform at least two essential actions. It must be able to identify or characterize an event, and it must also be able to determine when the particular event occurred in relation to other events. [...] "Meaning" requires that the unitary bits of information be kept in a proper sequence.
>
> EFRON, 1963A

In his 1963a paper, Efron points out the importance of event identification and event ordering for the construction of meaning. Given that we live and thrive in a multisensory environment, Efron's 'unitary bits' include information from multiple sensory modalities that may be complementary, contradictory or at some level of equivalence. The percept of these sensory 'bits' of information as synchronous and unified leads to the percept of a multisensory event rather than multiple, independent sensory events (Vatakis & Spence, 2010). Research to date has shown that the human perceptual system maintains the percept of synchrony between two sensory streams even though these streams may be processed/presented close but not exactly in time (Vatakis, 2013). The maintenance of a synchronous percept is accomplished through the hypothesized existence of a temporal window of integration (TWI; i.e., the interval in which no signal discrepancy is perceived, anything beyond this interval will normally be perceived as being desynchronized or asynchronous; e.g., King, 2005; Spence & Squire, 2003; Vatakis & Spence, 2010; Vroomen & Keetels, 2010). We, thus, have 'moments' (or 'functional moments' as per Wittmann, 2011) that have no perceivable duration and can be perceived as simultaneous or 'somewhat' simultaneous (yet order cannot be perceived) and 'moments' that are successive with specific and detectable order of presentation (e.g., Pöppel, 1985, 2004;

* MultiTime Lab, Cognitive Systems Research Institute, Athens, Greece and Department of Philosophy and History of Science, University of Athens, Greece.

** MultiTime Lab, Cognitive Systems Research Institute, Athens, Greece.

Wittmann, 2011). These 'moments' are integrated into 'event intervals' (or 'experienced moments'; Wittmann, 2011) that compose our percept of a continuously flowing multisensory event. But how these 'moments' and 'intervals' associate/interact so as to provide the continuously flowing in time experience of multisensory events?

The majority of clinical research in timing has focused on the 'event interval' level (e.g., Allman, 2011; Buhusi & Meck, 2005), while multisensory 'moments' and their integration have been understudied, thus, ignoring the potential link of the experience of intervals according to the synchrony and unity of moments. For example, it may be the case that high pacemaker rates may be associated with higher temporal resolution of the timing mechanism(s), which, in turn, should result in lower temporal discrimination thresholds (i.e., better performance and, thus, smaller TWIS; Rammsayer & Classen, 1997; Wenke & Haggard, 2009). Such associations have not been attempted partially due to that fact that researchers in the two areas of focus – synchrony perception and interval timing – have been working in isolation, but also because the mechanisms governing the integration of multisensory moments to event intervals have not been elucidated. Given, therefore, the previous focus on interval timing in clinical populations, in this chapter we will review the literature on multisensory temporal processing in various disorders/conditions in an attempt to draw some first conclusions and encourage future research on the association of multisensory temporal integration, interval timing, and event perception.

2 Multisensory Temporal Integration: A First Look

Our everyday and effortless (as it seems) experience of multisensory events, which are temporally and semantically unified, allows for faster and more accurate detection, discrimination, and localization of targets at hand. Thus, the multisensorial nature of our experiences, not only makes life more enjoyable, but also provides valuable behavioral and perceptual benefits (Calvert, Spence, & Stein, 2004). These benefits along with the altered percepts that are born out of crossmodal interactions (e.g., the McGurk effect, the visual influence on the perception of audiovisual speech; MacDonald & McGurk, 1978) have lead to a vigorous body of research on multisensory processing during the last 20–30 years. Given the 'young age' of this area, multisensory temporal integration have primarily been investigated in healthy participants, while the majority of studies with patients experiencing a disruption of audiovisual perception could potentially be interpreted as a consequence of other neurological deficits (e.g., Böhning, Campbell, & Karmiloff-Smith, 2002; Campbell et al.,

1990; Campbell et al., 1997). For example, Campbell and her colleagues have described two cases of patients with left and right hemispheric lesions who did not exhibit any susceptibility to the McGurk effect (Campbell et al., 1990). These patients, however, also suffered from aphasia, alexia, and prosopagnosia, which could also presumably have accounted for their inability to process and integrate auditory- and visual-speech signals. Similarly, a patient with bilateral lesions in V5/MT was found impaired in his ability to correlate visual lip-movements with the corresponding auditory-speech, which could, however, be due to akinetopsia (i.e., a motion processing deficit; Campbell et al., 1997). Williams syndrome patients have also been shown to exhibit a disrupted ability to integrate audiovisual speech, but this could probably be due to the visuospatial processing deficits in patients suffering from this syndrome (Böhning et al., 2002). Finally, neglect patients (i.e., inability to report, react, or search for stimuli located in the space contralateral to the patient's lesion) have shown evidence of a temporal distortion, which, however, is often confounded with the spatial deficits of this condition (Baylis et al., 2002; Becchio & Bertone, 2006; Rorden et al., 1997).

The difficulty to isolate disruptions specific to multisensory temporal integration, in addition to the malleable perception of synchrony (i.e., flexible mechanisms, percept that changes depending on stimulus qualities and measurements tasks; e.g., Vatakis, 2013), make the topic particularly challenging to study in clinical populations. In the following pages, we will review evidence that show cases of specific impairment in the perception of synchrony in the absence of any other major deficiencies as well as cases where multisensory temporal integration is greatly impaired as a consequence of other dysfunctions. We will also review cases where the research is as yet limited but potential impairment in multisensory temporal integration may be critical in terms of treatment, diagnosis or both. The reader must note that we mainly focused on reviewing multimodality and not unimodality. Thus, unimodal stimulation/experimentation is covered only when necessary.

3 Multisensory Temporal Integration: Specific Disruptions or Consequence of Other Deficits?

3.1 *Case Reports*

> I told my daughter her living room TV was out of sync. Then I noticed the kitchen telly was also dubbed badly. Suddenly I noticed that her voice was out of sync too. It wasn't the TV, it was me.
>
> PH'S experience as reported in NEW SCIENTIST, 2013

A specific disruption of the ability to integrate audiovisual speech has, in fact, never been reported until recently in two separate patient cases. First, in 2006, Hamilton, Shenton, and Coslett described a patient – AWF – who reported experiencing a temporal mismatch between the auditory- and visual-streams of a speech signal in the absence of any language or sensory impairment. AWF reported a difficulty in understanding face-to-face speech and described his experience as if "watching a movie with the audio out of sync" (Hamilton, Shenton, & Coslett, 2006). Imaging data from this patient revealed a biparietal hypoperfusion with prominence in the right region of the brain. In order to evaluate the deficit objectively, AWF was subjected to various behavioral tests, including a digit span task (with auditory-only or audiovisual presentations), a word-to-picture matching task (where auditory words were paired with congruent/incongruent facial movements, or no images), and a McGurk task. AWF exhibited a substantial impairment on a variety of these tasks, with the greatest impairments observed under conditions where multiple sensory inputs were presented, while no unimodal deficits were detected (through a visual/auditory temporal order judgment – TOJ – task using simple stimuli; i.e., judging the order of stimulus presentation). For instance, the patient's performance was lower in the digit span task in the presence of a face as compared to the no face condition and slower (but accurate) in the word-to-picture matching task when simultaneous audiovisual speech stimuli were presented. Thus, AWF appeared unable to efficiently integrate the auditory- and visual-streams of related multisensory events. The authors, therefore, focused on the integration abilities of AWF suggesting that the parietal cortical injury was potentially responsible for his desynchronized experiences without, however, providing much data on those experiences.

AWF's case (although insufficiently described) showed, for the first time, that the experience of desynchronized events is not improbable (Hamilton, Shenton, & Coslett, 2006), yet the direction (i.e., sensory stream leads/lags) and other details (e.g., width of TWI) of the experience were not reported, leaving us unable to paint a detailed picture and make specific hypothesis on the mechanisms governing audiovisual synchrony perception. While waiting for the next publication on AWF's case, another unique patient – PH – made his appearance in 2013, in an article by Freeman and colleagues. Specifically, the authors described PH, who had asynchronous speech experiences with the auditory stream (i.e., voices) preceding the visual (i.e., lip/facial movements). The medical history of PH included a surgery for treatment of pericarditis, a subsequently developed generalized myasthenia gravis, and, finally, 2–3 months later a sudden onset of timing distortions. Currently, there is no evidence linking myasthenia with timing. High-resolution magnetic resonance imaging (MRI) and diffusion tensor imaging (DTI) revealed that PH had lesions in pons and basal ganglia, both of which are areas associated with timing and multisensory processing. Behaviorally, PH was thoroughly tested so as to address whether or not the experienced asynchrony interrupted integration processes, thus, the authors examined the potential association or independence of the mechanisms governing event synchrony and unity (e.g., Spence & Squire, 2003; Vatakis & Spence, 2007).

PH along with a group of older and younger healthy controls (M = 65 and 22 years of age, respectively) was evaluated in a TOJ and identification task using both simple (stream-bounce illusion) and complex stimulus experimental set ups (e.g., McGurk effect; Freeman et al., 2013). PH's data verified his reported experiences. That is, his performance resulted in a point of subjective simultaneity (PSS) of 210ms auditory speech lagging visual speech, thus requiring visual speech to be presented first by a large interval for synchrony to be perceived. In speech identification, the opposite pattern was observed, with PH requiring 240ms auditory lead for optimal McGurk fusion. Comparison of PH to the controls showed larger deviations for the patient in terms of PSS and speech identification but much smaller just noticeable differences (JND) values as compared to the other two groups. No differences were noted in terms of the perception of synchrony for non speech stimuli. One particularly intriguing finding was the similar behavioral pattern exhibited by both PH and some controls, where the auditory/visual lead/lags noted in the TOJ task (synchrony perception) were inverted for the identification task (integration). These findings led the authors to suggest that it may be the case that the perception of synchrony and of integration depend on distinct timing mechanisms. Furthermore, they suggested the idea that the discrepancies in timing between mechanisms are not minimized (as previous claimed; e.g., Spence & Squire, 2003), but rather perceived relative to the average timing of these mechanisms. This idea was formulated in a theory, the 'temporal renormalization theory', which is one of the first to propose some kind of averaging between mechanisms for the percept of synchrony.

In 2008, we also worked with a patient – RW – whose case resembled those of AWF and PH (Vatakis, 2008). In particular, RW reported perceiving auditory-speech as occurring earlier than the corresponding visual-speech. This desynchronized experience was heightened when he watched television and, particularly, when he watched people singing. RW reported experiencing this desynchronization for 3 years prior to testing, had no history of neurological disease or any cognitive difficulties, and was not taking any medication. Initially, RW was evaluated through a simultaneity (SJ) task (i.e., judging if a stimulus was synchronous or not) for audiovisual speech (a female uttering /da/) and non speech (a complex harmonic tone paired with a white bar) stimuli. The results showed that RW indeed appeared to require a much larger auditory lead as compared to controls, while this was not the case for non speech stimuli. Given, however, that only a small number of stimuli were tested along with the fact that musical stimuli were not evaluated, RW was retested. Retesting included speech and musical stimuli in a TOJ task (see Vatakis & Spence, 2006; for design and stimuli). The reassessment showed no differences from healthy controls for both the musical and speech stimuli. Difficulty of access to RW did not allow further testing, the results, however, showed that RW did not have strong and consistent timing distortions (as he never reported problems in face-to-face communication) or it could also be the case that the experienced symptoms were no longer present at the time of testing.

Although such case studies are rare, they do show that impairments focused only in timing are probable. Thus, it could be the case that synchrony perception stems from a "renormalization" process (Freeman et al., 2013), however, further investigation is needed to explain: (a) why both RW and PH experienced auditory leads in their "out-of-synch" experiences? (b) why was speech the sole stimulus that was impaired? (c) could it be the case that predictive cue processing is also impaired in these patients, thus potentially affecting the renormalization process? (d) can synchrony and binding mechanisms truly be independent or rather in continuous "collaboration"?

3.2 *Schizophrenia*

> Time has stopped; there is no time… The past and future have collapsed into the present, and I can't tell them apart.
>
> The experience of an acute schizophrenic patient as reported in MELGES, 1982, P. XIX

Schizophrenia (SZ) is characterized by delusions, hallucinations, disorganized thinking, abnormal motor behavior (including catatonia), and/or negative symptoms (American Psychiatric Association, 2013). SZ has also been associated with a distorted sense of time continuity as reported by patients and clinicians (e.g., Gómez et al., 2014; Melges, 1982; Minkowski, 1933), as well as with behavioral and neuronal sensory processing abnormalities (e.g., Williams et al., 2010). The temporal and sensory implications in SZ have led to the investigation of potential impairments in multisensory integration (Tononi & Edelman, 2000), with the majority of research focusing on audiovisual (synchronous) speech (most often through the use of the McGurk effect; e.g., de Gelder et al., 2003; Pearl et al., 2009; Ross et al., 2007; White et al., 2014). A number of these studies have showed impaired speech integration, however, deficits have also been identified for simple stimuli. For instance, tasks such as backward masking (e.g., Saccuzzo & Schubert, 1981) have shown that schizophrenics require longer temporal separation of the target and mask in order to effectively locate, detect, or identify the given target.

Findings implicating larger temporal separation between stimuli in SZ (e.g., Saccuzzo & Schubert, 1981), led Foucher and colleagues (2007) to propose that integration per se may not be impaired but rather temporal processing. They examined this hypothesis by measuring simultaneity thresholds for unimodal (auditory, visual) and bimodal (audiovisual) stimuli in schizophrenics and healthy controls. Participants' performance validated the proposed hypothesis with patient data revealing wider temporal windows in all conditions as compared to the controls (Visual: 37 ± 15 vs. 27.5 ± 8ms, Auditory: 42.7 ± 16 vs. 30.5 ± 17ms, Audiovisual: 265 ± 86 vs. 203 ± 51ms, for patients vs. controls, respectively; but see Schwartz, Mallott, & Winstead, 1988). Further analysis showed that performance in all three conditions was highly correlated, thus verifying that the observed effects were not driven by a specific sensory impairment but distorted timing instead. The authors suggested that there might be a link between the extended temporal windows observed in their study and previously reported time estimation impairments in SZ. They further speculated that this association could be based on the disordered activation of the cerebellum (Andreasen, 1999; Minkowski, 1933) or the potential dysfunction of the basal ganglia (e.g., Lacruz et al., 1991), but as yet this has not been elucidated. It must be noted here that if synchrony and integration are two separate processes, this study shows differential temporal processing but does not address whether multisensory binding is intact or impaired in SZ.

Giersch and colleagues (2009) re-examined temporal discrimination thresholds in SZ by eliminating potential decisional (e.g., higher probability to assume synchrony in an SJ task) and attentional biases (e.g., attentional effects due to stimulus onset cues and spatial presentation). They, therefore, presented two target stimuli in different locations with synchronous or asynchronous luminance changes (see Elliott et al., 2007). Data from stabilized chronic outpatients and matched controls replicated Foucher et al.'s (2007) findings, with the patients exhibiting longer simultaneity thresholds as compared to controls, thus further supporting the idea that SZ is associated with a timing distortion.

The above-mentioned findings demonstrate the maintenance of event integration in SZ (but see e.g., Ross et al., 2007) over longer TWI. Clinical evidence, on the other hand, describe the experience of a fragmentation of events in time. This disparity was investigated by Lalanne, van Assche, and Giersch (2012a) using the Simon paradigm (Simon & Wolf, 1963). Specifically, they presented, within the TWI, two stimuli in or out of synchrony on the same or different locations on the screen, while participants responded using the left- and right-sided keys for simultaneity and asynchrony, respectively. It was expected that for same-side presentations, the participants should be biased towards the side of the stimulus (classical Simon effect), while for opposite side presentations, a bias was expected only for the asynchronous cases (although not perceived by the participants). Specifically, for the latter, if the first stimulus was detected as an isolated event on a particular location then responding should be biased to that location, while if a second stimulus was anticipated a bias should be expected to the location of the second stimulation (see also Lalanne et al., 2012b). Performance from stabilized outpatients (on medication) and healthy controls replicated, for one more time, the higher detection thresholds in SZ (Schmidt et al., 2011). Most importantly, it was shown that for shorter SOAs, the patients showed a clear bias towards the location of the first stimulus, while for larger SOAs both participant groups were biased towards the location of the second stimulus (Lalanne et al., 2012b). These findings were interpreted according to the predictive coding theory, with patients processing stimuli in a feed-forward way, having impaired prediction for the second stimulus, thus treating each stimulation as isolated rather that part of an event sequence.

The existence of an extended TWI implies higher levels of integration, however the Lalanne, van Assche, and Giersch's (2012a) findings of sensitivity to short asynchronies imply less integration (Vatakis & Spence, 2007; Welch & Warren, 1980). Given these conflicting views, Martin and colleagues (2013) studied the potential differences of implicit and explicit temporal processing and, thus, focusing on the relationship of timing and integration in SZ. Using the McGurk paradigm with stimuli in or out of synchrony, the participants were required to complete both an identification and SJ task. The identification task allowed for measuring integration (implicit timing), while the SJ task served as an explicit measure of timing. Such task comparisons have resulted in null effects in healthy participants (e.g., van Wassenhove et al., 2007; but see Freeman et al., 2013). Patients and controls showed no differences in the rates of McGurk fusion or combination and unimodal (visual and auditory) speech recognition. Thus, showing intact audiovisual speech integration in SZ (see also Surguladze et al., 2001). Differences, however, were found in the width of the TWI, with patients requiring smaller auditory leads for fusion stimuli and larger auditory lags for combination stimuli compared to controls. Thus, showing speech integration to be less tolerant in patients than in controls. This finding was in direct contrast with the SJ results, where patients showed TWIs larger (yet not significantly) than those of controls. These findings support previous reports of unimpaired integration and problematic timing, providing evidence for a potential dissociation of integration and synchrony timing mechanisms (as in Freeman et al., 2013).

The findings of distorted timing and intact integration capabilities are not limited to SJs but have also been observed in other temporal tasks. de Boer-Schellekens et al. (2013) utilized, for the first time, TOJs in a temporal ventriloquism paradigm (i.e., the presence of a beep before or after a flash alters the temporal percept of the flash as appearing earlier or later, respectively; Morein-Zamir, Soto-Faraco, & Kingstone, 2003). Adults with SZ (on medication; 7 on co-medication) and healthy controls completed a visual TOJ task, with white squares presented above or below the fixation with or without the presentation of auditory stimuli before or after the visual stimulation (0, 100, 200 or 300ms). Results showed worse JNDs for patients as compared to controls (28.4 vs. 21.6ms, respectively) for the smaller SOAs tested, while the presence of auditory stimulation lead to equal performance improvement for both groups. In terms of the PSS, the patient group demonstrated higher 'up-first' responses as compared to controls (10.1 vs. −5ms, respectively; negative values indicating 'down-first' presentations), while no effect of medication or symptom severity was found. The results of this study replicate those reporting an extended TWI but successful event integration. Additionally, they partly support the dysconnectivity hypothesis (i.e., impaired integration could lengthen neural conduction times and, in turn, lead to delayed action potentials and sensory streams not arriving in synchrony; e.g., Foucher et al., 2007) for the JND differences obtained at the lower SOA range tested, but not the null effects for the higher end of the SOA range.

Given the known TOJ and SJ differences in measuring synchrony perception (e.g., García-Pérez & Alcalá-Quintana, 2012; Vatakis et al., 2008) and the differential findings for longer vs. shorter intervals (e.g., de Boer-Schellekens et al., 2014; Lalanne, van Assche, & Giersch 2012a; Martin et al., 2013), Capa and colleagues (2014) compared patient performance in TOJs and SJs utilizing the same stimuli. This experimental set-up allowed them to address whether SZ is associated with distorted temporal ordering, in which case TOJs should be afflicted more than SJs, or with distortions only at brief time intervals, in which case both tasks should be afflicted at the same level. Data from stabilized patients (on medication) and healthy controls revealed major performance impairments in patients in the TOJ task, which was also maintained for larger SOAs. Additionally, the patients' error rate in the TOJ task (14 vs. 5% in controls) was higher than that predicted from their SJ performance (i.e., in larger asynchronies the patients could complete the SJ task but had difficulty in deciding order for the TOJ task). These results, therefore, point to a generalized distortion of ordering in time rather than one limited to very brief intervals.

The consistent finding of an extended TWI for SZ is equivalent to an extended experience of simultaneity, which, in turn, can be interpreted as an extended integration of past and present events and, thus, an altered structure of conscious experience (Husserl, 1991). Schmidt and colleagues (2011) examined whether or not these distorted experiences are specific to SZ or rather general to psychosis. They, thus, tested chronic schizophrenic (on medication) and first-episode psychosis (FEP of various diagnoses; on medication) patients along with healthy controls on an SJ task (as in Elliott et al., 2007, using a staircase procedure). The results of this study replicated the existence of an extended TWI for schizophrenics but also demonstrated – for the first time – that FEP patients exhibit higher simultaneity thresholds as compared to controls. These findings (although with a small N) point out the potential of altered timing in psychosis in general, rather to a specific disorder, thus putting forward the need for comparisons of different clinical populations or of studies with different populations but similar experimental set-ups.

Overall, there is a consistency in the findings pertaining to SZ and multisensory temporal processing. That is, schizophrenics maintain their ability to integrate audiovisual (as no study to date have examined other modalities) events but this integration happens over an extended TWI. Given that simultaneity and order are necessary for the perception of an event and that of a series of events linked in time, the timing distortion in SZ may also be associated with the disorganized speech and action understanding/execution impairments reported in SZ (e.g., de Gelder et al., 2003; Delevoye-Turrell et al., 2007). All of the experimentation, however, has focused on simple stimuli, thus it would be interesting to see how multisensory speech or other stimuli are integrated. Additionally, it would be interesting to see how this extended TWI affects both time estimation and sequencing tasks for simple and complex actions in SZ.

3.3 *Autism Spectrum Disorders*

> Sometimes the channels get confused… Sometimes I know that something is coming in somewhere, but I can't tell right away what sense it's coming through.
>
> CESARONI & GARBER, 1991, P. 305

Autism spectrum disorder (ASD) is characterized by impaired social communication and interaction and restricted, repetitive patterns of behavior, interests, and/or activities (American Psychiatric Association, 2013). Temporal processing distortions have also been reported in ASD, however, the empirical literature is limited with the majority of research been initiated during the last few years. This research activity has shown that children with ASD experience difficulties with mental time travel, time estimation, and time continuity (e.g., Allman, DeLeon, & Wearden, 2011; Szelag et al., 2004), potentially due to a hypothesized critical deficit in the temporal synchronization between local and distributed neural networks (Brock et al., 2002).

ASD has also been associated both through autobiographical and caretaker reports (e.g., Cesaroni & Garber, 1991; Iarocci & McDonald, 2005) with sensory abnormalities such as: hyper/hyposensitivity to sensory stimulation, sensory distortions (e.g., Rogers, Hepburn, & Wehner, 2003; Waterhouse, Fein, & Modahl, 1996), problematic processing of multiple sensory inputs simultaneously (e.g., Brandwein et al., 2013; Bryson, 1970, 1972; Collignon et al., 2013; Foxe et al., 2013; Lovaas & Schreibman, 1971; Lovaas et al., 1971; but see Haviland et al., 1996; Loveland et al., 1995; Walker-Andrews et al., 1994), and difficulties identifying the sensory origin of an input (e.g., Cesaroni & Garber, 1991; O'Neill & Jones, 1997). Yet, the actual validity or importance of sensory disturbances in ASD is often put in question (e.g., Ashwin et al., 2009; Bonnel et al., 2003; Magnee et al., 2008; Mongillo et al., 2008), as evidenced from the removal of this issue from the DSM versions following DSM III (American Psychiatric Association, 1980) and its reintroduction in the DSM-5 (American Psychiatric Association, 2013). The wavering noted in the DSM guidelines has also been noted in research where the potential impaired sensory and multisensory processing hypothesis has not always been verified, in addition to the limited generalization of past findings due to small population sizes, absence of comparison groups, variable or unreported symptom severity, task difficulty, and variable – in terms of age and diagnosis-population groups (e.g., Bedko et al., 2006; Foxe et al., 2013; Kolko, Anderson, & Campbell, 1980; Lovaas et al., 1971; Woynaroski et al., 2013).

A number of studies have looked at multisensory temporal processing in ASD utilizing simple, well-studied paradigms such as the sound induced flash illusion (SIFI; i.e., a brief flash paired with two auditory beeps is actually perceived as two distinct flashes; e.g., Shams, Kamitani, & Shimojo, 2000). First, van der Smagt, van Engel, and Kemner (2007) examined low-level audiovisual integration in high-functioning adults with ASD and healthy controls (IQ matched) through the SIFI. Participants were presented with a series of auditory (beeps) and visual stimuli in 3 visual conditions (1, 2 or 3 flashes) combined with 4 auditory conditions (0, 1, 2 or 3 beeps), and, thus, including consistent and inconsistent (i.e., different numbers of beeps/flashes) stimulus pairings. The performance of flash detection showed that the number of concurrently presented sounds influenced the perceived number of disks similarly for both participant groups. Thus, indicating intact integration of auditory and visual information at an early level of processing in ASD (but see Collignon et al., 2013). Given, however, that SIFI is highly dependent on the presentation timing of the auditory stimuli, Foss-Feig and colleagues (2010) conducted the exact same study with the addition of a timing manipulation. They tested children with ASD and typically-developing (TD) children matched in verbal/nonverbal cognitive abilities, reading level, and gender. Analysis of the participant reports on the number of flashes perceived showed that the ASD group experienced the illusion over a wider range of SOAs as compared to the TD group, with the TWI in ASD being double the size of that observed in TD children (Kwakye et al., 2011). The results from these two studies, therefore, support that integration is possible in ASD for simple stimuli but temporal processing seems to be impaired.

Shifting to more complex and social stimuli, Bebko and colleagues (2006) focused on defining whether or not people with ASD have impaired multisensory temporal integration capabilities when confronted with speech or action-related stimuli. They utilized a preferential looking paradigm so as to avoid instructions and/or task demands influencing participants' performance. They tested 3 groups of participants: children with ASD, children with other developmental disabilities, and TD children, using three different types of audiovisual stimuli: (a) non linguistic (segment of a children's game with a ball moving through a series of plastic ramps and cliffs along with the respective sounds), (b) simple linguistic (a woman counting at approximately one number per 1.5 s), and (c) complex linguistic (a woman telling a story). The analysis for looking away behavior showed more looking away time for simple linguistic as compared to non linguistic stimuli but no group interaction was noted. In terms of response to the asynchrony presented, ASD children showed no difference from chance performance in preference of synchronous versus asynchronous displays. Additionally, although there was a preference for synchronous displays of non linguistic stimuli in ASD, no interaction of stimulus type with participant group was obtained, instead only a main effect of stimulus type was noted. Thus, although this study utilized a low in demands task and a balanced group of participants, the results did not provide a clear picture of the multisensory temporal processing capabilities in ASD.

Further investigations utilizing complex stimuli with a focus on speech extended on the work first conducted by Bebko et al. (2006), resulting in a still conflicting picture. For example, Grossman, Schneps, and Tager-Flusberg (2009) examined whether individuals with ASD can integrate visual- and auditory-speech information presented in a meaningful linguistic context at normal speaking rate. Adolescents with ASD and TD controls matched in age, verbal IQ, and receptive vocabulary were tested. Participants completed an SJ task for a 3-s video of a woman describing the baking of various desserts with the audiovisual streams presented in: synchrony, high slip rates (10, 12 or 14 frames), or low slip rates (4, 6 or 8 frames; 4 frames~120ms). The results showed that ASD participants were as accurate as their TD peers at all slip rates, thus showing normal integration of speech stimuli in ASD. It must be noted, however, that the null results could have been driven by: (a) the potential adaptation to the asynchrony due to the long video duration, (b) the hypothesized developmental "catch-up" noted in ASD, and/or (c) the mild severity of the symptoms of the participants tested. Regarding the latter issue, Woynaroski and colleagues (2013) examined unisensory and multisensory processing capabilities in ASD along with symptom severity. ASD and TD participants were presented with McGurk stimuli in matched or mismatched formats and with the sensory streams either in synchrony or asynchronously. The analysis revealed intact auditory but impaired visual and audiovisual speech perception in ASD in comparison to the TD group. Deficits were also noted in the case of non-verbal responding with poor performance in visual-only and audiovisual matched stimulus presentations. Additionally, a wider TWI was observed for the case of mismatched stimulus presentations with visual speech influencing mismatched speech reports more in ASD than in TD. Finally, in terms of symptom severity, it was found that performance in matched speech was highly correlated with communication-related symptoms and fusion percepts were highly correlated with sensory processing symptoms. Overall, therefore, this study showed a wide range of deficits for visual and multisensory temporal processing in ASD, which was associated with some ASD symptoms.

In an attempt to further elucidate multisensory temporal integration in complex stimuli not limited to speech, de Boer-Schellekens, Eussen, and Vroomen (2013) conducted a study utilizing TOJs and a range of audiovisual stimuli from simple (flash-beep) to more complex ones with or without social aspects (i.e., handclap and speech – a Dutch female uttering/bi/). High-functioning adolescents in the autism spectrum (patients with autism disorder, Asperger's syndrome, and pervasive developmental disorder not otherwise specified, PDD-NOS) and TD controls (age-, gender-, and IQ-matched) completed the TOJ task revealing that the ASD group had larger JNDs than the control group for all conditions tested. Thus, demonstrating that ASD individuals suffer from a generalized impairment in audiovisual temporal processing, which is not limited to speech. Overall, all groups performed best for the hand clapping condition and, subsequently, for the simple flash-beep and speech stimulation. Based on the fact that both groups performed best with the clapping hands, a condition that involves anticipatory visual information, suggests that individuals with ASD can use predictive information. Given that this study had a very specific set of patients and limited population size, Stevenson et al. (2014) conducted a similar study with a large sample of ASD (Autism = 25%, Asperger = 66%, and PDD = 9%) and TD age-matched individuals. The participants were presented with audiovisual stimuli of simple flashes and beeps, dynamic handheld tools, and single syllable utterances (/ba/ and /ga/) and completed an SJ task for all conditions, a McGurk task for the speech stimuli, and unimodal (auditory and visual) TOJ tasks for the simple stimulus conditions. Participants' performance revealed a strong link between multisensory temporal processing and speech perception in ASD. Specifically, the poorer an individual's temporal acuity across vision and audition (i.e., larger TWI), the weaker their ability to bind auditory and visual speech to create a robust McGurk percept and to integrate multisensory stimuli in general. The better an individual's multisensory temporal acuity, the more reliable perceived synchrony was as a predictor of integration (i.e., whether or not two sensory signals originated from a single external event simultaneously and should, thus, be perceptually bound). The width of the TWI for both participant groups was found to be dependent on the complexity of the stimuli presented, with the largest widths obtained for speech (e.g., Vatakis & Spence, 2006). In ASD, there was a particularly enlarged window for speech as compared to the rest of the stimuli (Bebko et al., 2006). Finally, there was also a marginal effect of diagnosis (collapsed across stimulus types), suggesting impairments for the cases of simple stimulus presentations across the different groups in ASD (de Boer-Schellekens et al., 2013; Foss-Feig et al., 2010; Kwakye et al., 2011). Although this study tested participants from 6 to 18 years of age, no separate analysis by age group was conducted. The authors interpreted their speech findings as an effect driven by either long processing times for speech perception or extended processing times within each sensory stream of the speech signal, and, thus, interpret these as supporting the temporal binding hypothesis of autism (Brock et al., 2002; but see Chan & Naumer, 2014).

Overall, the research conducted to date shows that ASD is associated with an extended TWI, but it remains to be clarified whether this extended window is limited to speech or generalized to all types of stimuli. One way to account for this debate in the literature is to further experiment with stimuli of different modalities and modality combinations using implicit or explicit paradigms (e.g., Wada et al., 2014). Additionally, implementation of complex stimulation should take into consideration issues related to semantics and emotional content and, thus, experiment with stimulation that is controlled for social-related activities (e.g., Bakirtzi & Vatakis, 2014). Further experimentation should also consider in detail sample size and population selection both for the ASD and the TD group. For instance, participants in the ASD group should also be screened for undiagnosed attentional deficits, while the TD group should be screened for ASD-related symptoms. On the latter topic, Donohue, Darling, and Mitroff (2012) conducted a study on multisensory processing and ASD-related symptomology in an adult, *non-clinical* population. Specifically, they tested 101 participants on the Autism Spectrum Quotient (ASQ) questionnaire (commonly used as a screening measure; covers issues under the categories of social skill, attention switching, attention to detail, communication, and imagination) and a multisensory SJ task. Analyses revealed that the PSS correlated significantly with the scores on the ASQ, with higher ASQ scores (i.e., higher ASD symptomatology) being associated with an auditory-first bias in responding (Russo et al., 2012). This association could be related to higher auditory processing sensitivity and/or auditory effects on attention and, subsequently, to a potentially larger TWI that is auditory shifted.

3.4 *Attention Deficit Hyperactivity Disorder*

Attention deficit hyperactivity disorder (ADHD) is prevalent in children and it is characterized by a persistent pattern of inattention (i.e., being disorganized, wandering off, difficulty in sustaining focus independent of disobedience or lack of understanding) and/or hyperactivity (i.e., excessive motor activity) and impulsivity (i.e., actions without forethought, potentially harming, that result in immediate rewards) that interferes with functioning and/or development (American Psychiatric Association, 2013). Timing distortions have also been noted in ADHD but mainly in terms of time estimation (e.g., Seri et al., 2002; Smith et al., 2002; Wittmann et al., 2011) with the majority of the findings showing impulsivity to result to poor motor inhibition and, thus, interval underestimation (Brown & Vickers, 2004; Smith et al., 2002). Such distortions have been associated with basal ganglia dysfunction (Meck, 1996), but alternative or complementary accounts put forward a potential working memory deficit (Brown & Vickers, 2004). More recently, the association of hyper-responsive superior colliculus (SC) with increased distractibility makes for an alternative model for ADHD pathology (Overton, 2008).

Potential distortions in temporal processing in ADHD have rarely been investigated. The only investigation on the topic is the study by Brown and Vickers (2004) using simple visual stimulation in adolescent ADHD patients. Specifically, they investigated the perceptual nature of the temporal deficits previously reported in ADHD by minimizing memory and motor-related demands, as well as task complexity. ADHD patients (tested on- and two days off-medication) and healthy controls (age- and gender-matched) completed a verbal SJ task (in a staircase paradigm) for pairs of red light emitting diodes (LEDs). The LEDs were presented bilaterally and unilaterally over a set of different SOAs.

Participants' performance revealed no temporal processing differences in ADHD participants (on- or off-medication) as compared to controls (mean thresholds: 30 and 31ms, respectively). Given these null results as compared to the time estimation studies, the authors argued against the basal ganglia dysfunction account for ADHD. It must be noted however that the small sample tested, the absence of younger population (as it is usually the case for ADHD), and the unimodal nature of the study render these results preliminary.

Given the limited amount of research on the topic, the involvement of attention in multisensory temporal integration (Talsma et al., 2010), and the recent proposal of SC involvement in ADHD (an area known for its involvement in multisensory integration; Stein & Stanford, 2008), it is pertinent to investigate more thoroughly whether or not ADHD patients do exhibit multisensory temporal integration deficits (Panagiotidi, Stafford, & Overton, 2014) and how these potential deficits associate with those reported in time estimation.

3.5 *Aphasia*

> A word is a sequence of auditory events (phonemes)... suppose that an individual loses the capacity for identifying the correct order of phonemes. The word 'electrician' 'i lek' tris' an' might as easily be interpreted as 'tris' lek' an ' or as'an i tris' lek'.
>
> EFRON, 1963A, P. 404

Aphasia is associated with the loss of one's ability to speak and/or comprehend language along with the inability to read and write, usually caused by stroke or brain tumor. Research on aphasia has focused on the identification of the disrupted cognitive processes underlying language, along with the classification of aphasia depending on the symptoms and the brain areas affected (e.g., Charidimou et al., 2014). Efron (1963a), however, proposed that aphasia may be related to impaired temporal processing emphasizing the importance of correct phoneme ordering in language and the aphasic symptoms of jargon, neologisms, and "word-salad" in speech. In addition, data from Efron and others (e.g., Efron, 1963b,c,d; Nicholls, 1996) support that simultaneity and temporal order are processed in the hemisphere dominant for speech (Wittmann et al., 2004).

Based on this hypothesis, Efron, in his 1963a study, examined patients with lesions of the left hemisphere (LH; with some degree of aphasia) or the right hemisphere (RH) in the same general areas, and patients with no aphasia (matched in age and general alertness). The patients were asked to complete a visual and an auditory tertiary task (i.e., reporting synchronous or asynchronous presentations and in the case of asynchrony reporting of order; the "I do not know" option was also available) using simple stimulation. Analysis revealed order distortions for the patients with LH lesions (and some degree of aphasia) only, thus supporting the association of hemispheric dominance and aphasia (but see Sherwin & Efron, 1980) with timing. Results also showed that 'expressive' aphasics (i.e., clear understanding of speech but no standard intonation, speech is slow and staggered with long breaks) had worse auditory discrimination performance as compared to visual, while "receptive" aphasics (i.e., inability to understand normal speech, generation of speech with the right intonation and speed is possible, but words are not meaningfully combined) showed the reverse pattern. Although one of the first studies to explore the link of timing and aphasia, the sample size tested was quite small, thus making the generalization of the results and the mainstreaming of the idea difficult. Subsequent investigations, however, replicated Efron's findings in terms of poor auditory discrimination thresholds in aphasia (e.g., Edwards & Auger, 1965; Holmes, 1965), but no such replications were reported for vision (Holmes, 1965; Jerger et al., 1969).

Given the methodological differences of the previously conducted studies (e.g., spatial location, different stimulus types, no stimulus control for intensity), Swisher and Hirsh (1972) conducted a study to clarify whether the previously noted temporal distortions in aphasia were valid or rather a methodological artifact. They, therefore, presented all possible stimulus configurations (i.e., visual/auditory: different color/pitch in the same or different location/ear with a color/pitch response, same color/click in different location/ear with a location response) to different patient (LH damage with receptive aphasia, LH damage with expressive aphasia and hemiplegia, RH hemiplegia with no aphasia) and healthy control (young and older participant groups) groups. Their results were similar to those obtained by Efron (1963a), with the presentation of different visual stimuli in the same location resulting to larger discrimination thresholds for the LH impaired as compared to the RH impaired patients. No such differences were noted for visual presentations in different locations. For auditory presentations, the deficits were greater in comparison to those for the visual modality with the receptive aphasics showing larger discrimination thresholds (see also von Steinbüchel et al., 1999) for presentations in the same as compared to different ears. Stimulation of different type and location improved performance for the LH patient groups. The greater effect observed in receptive aphasics for audition could be driven by a number of factors such as symptom severity as well as specific brain regions afflicted, yet it still supports Efron's proposal of a close association of language function and temporal processing (see also von Steinbüchel & Wittmann, 1997; Wittmann et al., 2004). This close association is as yet undefined given that limited research conducted in recent years and the absence of experimentation with multisensory stimuli.

3.6 *Dyslexia and Specific Language Impairments*

Developmental dyslexia has been repeatedly associated with impaired or slowed temporal processing (e.g., Farmer & Klein, 1995), a direct, however, association has not been accepted as yet (e.g., Protopapas, 2014). Unimodal visual and auditory impairments have lead to the hypotheses of impaired functioning of a specific modality, a general perceptual impairment or a speech-specific dysfunction that is associated with phonological processing (e.g., Laasonen et al., 2000). The use of unimodal as well as multimodal stimulation of nonlinguistic stimuli can provide insight as to the potential impairments in multisensory temporal processing, which is critical in the development of reading and language acquisition (e.g., Bremner, Lewkowicz, & Spence, 2012).

One of the first studies to examine the potential association of dyslexia and temporal processing both unimodally (i.e., clicks in the auditory, flashes in the visual, and indentations of the skin in the tactile modality) and multimodally was conducted by Laasonen and colleagues (2000). Specifically, they tested developmentally dyslexic readers and control readers through a within- and between-modality SJ task (using stimulus pulses in two trains for a constant interval on distinct spatial locations) along with a series of other tests. The data, overall, showed that the dyslexic group segregated spatiotemporal perceptual information at a slower rate than the control group. These differences were noted not only for unimodal presentations but also and more pronounced for bimodal ones with the highest differences noted for the audiovisual combinations presented. A subsequent replication (Laasonen, Service, & Virsu, 2002; Laasonen et al., 2001) utilizing adult participants and an additional TOJ task measure continued to show impairments but at a different level for each modality. Both tasks showed poor discrimination thresholds for the dyslexic group for all modalities and their combinations, but with auditory, tactile, and audiotactile stimulation measuring at the lowest performance as compared to the control group. Dyslexia, therefore, was associated with a generalized temporal impairment in perceiving rapidly presented sequential non speech events, which was found to deteriorate even more with age (Virsu, Lahti-Nuuttila, & Laasonen, 2003).

The above-mentioned studies provide the first evidence of a multimodal temporal processing impairment in dyslexia. It is as yet unknown, however, whether or not multisensory temporal integration is also affected, given that a potentially extended TWI could interfere with processes that require rapid and accurate integration of sensory inputs (e.g., reading; King, Wood, & Faulkner, 2007). A first investigation on this issue was undertaken by Hairston and colleagues (2005) using simple stimuli in a temporal ventriloquism paradigm (e.g., Morein-Zamir et al., 2003). Specifically, they tested dyslexic and typical-reading participants (IQ above 70) in a visual TOJ task (staircase procedure) and, subsequently, a visual TOJ task in the presence of synchronous or asynchronous auditory stimuli. The results showed impaired visual TOJs in dyslexia, as per previously mentioned studies. In terms of integration, in the presence of synchronous auditory stimulation, the dyslexic group performed significantly better compared to the unimodal performance, while no such difference was obtained for the control group. When auditory delays were introduced, the control group showed benefits for delays over 100ms and up to 250ms, while the dyslexic group benefited for delays as low as 50ms and as high as 350ms. Thus, exhibiting an extended TWI as compared to the control group. Similar results were obtained for audiovisual speech stimuli for children with specific language impairment (SLI; Pons et al., 2012), where auditory speech delays of up to 666ms went unnoticed when compared to a control group. These latter results could be attributed to dysfunctional predictive mechanisms of the sensory inputs for the percept of unified multisensory speech and/or extended TWIs.

The above-mentioned deviations of temporal processing and multisensory integration in dyslexia and SLI could be driven by an attentional impairment (e.g., spatial allocation of attention), which has also been under-researched (e.g., Hood & Conlon, 2004; Landerl & Willburger, 2010; Pons et al., 2012; Talcott et al., 2002). Further research is, thus, needed in the potential associations of timing, modality, and attention (Harrar et al., 2014) in dyslexia and SLI utilizing both simple and complex stimuli (Kronschnabel et al., 2014, for use of complex stimulation in terms of congruency but not timing).

3.7 *Parkinson's Disease*

Parkinson's disease (PD) is a neurodegenerative disease characterized by motor-related impairments (e.g., bradykinesia, rigidity, postural instability), as well as cognitive deficits, including timing distortions, which may be associated with executive dysfunction affecting approximately 30% of PD patients (Parker et al., 2013). The timing distortions noted in PD relate to time estimation, while data on temporal discrimination is sparse. Additionally, past studies have shown a large percentage of PD patients to exhibit primary sensory impairments (e.g., Koller, 1984; Snider et al., 1976). It is, however, unclear why these sensory and timing deficits occur. Some researchers promote basal ganglia as a key contributor in these PD disturbances given its primary role in motor control and its involvement in sensory (Graziano & Gross, 1993; Snider et al., 1976) and sensorimotor integrative function (Artieda et al., 1992), but as yet no concrete conclusions have been reached.

A first set of investigations focusing on temporal discrimination using tactile (Lacruz et al., 1991) or proprioceptive (Fiorio et al., 2007) stimuli showed PD impairments, with discrimination thresholds being larger for the patients as compared to controls. As revealed by these studies, the stimulation location is critical since larger thresholds were noted for stimulation at the hand contralateral to the lesion or the affected arm in the case of unilateral bradykinetic rigid PD patients (Fiorio et al., 2007; Lacruz et al., 1991). A direct comparison of different modalities was conducted by Artieda et al. (1992) by measuring temporal discrimination thresholds for somaesthetic, auditory, and visual stimuli in PD (grouped by severity: mild, moderate, severe) and healthy controls. Utilizing the method of limits, the authors asked participants to detect the number of sensory inputs (1 vs. 2) they were receiving. Analysis of these reports showed that PD patients were deficient in all three modalities and the degree of the deficit increased with symptom severity. This deficit was expressed with significantly extended TWI as compared to controls. Thus, in addition to other timing deficits noted in PD (e.g., time estimation, sensorimotor synchronization), the authors proposed that these effects may be due to slow running internal clocks, which, in turn, impaired patients' temporal resolution for simple stimuli.

Results, although limited, support impaired temporal discrimination thresholds in PD when the stimulus input is presented unimodally. It is as yet unknown how PD patients perform in the presence of multisensory and complex stimulation. Given the data supporting the involvement of basal ganglia in multisensory input and timing (e.g., Graziano & Gross, 1993), it may be the case that the indirect basal ganglia circuit inhibiting neurons from thalamus to cortex is impaired given research suggesting dissociation of motor and temporal discrimination and direct basal ganglia circuits for movement (Lucas et al., 2013; Widnell, 2005).

Similar to PD, Dystonia is a neurological condition associated with impaired functioning of the basal ganglia and characterized by motor disturbances (repetitive movements, abnormal postures) in many different (generalized dystonia) or specific body parts (focal dystonia; e.g., focal hand dystonia, cervical dystonia) along with abnormal somatosensory processing (e.g., Tinazzi et al., 2004). A small number of psychophysical studies have shown that dystonic patients demonstrate distorted temporal discrimination with tactile and even visual stimuli (e.g., Aglioti et al., 2003; Fiorio et al., 2003). In terms of multisensory temporal integration, recent studies have demonstrated deficits in visuo-tactile integration in generalized and focal hand dystonia, while visual temporal discrimination was found to be impaired only for generalized dystonia. Similarly, Tinazzi and colleagues found patients with cervical dystonia to exhibit poor tactile and visuo-tactile discrimination as compared to healthy controls, yet visual thresholds were normal. These differences may imply that each type of dystonia is governed by slightly different impairments localized at different neural circuits composing the basal ganglia. Thus, further understanding of basal ganglia dysfunction and the commonalities and differences of the various types of dystonia will allow us to also understand PD impairments.

4 Looking Forward to Further Experimentation

The perception of order and simultaneity are fundamental in the perception of unified multisensory events. We suggest that distortion at any level of the timing experience (simultaneity, order, duration) leads to a disruption to all other levels of our experience and collectively this disruption may be associated with an altered clinical picture. Current investigations, however, focus only on one of those levels of the timing experience, thus potentially ignoring associations/interactions that may be important for understanding a specific clinical case, a collection of symptoms, and/or the implementation of a particular intervention. Time estimation has been investigated more vigorously than multisensory temporal processing in the clinical domain. The few and recent studies focusing on the latter have shown, for the first time, a potential distortion of timing in the absence of any other impairments (see Section 3.1 of this chapter), timing being the critical component of a disorder (see Section 3.2 of this chapter), synchrony and binding being potentially differentiated in terms of timing mechanisms (see Sections 3.1 and 3.2 of this chapter), as well as timing distortions being co-present with other impairments, with the latter potentially causing the timing distortions noted (see Section 3.7 of this chapter).

In this chapter, we explored a range of different clinical/neurological cases that experimentation lead to findings of impaired sensory discrimination and mostly extended TWIs (with some tendency for the auditory stream to be leading), while in a few studies we also obtained data regarding multisensory binding although concrete conclusions were not reached (see Table 1.1). The literature covered here may tempt one to claim that there is a single, common mechanism that governs all these distortions in timing and integration. This however will be premature, simplistic, and potentially wrong (e.g., Freeman et al., 2013; Martin et al., 2013, showing evidence of different timing mechanisms). For instance, the

TABLE 1.1 *A brief overview of all the studies reviewed in this chapter covering issues related to multisensory temporal processing.*

	Behavioral task	Stimulus type			Modality	Distortions
		Abstract	Speech	Action		
Case study						
Hamilton et al. (2006)	Match detection, Digit span, and others	✓	✓		A, V, AV	Reported timing deficit but not tested; AV binding impaired
Freeman et al. (2013)	TOJ, Identification	✓	✓		A, V, AV	Impaired synchrony perception requiring a V lead of 210ms for speech; Opposite pattern for binding with PSS measured at 240ms of A leading speech
Vatakis (2008)	TOJ		✓	✓	AV	Impaired synchrony perception requiring a V lead of 210ms for speech; Opposite pattern for binding with PSS measured at 240ms of A leading speech
ADHD						
Brown & Vickers (2004)	SJ	✓			V	No deficit detected
Aphasia						
Efron (1963a)	Ternary	✓			A, V	Expressive: Poor A discrimination; Receptive: Poor V discrimination
Swisher & Hirsh (1972)	TOJ	✓			A, V	Expressive: Poor V discrimination; Both types: Poor A discrimination

von Steinbüchel et al. (1999)	Ternary	✓		A	Receptive: Poor A discrimination
Parkinson's disease					
Artieda et al. (1992)	SJ modified	✓		A, V, P	Poor discrimination thresholds for all modalities tested
Fiorio et al. (2007)	Motion detection	✓		P	Poor discrimination thresholds
Schizophrenia					
Foucher et al. (2007)	SJ	✓		A, V, AV	Extended TWI
Giersch et al. (2009)	Change detection	✓		V	Extended TWI
Schmidt et al. (2011)	Change detection	✓		V	Extended TWI
Lalanne et al. (2012b)	Simon effect	✓		V	Extended TWI; Short SOAs: First stimulus presentation bias
Martin et al. (2013)	SJ, Identification		✓	A, V, AV	Identification: Window width with smaller A leads for fusion and larger A lags for combination stimuli; S J: Extended TWI; Normal binding
de Boer-Schellekens et al. (2014)	TOJ	✓		V, AV	Poor discrimination; PSS differences; Normal AV integration
Capa et al. (2014)	TOJ, SJ	✓		V	Poor order discrimination for short SOAs

TABLE 1.1 *A brief overview of all the studies reviewed in this chapter covering issues related to multisensory temporal processing* (cont.)

	Behavioral task	Stimulus type			Modality	Distortions
		Abstract	Speech	Action		
Dystonia						
Tinazzi et al. (2004)	SJ	✓			V, T, VT	Poor T and VT discrimination
Autism						
Bebko et al. (2006)	Preferential looking		✓	✓	AV	Unpaired synchrony perception for speech stimuli
van der Smagt, van Engel, and Kemner (2007)	V detection	✓			V, AV	Normal AV integration
Grossman, Schneps, & Tager-Flusberg (2009)	SJ		✓		AV	Normal timing; AV integration normal
Foss-Feig et al. (2010)	V detection	✓			VA-V	Extended TWI; Persistence of binding
Kwakye et al. (2011)	TOJ				A, V, AV	V: Normal, A: Poor discrimination, Extended TWI; Persistence of binding
de Boer, Eussen, & Vroomen (2013)	TOJ	✓	✓	✓	AV	Poor discrimination thresholds

Woynaroski et al. (2013)	Recognition		✓		A, V, AV	Poor V and AV discrimination thresholds; Poor binding; Speech congruence positively correlated with communication related symptoms; McGurk fusions positively correlated with sensory processing symptoms
Stevenson et al. (2014)	SJ, TOJ, Identification	✓	✓	✓	A, V, AV	Poor V and A temporal acuity; Weaker binding
Dyslexia						
Hairston et al. (2005)	TOJ	✓			V, AV	Poor V order; Extended binding
Laasonen et al. (2000)	RIS	✓			A, V, T	Temporal impairment of uni/multisensory information
Laasonen et al. (2001)	RIS, TOJ	✓			A, V, T	Pansensory perceptual temporal acuity impairment
Laasonen et al. (2002)	RIS, TOJ				A, V, T	TOJ: Poor discrimination thresholds for A-T; RIS: Poor discrimination thresholds for A-T, V-T
Specific Language Impairment						
Pons et al. (2012)	Preferential looking		✓		AV	Poor detection of large A delays

Proprioreception (P); Auditory (A); Visual (V); Tactile (T); Temporal order judgment (TOJ); Simultaneity judgment (SJ); Stimulus onset asynchrony (SOA); Temporal window of integration (TWI); Rate of information segregation (RIS) enlarged windows noted in dyslexia, schizophrenia, and ASD cannot be governed by the same mechanisms. That is, dyslexics have been shown to exhibit impairments in auditory and visual temporal processing (e.g., Hairston et al., 2005; Laasonen et al., 2001), while investigations with ASD have resulted in unclear findings with some studies showing auditory impairments only (e.g., Kwakye et al., 2011). These findings may be related to different types or levels of disruption in multisensory temporal processing. Thus, for instance, ASD has been associated with local hyperconnectivity and long-range hypoconnectivity, but dyslexia with proximal hypoconnectivity but distal hyperconnectivity (e.g., Williams & Casanova, 2010), while both disorders have been associated with cerebellum impairments (e.g., Hodge et al., 2010). On the other hand (see Section 3.2 of this chapter), patients suffering from SZ do exhibit an extended TWI but recent behavioral evidence support the idea of patients getting "stuck" on the first stimulus presentation and, thus, leading to impaired discrimination performance.

The recent hypothesis that multisensory temporal processing may be governed by separate mechanisms for synchrony and integration (e.g., Freeman et al., 2013) is also supported by recent imaging studies. Specifically, Stevenson et al. (2010, 2011) in an fMRI study utilizing audiovisual speech in or out of synchrony identified two subregions of the multisensory superior temporal cortex (STC), one responding to synchrony independent of integration and another responding to integration regardless of the timing of the sensory streams. A limitation was also noted with increased synchrony leading to higher probability of integration. Thus, although separate, the mechanisms do 'collaborate' for the achievement of a unified moment and, thus, 'meaningful' interval. Given the distortions of temporal integration of multisensory events in many of the disorders reviewed in this chapter (see also Chapters 3 and 5 of this book) it may be important to further examine STC in terms of integration (e.g., Pekkola et al., 2006), timing, and anatomical abnormalities (e.g., Boddaert et al., 2004). Parietal regions should also be studied further given that in SZ functional differences are also noted in regions that project directly to STC such as parietal regions in the dorsal visual stream (Doniger et al., 2002). These latter regions were also implicated in the case of the patient AWF (Hamilton et al., 2006). Thus, it may be the case that STC is core in the unified percept of synchronous or near-synchronous moments (Stevenson et al., 2011).

Overall, it is pertinent to better understand the role of timing in the clinical/neurological patient. Such understanding will allow for the study of timing and binding mechanisms, but most importantly will allow for the development of diagnostic and/or rehabilitation tools. In terms of the latter, given that a number of disorders have been associated with an enlarged TWI, Powers, Hillock, and Wallace (2009) showed that perceptual training can change perceived SJs with the change depicting a true perceptual change rather that a short-lived training effect (as shown by a 1 week after-training post-test; see also Powers et al., 2012; see also Schlesinger et al., 2014, for narrowing of the TWI with feedback training). Given that we support an association of the different levels of the timing experience, changes of the width of the TWI will also lead to temporal discrimination and potentially time estimation changes. Further research looking at the different levels of the timing in the same paradigm, with identical multisensory stimulation, and the same group of individuals (healthy or not) will allow for a unified model of our everyday unified and continuously flowing in time experience.

Acknowledgments

This work has been supported by the European project COST ISCH Action TD0904 (<www.timely-cost.eu>) "Time In MEntaL activitY: theoretical, behavioral, bioimaging and clinical perspectives (TIMELY)" and co-financed by the European Union (European Social Fund, ESF) and Greek national funds through the Operational Program "Education and Lifelong Learning" of the National Strategic Reference Framework (NSRF) – Research Funding Program: THALIS-UOA-375737.

References

Aglioti, S.M. et al. 2003. "Temporal Discrimination of Cross-Modal and Unimodal Stimuli in Generalized Dystonia" *Neurology* 60: 782–85.

Allman, Melissa J. 2011. "Deficits in Temporal Processing Associated with Autistic Disorder" *Frontiers in Integrative Neuroscience* 5(2): 1–2.

Allman, Melissa J., Iser G. DeLeon, and John H. Wearden. 2011. "Psychophysical Assessment of Timing in Individuals with Autism" *American Journal on Intellectual and Developmental Disabilities* 116: 165–78.

American Psychiatric Association. 1980. *Diagnostic and Statistical Manual of Mental Disorders* (3rd ed.). Arlington, VA: American Psychiatric Publishing.

American Psychiatric Association. 2013. *Diagnostic and Statistical Manual of Mental Disorders* (5th ed.). Arlington, VA: American Psychiatric Publishing.

Andreasen, Nancy C. 1999. "A Unitary Model of Schizophrenia: Bleuler's 'Fragmented Phrene' as Schizencephaly" *Archives of General Psychiatry* 56(9): 781–87.

Artieda J. et al. 1992. "Temporal Discrimination is Abnormal in Parkinson's Disease" *Brain* 115(1): 199–210.

Ashwin, Emma et al. 2009. "Eagle-Eyed Visual Acuity: An Experimental Investigation of Enhanced Perception in Autism" *Biological Psychiatry* 65: 17–21.

Bakirtzi, Venetia, and Argiro Vatakis. 2014. "The Perception of Integrated Events in Autism Spectrum Disorders: The Role of Semantic Relatedness and Timing" *Procedia – Social and Behavioral Sciences* 126: 212–13.

Baylis, Gordon C. et al. 2002. "Visual Extinction with Double Simultaneous Stimulation: What is Simultaneous?" *Neuropsychologia* 40: 1027–34.

Bebko, James M. et al. 2006. "Discrimination of Temporal Synchrony in Intermodal Events by Children with Autism and Children with Developmental Disabilities without Autism" *Journal of Child Psychology and Psychiatry* 47(1): 88–98.

Becchio, Cristina, and Cesare Bertone. 2006. "Time and Neglect: Abnormal Temporal Dynamics in Unilateral Spatial Neglect" *Neuropsychologia* 44: 2775–82.

Boddaert, N. et al. 2004. "Superior Temporal Sulcus Anatomical Abnormalities in Childhood Autism: A Voxel-Based Morphometry MRI Study" *NeuroImage* 23: 364–69.

Böhning, M., R. Campbell, and A. Karmilov-Smith. 2002. "Audiovisual Speech Perception in Williams Syndrome" *Neuropsychologia* 40: 1396–406.

Bonnel, Anna et al. 2003. "Enhanced Pitch Sensitivity in Individuals with Autism: A Signal Detection Analysis" *Journal of Cognitive Neuroscience* 15: 226–35.

Brandwein, Alice B. et al. 2013. "The Development of Multisensory Integration in High-Functioning Autism: High-Density Electrical Mapping and Psychophysical Measures Reveal Impairments in the Processing of Audiovisual Inputs" *Cerebral Cortex* 23: 1329–41.

Bremner, Andrew J., David J. Lewkowicz, and Charles Spence. *Multisensory Development.* Oxford, UK: Oxford University Press, 2012.

Brock, Jon et al. 2002. "The Temporal Binding Deficit Hypothesis of Autism" *Development and Psychopathology* 14: 209–24.

Brown, Lenora, and Joan Vickers. 2004. "Temporal Judgments, Hemispheric Equivalence, and Interhemispheric Transfer in Adolescents with Attention Deficit Hyperactivity Disorder" *Experimental Brain Research* 154(1): 7684.

Bryson, Carolyn Q. 1970. "Systematic Identification of Perceptual Disabilities in Autistic Children" *Perception and Motor Skills* 31: 239–46.

Bryson, Carolyn Q. 1972. "Short-Term Memory and Crossmodal Information Processing in Autistic Children" *Journal of Learning Disabilities* 5: 81–91.

Buhusi, Catalin V., and Warren H. Meck. 2005. "What Makes Us Tick? Functional and Neural Mechanisms of Interval Timing" *Nature Reviews Neuroscience* 6(10): 755–65.

Calvert, Gemma A., Charles Spence, and Barry E. Stein. *The Handbook of Multisensory Processes.* Cambridge, MA: The MIT Press, 2004.

Campbell, Ruth et al. 1990. "Neuropsychological Studies of Auditory-Visual Fusion Illusions. Four Case Studies and their Implications" *Neuropsychologia* 28: 787–802.

Campbell, R. et al. 1997. "Speechreading in the Akinetopsic Patient (LM)" *Brain* 120: 1793–803.

Capa, Rémi L. et al. 2014. "Patients with Schizophrenia Selectively Impaired in Temporal Order Judgments" *Schizophrenia Research* 156: 51–55.

Cesaroni, L., and M. Garber. 1991. "Exploring the Experience of Autism through Firsthand Accounts" *Journal of Autism and Developmental Disorders* 21: 303–14.

Chan, Jason S., and Marcus J. Naumer. 2014. "Explaining Autism Spectrum Disorders: Central Coherence Versus Predictive Coding Theories" *Journal of Neurophysiology*.

Charidimou, Andreas et al. 2014. "Why is it Difficult to Predict Language Impairment and Outcome in Patients with Aphasia after Stroke?" *Journal of Clinical Neurology* 10(2): 75–83.

Collignon, Olivier et al. 2013. "Reduced Multisensory Sacilitation in Persons with Autism" *Cortex* 49: 1704–10.

de Boer-Schellekens, Liselotte, Mart Eussen, and Jean Vroomen. 2013. "Diminished Sensitivity of Audiovisual Temporal Order in Autism Spectrum Disorder" *Frontiers in Integrative Neuroscience* 7(8).

de Boer-Schellekens, Liselotte et al. 2014. "Sound Improves Diminished Visual Temporal Sensitivity in Schizophrenia" *Acta Psychologica* 147: 136–42.

de Gelder, Beatrice et al. 2003. "Audio-Visual Integration in Schizophrenia" *Schizophrenia Research* 59: 211–18.

Delevoye-Turrell, Yvonne et al. 2007. "Motor Fluency Deficits in the Sequencing of Actions in Schizophrenia" *Journal of Abnormal Psychology* 116: 56–64.

Doniger, Glen M. et al. 2002. "Impaired Visual Object Recognition and Dorsal/Ventral Stream Interaction in Schizophrenia" *Archives of General Psychiatry* 59(11): 1011–20.

Donohue, Sarah E., Elise F. Darling, and Stephen R. Mitroff. 2012. "Links between Multisensory Processing and Autism" *Experimental Brain Research* 222(4): 377–87.

Edwards, Allen E., and Richard Auger. 1965. "The effect of aphasia on the perception of precedence" *Proceedings of the Annual Convention of the American Psychological Association* 207–08.

Efron, Robert. 1963a. "Temporal Perception, Aphasia and déjà vu" *Brain* 86: 403–24.

Efron, Robert. 1963b. "The Effect of Handedness on the Perception of Simultaneity and Temporal Order" *Brain* 86: 261–84.

Efron, Robert. 1963c. "The Effects of Stimulus Intensity on the Perception of Simultaneity in Right- and Left-Handed Subjects" *Brain* 86: 285–94.

Efron, Robert. 1963d. "An Extension of the Pulfrich Stereoscopic Effect" *Brain* 86: 295–300.

Elliott, Mark A., Zhuanghua Shi, and Fatma Surer. 2007. "The Effects of Subthreshold Synchrony on the Perception of Simultaneity" *Psychological Research* 71: 687–93.

Farmer, Mary E., and Raymond M. Klein. 1995. "The Evidence for a Temporal Processing Deficit Linked to Dyslexia: A Review" *Psychonomic Bulletin & Review* 2(4): 460–93.

Fiorio, Mirta et al. 2003. "Temporal Processing of Visuotactile and Tactile Stimuli in Writer's Cramp" *Annual Neurology* 53: 630–35.

Fiorio, Mirta et al. 2007. "Defective Temporal Discrimination of Passive Movements in Parkinson's Disease" *Neuroscience Letters* 417: 312–15.

Foss-Feig, Jennifer H. et al. 2010. "An Extended Multisensory Temporal Binding Window in Autism Spectrum Disorders" *Experimental Brain Research* 203: 381–9.

Foucher, J.R. et al. 2007. "Low Time Resolution in Schizophrenia: Lengthened Windows of Simultaneity for Visual, Auditory and Bimodal Stimuli" *Schizophrenia Research* 97: 118–27.

Foxe, John J. et al. 2013. "Severe Multisensory Speech Integration Deficits in High-Functioning School-Aged Children with Autism Spectrum Disorder (ASD) and Their Resolution during Early Adolescence" *Cerebral Cortex* 25(2): 298-312.

Freeman, Elliot D. et al. 2013. "Sight and Sound Out of Synch: Fragmentation and Renormalisation of Audiovisual Integration and Subjective Timing" *Cortex* 49(10): 2875–87.

Fuchs, Thomas. 2007. "The Temporal Structure of Intentionality and Its Disturbance in Schizophrenia" *Psychopathology* 40: 229–35.

García-Pérez, Miguel A., and Rocío Alcalá-Quintana. 2012. "On the Discrepant Results in Synchrony Judgment and Temporal-Order Judgment Tasks: A Quantitative Model" *Psychonomic Bulletin Review* 19, 820–46.

Giersch, Anne et al. 2009. "Extended Visual Simultaneity Thresholds in Patients with Schizophrenia" *Schizophrenia Bulletin* 35(4): 816–25.

Gómez, Jose et al. 2014. "Time Perception Networks and Cognition in Schizophrenia: A Review and a Proposal" *Psychiatry Research.*

Graziano, Michael S.A., and Charles G. Gross. 1993. "A Bimodal Map of Space: Somatosensory Receptive Fields in the Macaque Putamen with Corresponding Visual Receptive Fields" *Experimental Brain Research* 97: 96–109.

Grossman, Ruth B., Matthew H. Schneps, and Helen Tager-Flusberg. 2009. "Slipped Lips: Onset Asynchrony Detection of Auditory-Visual Language in Autism" *Journal of Child Psychology and Psychiatry* 50(4): 491–97.

Hairston, W. David et al. 2005. "Altered Temporal Profile of Visual-Auditory Multisensory Interactions in Dyslexia" *Experimental Brain Research* 166: 474–80.

Hamilton, Roy H., Jeffrey T. Shenton, and H. Branch Coslett. 2006. "An Acquired Deficit of Audiovisual Speech Processing" *Brain and Language* 98: 66–73.

Harrar, Vanessa et al. "Multisensory Integration and Attention in Developmental Dyslexia" *Current Biology* 24(5): 531–35.

Haviland, Jeannette M. et al.. 1996. "Intermodal Perception of Emotional Expressions by Children with Autism" *Journal of Developmental and Physical Disabilities* 8: 77–88.

Hodge, Steven M. et al. 2010. "Cerebellum, Language, and Cognition in Autism and Specific Language Impairment" *Journal of Autism and Developmental Disorders* 40(3): 300–16.

Holmes, Helen L. 1965. *Disordered Perception of Auditory Sequences in Aphasia.* Doctoral thesis, Harvard.

Hood, Michelle, and Elizabeth Conlon. 2004. "Visual and Auditory Temporal Processing and Early Reading Development" *Dyslexia* 10: 1–19.

Husserl, E. *Zur Phänomenologie des inneren Zeitbewußtseins (1983–1917)*. Ed. R. Boehm. The Hague: Martinus Nijhoff, 1966; On the Phenomenology of the Consciousness of Internal Time (1983–1917). Trans. J. Brough. Dordrecht: Kluwer Academic Publishers, 1991.

Iarocci, Grace, and John McDonald. 2005. "Sensory Integration and the Perceptual Experience of Persons with Autism" *Journal of Autism and Developmental Disorders* 36(1): 77–90.

Jerger, J., N.J. Weikers, F.W. Sharbrough, and S. Jerger. 1969. "Bilateral Lesions of the Temporal Lobe. A Case Study." *Acta Otolaryngologica Supplementum* 258: 1–51.

King, Andrew J. 2005. "Multisensory Integration: Strategies for Synchronization" *Current Biology* 15: R339–41.

King, Bernardine, Clare Wood, and Dorothy Faulkner. 2007. "Sensitivity to Visual and Auditory Stimuli in Children with Developmental Dyslexia" *Dyslexia* 14 (2): 116–41.

Kolko, David J., Lowell Anderson, and Magda Campbell. 1980. "Sensory Preference and Overselective Responding in Autistic Children" *Journal of Autism and Developmental Disorders* 10: 259–71.

Koller, William C. 1984. "Sensory Symptoms in Parkinson's Disease" *Neurology* 34(7): 957.

Kronschnabel, Jens et al. 2014. "The Level of Audiovisual Print-Speech Integration Deficits in Dyslexia" *Neuropsychologia* 62: 245–61.

Kwakye, Leslie D. et al. 2011. "Altered Auditory and Multisensory Temporal Processing in Autism Spectrum Disorders" *Frontiers in Integrative Neuroscience* 4(129).

Laasonen, Marja et al. 2000. "Rate of Information Segregation in Developmentally Dyslexic Children" *Brain* 75: 66–81.

Laasonen, Marja, Elisabet Service, and Veijo Virsu. 2001. "Temporal Order and Processing Acuity of Visual, Auditory, and Tactile Perception in Developmentally Dyslexic Young Adults" *Cognitive, Affective, & Behavioral Neuroscience* 1(4): 394–410.

Laasonen, Marja, Elisabet Service, and Veijo Virsu. 2002. "Crossmodal Temporal Order and Processing Acuity in Developmentally Dyslexic Young Adults" *Brain and Language* 80: 340–54.

Lacruz, F. et al. 1991. "The Anatomical Basis of Somaesthetic Temporal Discrimination in Humans" *Journal of Neurology, Neurosurgery, and Psychiatry* 54: 1077–81.

Lalanne, Laurence, Mitsouko van Assche, and Anne Giersch. 2012a. "When Predictive Mechanisms Go Wrong: Disordered Visual Synchrony Thresholds in Schizophrenia" *Schizophrenia Bulletin* 38(3): 506–13.

Lalanne, Laurence et al. 2012b. "Looking Forward: An Impaired Ability in Patients with Schizophrenia?" *Neuropsychologia* 50: 2736–44.

Landerl, Karin, and Edith Willburger. 2010. "Temporal Processing, Attention, and Learning Disorders" *Learning and Individual Differences* 20: 393–401.

Lovaas, Ivar O., and Laura Schreibman. 1971. "Stimulus Overselectivity of Autistic Children in a Two Stimulus Situation" *Behavioral Research and Therapy* 9: 305–10.

Lovaas, Ivar O. et al. 1971. "Selective Responding by Autistic Children to Multiple Sensory Input" *Journal of Abnormal Psychology* 77(3): 211–22.

Loveland, Katherine A. et al. 1995. "Intermodal Perception of Affect in Persons with Autism or Down Syndrome" *Developmental Psychopathology* 7: 409–18.

Lucas, Marina et al. 2013. "Time Perception Impairs Sensory-Motor Integration in Parkinson's Disease" *International Archives of Medicine* 6(39).

MacDonald, John, and Harry McGurk. 1978. "Visual Influences on Speech Perception Process" *Perception & Psychophysics* 24: 253–57

Magnee, Maurice J.C.M. et al. 2008. "Audiovisual Speech Integration in Pervasive Developmental Disorder: Evidence from Event-Related Potentials" *Journal of Child Psychology Psychiatry* 49(9): 995–1000.

Martin, Brice et al. 2013. "Temporal Event Structure and Timing in Schizophrenia: Preserved Binding in a Longer 'Now'" *Neuropsychologia* 51: 358–71.

Meck, Warren H. 1996. "Neuropharmacology of Timing and Time Perception" *Brain Research. Cognitive Brain Research* 3(3–4): 227–42.

Melges, Frederick T. 1982. *Time and the Inner Future: A Temporal Approach to Psychiatric Disorders.* New York: Wiley & Sons.

Minkowski, Eugene. 1933. *Lived Time: Phenomenological and Psychopathological Studies.* Transl. by Nancy Metzel, Northwestern University Press, Evanston. 1970.

Mongillo, Elizabeth A. et al. 2008. "Audiovisual Processing in Children with and without Autism Spectrum Disorders" *Journal of Autism and Developmental Disorders* 38: 1349–58.

Morein-Zamir, Sharon, Salvador Soto-Faraco, and Alan Kingstone. 2003. "Auditory Capture of Vision: Examining Temporal Ventriloquism" *Cognitive Brain Research* 17: 154–63.

New Scientist. 2013. *Mindscapes: First Man to Hear People Before they Speak.* Coverage of Freeman, Ipser et al. (2013) by Helen Thomson. *New Scientist*, July 4, 2013.

Nicholls, Michael E.R. 1996. "Temporal Processing Asymmetries between the Cerebral Hemispheres: Evidence and Implications" *Laterality: Asymmetries of Body, Brain and Cognition* 1(2): 97–138.

O'Neill, Meena, and Robert S.P. Jones. 1997. "Sensory-Perceptual Abnormalities in Autism: A Case for More Research?" *Journal of Autism and Developmental Disorders* 27(3): 283–93.

Overton, Paul G. 2008. "Collicular Dysfunction in Attention Deficit Hyperactivity Disorder" *Medical Hypotheses* 70(6): 1121–27.

Panagiotidi, Maria, Tom Stafford, and Paul G. Overton. 2014. "Multisensory Integration Abnormalities and ADHD: Evidence from a Study on a Subclinical Population" Poster Presented at the 15th International Multisensory Research Forum, Amsterdam, Netherlands.

Parker, Krystal et al. 2013. "Executive Dysfunction in Parkinson's Disease and Timing Deficits" *Frontiers in Integrative Neuroscience* 7(75).

Pearl, Doron et al. 2009. "Differences in Audiovisual Integration, as Measured by McGurk Phenomenon, among Adult and Adolescent Patients with Schizophrenia and Age-Matched Healthy Control Groups" *Comprehensive Psychiatry* 50: 186–92.

Pekkola, Johanna et al. 2006. "Perception of Matching and Conflicting Audiovisual Speech in Dyslexic and Fluent Readers: An fMRI Study at 3 T" *NeuroImage* 29: 797–807.

Pons, Ferran et al. 2012. "Perception of Audio-Visual Speech Synchrony in Spanish-Speaking Children with and without Specific Language Impairment" *Journal of Child Language* 40(3): 687–700.

Pöppel, Ernst. 1985. *Grenzen des bewußtseins: Uber wirklichkeit und welterfahrung.* Stuttgart: Deutsche Verlags-Anstalt GmbH.

Pöppel, Ernst. 2004. "Lost in Time: A Historical Frame, Elementary Processing Units and the 3-Second Window" *Acta Neurobiologia Experimentalis* 64: 295–301.

Powers, Albert R. III, Andrea R. Hillock, and Mark T. Wallace. 2009. "Perceptual Training Narrows the Temporal Window of Multisensory Binding" *Journal of Neuroscience* 29(39): 12265–74.

Powers, Albert R. III, Matthew A. Hevey, and Mark T. Wallace. 2012. "Neural Correlates of Multisensory Perceptual Learning" *Journal of Neuroscience* 32(18): 6263–74.

Protopapas, A. 2014. "From Temporal Processing to Developmental Language Disorders: Mind the Gap" *Philosophical Transactions of the Royal Society B* 369: 20130090.

Rammsayer, Thomas, and Wilhelm Classen. 1997. "Impaired Temporal Discrimination in Parkinson's Disease: Temporal Processing of Brief Durations as an Indicator of Degeneration of Dopaminergic Neurons in the Basal Ganglia" *International Journal of Neuroscience* 91: 45–55.

Rogers, Sally J., Susan Hepburn, and Elizabeth Wehner. 2003. "Parent Reports of Sensory Symptoms in Toddlers with Autism and Those with Other Developmental Disorders" *Journal of Autism & Developmental Disorders* 33: 631–42.

Rorden, Chris et al. 1997. "Visual Extinction and Prior Entry: Impaired Perception of Temporal Order with Intact Motion Perception after Unilateral Parietal Damage" *Neuropsychologia*, 35: 421–33.

Ross, Lars A. et al. 2007. "Impaired Multisensory Processing in Schizophrenia: Deficits in the Visual Enhancement of Speech Comprehension under Noisy Environmental Conditions" *Schizophrenia Research* 97: 173–83.

Russo, N. et al. 2012. "Parameters of Semantic Multisensory Integration Depend on Timing and Modality Order among People on the Autism Spectrum: Evidence from Event-Related Potentials" *Neuropsychologia* 50: 2131–41.

Saccuzzo, Dennis P., and Donald L. Schubert. 1981. "Backward Masking as a Measure of Slow Processing in Schizophrenia Spectrum Disorders" *Journal of Abnormal Psychology* 90(4): 305–12.

Schlesinger, Joseph J. et al. 2014. "Improving Pulse Oximetry Pitch Perception with Multisensory Perceptual Training" *Anesthesia & Analgesia* 118(6): 1249–53.

Schmidt, Heike et al. 2011. "Low-Level Temporal Coding Impairments in Psychosis: Preliminary Findings and Recommendations for Further Studies" *Journal of Abnormal Psychology* 120(2): 476–82.

Schwartz, Barry D., David B. Mallott, and Daniel K. Winstead. 1988. "Preattentive Deficit in Temporal Processing by Chronic Schizophrenics" *Biological Psychiatry* 23: 664–69.

Seri, Yifat, Ora Kofmar, and Lidor Shay. 2002. "Time Estimation Could Be Impaired in Male, But Not Female Adults with Attention Deficits" *Brain and Cognition* 48(2–3): 553–58.

Shams, Ladan, Yukiyasu Kamitani, and Shinsuke Shimojo. 2000. "What You See is What You Hear" *Nature* 408: 788.

Sherwin, Ira, and Robert Efron. 1980. "Temporal Ordering Deficits Following Anterior Temporal Lobectomy" *Brain and Language* 11: 195–203.

Simon, J. Richard, and James D. Wolf. 1963. "Choice Reaction Times as a Function of Angular Stimulus–Response Correspondence and Age" *Ergonomics* 6: 99–105.

Smith, Anna et al. 2002. "Evidence for a Pure Time Perception Deficit in Children with ADHD" *Journal of Child Psychology and Psychiatry* 43(4): 529–42.

Snider, Stuart R. et al. 1976. "Primary Sensory Symptoms in Parkinsonism" *Neurology* 26: 423–29.

Spence, Charles, and Sarah Squire. 2003. "Multisensory Integration: Maintaining the Perception of Synchrony" *Current Biology* 13(13): R519–21.

Stein, Barry E., and Terrence R. Stanford. 2008. "Multisensory Integration: Current Issues from the Perspective of the Single Neuron" *Nature Reviews Neuroscience* 9: 255–66.

Stevenson, Ryan A. et al. 2010. "Neural Processing of Asynchronous Audiovisual Speech Perception" *Neuroimage* 49: 3308–18.

Stevenson, Ryan A. et al. 2011. "Discrete Neural Substrates Underlie Complementary Audiovisual Speech Integration Processes" *NeuroImage* 55: 1339–45.

Stevenson, Ryan A. et al. 2014. "Multisensory Temporal Integration in Autism Spectrum Disorders" *Journal of Neuroscience* 34(3): 691–97.

Surguladze, Simon A. et al. 2001. "Audio-Visual Speech Perception in Schizophrenia: An fMRI Study" *Psychiatry Research: Neuroimaging* 106: 1–14.

Swisher, Linda, and Ira J. Hirsh. 1972. "Brain Damage and the Ordering of Two Temporally Successive Stimuli" *Neuropsychologia* 10: 137–52.

Szelag, Elzbieta et al. 2004. "Temporal Processing Deficits in High-Functioning Children with Autism" *British Journal of Psychology* 95: 269–82.

Talcott, Joel B. et al. 2002. "On the Relationship between Dynamic Visual and Auditory Processing and Literacy Skills: Results from a Large Primary-School Study" *Dyslexia* 8: 204–25.

Talsma, Durk et al. 2010. "The Multifaceted Interplay Between Attention and Multisensory Integration" *Trends in Cognitive Science* 14(9): 400–10.

Tinazzi, Michele et al. 2004. "Timing of Tactile and Visuo-Tactile Events is Impaired in Patients with Cervical Dystonia" *Journal of Neurology* 251: 85–90.

van der Smagt, Maarten J., Herman van Engeland, and Chantal Kemner. 2007. "Brief Report: Can You See What is Not There? Low-Level Auditory-Visual Integration in Autism Spectrum Disorder" *Journal of Autism and Developmental Disorders* 37: 2014–19.

van Wassenhove, Virginie, Ken W. Grant, and David Poeppel. 2007. "Temporal Window of Integration in Auditory-Visual Speech Perception" *Neuropsychologia* 45: 598–607.

Vatakis, Argiro. 2008a. "Examining the Possibility of an Acquired Deficit in Audiovisual Temporal Perception for Speech and Musical Events" *Annals of General Psychiatry* 7(1): S137.

Vatakis, Argiro et al. 2008b. "Audiovisual Temporal Adaptation of Speech: Temporal order Versus Simultaneity Judgments" *Experimental Brain Research* 185, 521–29.

Vatakis, Argiro. 2013. "The Role of Stimulus Properties and Cognitive Processes in the Quality of the Multisensory Perception of Synchrony" In L. Albertazzi (Ed.), *The Wiley-Blackwell Handbook of Experimental Phenomenology.* Subtitle: Visual Perception of Shape, Space and Appearance. Wiley-Blackwell.

Vatakis, Argiro, and Charles Spence. 2006. "Audiovisual Synchrony Perception for Music, Speech, and Object Actions" *Brain Research* 1111: 134–42.

Vatakis, Argiro, and Charles Spence. 2007. "Crossmodal Binding: Evaluating the 'Unity Assumption' Using Audiovisual Speech Stimuli" *Perception & Psychophysics* 69(5): 744–56.

Vatakis, Argiro, and Charles Spence. 2010. "Audiovisual Temporal Integration for Complex Speech, Object-Action, Animal Call, and Musical Stimuli" In M.J. Naumer & J. Kaiser (Eds.), *Multisensory Object Perception in the Primate Brain.* Springer-Verlag: Berlin Heidelberg.

Virsu, Veijo, Pekka Lahti-Nuuttila, and Marja Laasonen. 2003. "Crossmodal Temporal Processing Acuity Impairment Aggravates with Age in Developmental Dyslexia" *Neuroscience Letters* 336: 151–54.

von Steinbüchel, Nicole, and Marc Wittmann. 1997. "Temporal Order Threshold in Patients with Brain Injury" *International Journal of Psychophysiology* 25: 80–81.

von Steinbüchel, Nicole, Marc Wittmann, Hans Strasburger, and Elzbieta Szelag. 1999. "Auditory Temporal-Order Judgement is Impaired in Patients with Cortical Lesions in Posterior Regions of the Left Hemisphere" *Neuroscience Letters* 264(1–3): 168–71.

Vroomen, Jean, and Mirjam Keetels. 2010. "Perception of Intersensory Synchrony: A Tutorial Review" *Attention, Perception, & Psychophysics* 72(4): 871–84.

Tononi, Giulio, and Gerald M. Edelman. 2000. "Schizophrenia and the Mechanisms of Conscious Integration" *Brain Research Reviews* 31(2–3): 391–400.

Wada, Makoto et al. 2014. "Spatio-Temporal Processing of Tactile Stimuli in Autistic Children" *Scientific Reports* 4: 5985.
Walker-Andrews, A.S. et al. 1994. "Brief Report: Preferential Looking in Intermodal Perception by Children with Autism" *Journal of Autism and Developmental Disorders* 24: 99–107.
Waterhouse, Lynn, Deborah Fein, and Charlotte Modahl. 1996. "Neurofunctional Mechanisms in Autism" *Psychological Review* 103: 457–89.
Welch, Robert B., and David H. Warren. 1980. "Immediate Perceptual Response to Intersensory Discrepancy" *Psychological Bulletin* 88: 638–67.
Wenke, Dorit, and Patrick Haggard. 2009. "How Voluntary Actions Modulate Time Perception" *Experimental Brain Research* 196: 311–18.
White, Thomas P. et al. 2014. "Eluding the Illusion? Schizophrenia, Dopamine and the McGurk Effect" *Frontiers in Human Neuroscience* 8: 565.
Widnell, Katherine. 2005. "Pathophysiology of Motor Fluctuations in Parkinson's Disease" *Movement Disorders* 20: S17–22.
Williams, Emily L., and Manuel F. Casanova. 2010. "Autism and Dyslexia: A Spectrum of Cognitive Styles as Defined by Minicolumnar Morphometry" *Medical Hypotheses* 74(1): 59–62.
Williams, Lisa E. et al. 2010. "Reduced Multisensory Integration in Patients with Schizophrenia on a Target Detection Task" *Neuropsychologia* 48: 3128–36.
Wittmann, Marc. 2011. "Moments in Time" *Frontiers of Integrative Neuroscience* 5, 66.
Wittmann, Marc et al. 2004. "Effects of Brain-Lesion Size and Location on Temporal-Order Judgment" *NeuroReport* 15(15): 2401–05.
Wittmann, Marc et al. 2011. "Neural Substrates of Time Perception and Impulsivity" *Brain Research* 1406: 43–58.
Woynaroski, Tiffany G. et al. 2013. "Multisensory Speech Perception in Children with Autism Spectrum Disorders" *Journal of Autism and Developmental Disorders* 43: 2891–902.

CHAPTER 2

Abnormal Timing and Time Perception in Autism Spectrum Disorder? A Review of the Evidence

Melissa J. Allman and Christine M. Falter***

1 Introduction

Autism spectrum disorders (ASD) is a cluster term for impairment in areas such as communication, social interaction, and imagination, and restricted and repetitive behaviors. Affected individuals, who grow up with this disorder, appear to perceive the world in profoundly different ways, and this may ultimately underlie the manifestation of diagnostic behavioral phenotypic characteristics (as for other forms of psychological disorder; e.g., schizophrenia, ADHD, other developmental disorders). The question as to whether some of these differences are attributable to pathophysiological differences in how the brain keeps time has recently become a focus of autism research. So too has the nature of the relationship between timing and time processing and other cognitive abilities that are compromised in ASD (e.g., perception of temporal structure and social reciprocity, such as to and fro- in social exchanges; see Chapter 3 of this book). As evidenced in this book, the interest in the potential correspondence of distortions in psychological time with diagnostic features of certain disorders is burgeoning in clinical populations often co-morbid to ASD including ADHD (Barkley et al., 1997; Noreika, Falter, and Rubia, 2013; Rubia et al., 1999; Toplack, Dockstader, and Tannock, 2006; Yang et al., 2007) and schizophrenia (Penney et al., 2005; see also Allman et al., 2014), however relatively little is known about timing abilities in individuals with ASD. This is despite the fact that it is widely acknowledged that these individuals have a poor sense of time at the intuitive level *"to a degree that is markedly discrepant with their level of intelligence"* (Wing, 1996: 89). There are at least three different ways in which time processing may be related to cognitive development, including a direct causal relationship or an associative interaction with other functional deficits (Falter and Noreika, 2011). The mechanisms by which a primary deficit of interval timing might contribute to diagnostic impairments

* Department of Psychology, Michigan State University, East Lansing, MI, USA.
** Department of Psychiatry, University Hospital Cologne, Cologne, Germany.

 | DOI 10.1163/9789004230699_003

have been speculated on in a temporal deficit hypothesis of autism (Allman, 2011; Allman and DeLeon, 2009). By this form of account, a fundamental building block of timing ability is required for cognitive development and psychophysiological differences in timing ability may account for diagnostic features of the disorder (see also Allman et al., 2012; Boucher, 2001).

The goal of this chapter is to outline the current literature on the state-of-the-field of autism timing research. The research on timing in ASD is like a jigsaw puzzle with many pieces missing, and there are clusters of findings related to islets of timing ability. Typically any findings of interest are confounded by other studies showing no effect. We begin by providing a general discussion of the relevance of timing and time processing to improving our understanding of ASD, which includes mention of certain studies (not included in Table 2.1). We then summarize research studies examining time processing in this population, which we have chosen to organize by duration range, and conclude by couching these findings in the "so what" for autism research.

2 Background

Autism is a spectrum developmental disorder, meaning that diagnosis of the condition falls along a *spectrum* of impairment/degree of severity. Thus, two children who are both diagnosed with ASD may present with behavioral symptoms that are quite different to the extent you might consider that they had different disorders; for example, one individual may be non-verbal; another may talk constantly about their fascination with trains, another may be an undergraduate or graduate student at a university. Of course, comparing the mental capabilities of such a broad range of individuals, who although they share the same diagnosis may nevertheless *appear* very different, is a challenge to psychological theories underlying this complex disorder. Individuals with ASD tend to present with differences in auditory and tactile perception. Sensory problems are based on how the brain processes incoming sensory input from multiple modalities across each of the five senses (vision, hearing, tactile, smell, and taste) that is likely intimately connected with the brain's sense of time, and ability to perceive the temporal structure of and between events.

It would be fair to say there are currently three dominant neuropsychological theories of ASD: First, the notion that there is under-connectivity between long-distance cortical regions and over-connectivity within regions (e.g., Frith, 2004). According to this way of thinking, affected individuals become overly concerned with local-level features at sometimes reduced processing of global-level events (i.e., weak central coherence; Happe and Frith, 2006). A second

TABLE 2.1 *Details and results of the studies on timing studies in ASD and healthy controls.*

Study	Paradigm	Modality	N (ASD/Con)	Mean Age ASD/Con	% Males ADHD/Con	Mean IQ ASD/Con	% Diagnosis (% comorbidities)	Significant group differences[1]
Mostofsky et al., 2000	Unfilled interval discrimination thresholds (550ms)	aud	11/17	13.3/12.5	55/35	FIQ 101/105	100 Autism	n.s.
Szelag et al., 2004	Duration reproduction (1000–5500ms)	vis aud	7/7	12.6/n.r.	57/57	PIQ 82–102/95–145	100 Autism	– Overestimation > 3500ms – Underestimation < 3500ms – CV
Gowen and Miall, 2005	Synchronized button pressing with beeps (400–800ms)	aud	12/12	27.4/28.1	67/67	FIQ 104/112	100 AS	– ADS – SD
	Continued button pressing after beeps (400–800ms)							SD

1 Only conditions and group comparisons yielding significant differences are reported in this column. If not further specified then all conditions yielded a significant group difference favouring Con.

TABLE 2.1 *Details and results of the studies on timing studies in asd and healthy controls* (cont.)

Study	Paradigm	Modality	N (ASD/ Con)	Mean Age ASD/Con	% Males ADHD/ Con	Mean IQ ASD/ Con	% Diagnosis (% comorbidities)	Significant group differences[1]
Lepisto et al., 2005	Duration MMN (100/33ms)	aud	15/15	9.4/9.4	87/87	PIQ 95/115 VIQ 59/107	100 Autism (13 ADHD)	MMN amplitude
Bebko et al., 2006	Aud-vis asynchrony detection in non-linguistic stimuli	aud-vis	16/16 15 DD	5.4/2.4 DD 4.9	n.r.	FA 2.3/n.r. DD 2.1	n.r. Autism or PPD	n.s.
	Aud-vis asynchrony detection in linguistic stimuli							Preferential looking (ASD < DD)
Lepisto et al., 2006	Duration MMN (100/33ms)	aud	10/10	8.1/8.1	80/80	PIQ 112/114 VIQ 108/107	100 AS (20 ADHD)	MMN amplitude
Boucher et al., 2007	Diachronic thinking test	–	23/13 LD + 10Con	12.7/12.3	83/78	RPM 29/27 BPVS 90/88	n.r. Autism or AS	– Diachronic thinking tendency – Temporal transformation – Temporal synthesis

			15/15 (LD)	14.3/14.7	80/87	RPM 26/24 BPVS 83/90	n.r. Autism or AS	– Diachronic thinking tendency – Temporal transformation – Temporal synthesis
Wallace and Happe, 2008	Duration estimation (2–45 s)	–	25/25	14.1/13.8	100/92	FIQ 96/100	n.r. Autism or AS	n.s.
	Duration reproduction (2–45 s)							ACC[2]
	Duration production (2–45 s)							n.s.
Jones et al., 2009	Duration discrimination thresholds (640ms)	aud	72/26 22 SEN	15.6/15.6	92/96	FIQ 88/89 PIQ 93/93 VIQ 84/87	54 Autism; 4 Atypical Autism; 39 other PDD; 3 PDD unspecified	n.s.
Foss-Feig et al., 2010	Flash-beep illusion	aud-vis	29/17	12.6/12.1	59/82	FIQ 108/107 PIQ 110/103 VIQ 105/109	n.r. Autism, ASD	Higher probability of illusion within wider range of SOAs

2 Superior performance in the ASD group.

TABLE 2.1 *Details and results of the studies on timing studies in asd and healthy controls* (cont.)

Study	Paradigm	Modality	N (ASD/Con)	Mean Age ASD/Con	% Males ADHD/Con	Mean IQ ASD/Con	% Diagnosis (% comorbidities)	Significant group differences[1]
Martin et al., 2010	Duration reproduction (0.5–4.1 s)	aud	20/20	36/35	75/65	FIQ 106/108 PIQ 105/106 VIQ 107/108	100 ASD	– ADS – Overestimation > 2300ms – Underestimation < 2300ms – CV
Allman et al., 2011	Temporal bisection task (1/4 s; 2/8 s)	vis	13/12	10.3/10.3	100/75	VIQ 100/n.r.	100 Autism	– ACC (1/4 s) – shorter BP – WR (2/8 s)
Kwakye et al., 2011	Temporal order judgment (0–500ms)	vis aud aud-vis	35/27	12.2/11.7	86/81	FIQ 103/110	n.r. Autism, AS, PDD unspecified	Temporal order thresholds (aud, aud-vis)
Maister and Plaisted, 2011	Duration reproduction (0.5–45 s)	vis	21/21	11.3/10.7	95/62	RPM 40/38 BPVS 102/111	100 Autism	– PDS (0.5, 1, 2, 45 s) – Intraindividual variability
			15/15	11.8/11.2	100/73	RPM 40/40 BPVS 106/116	100 Autism	PDS (0.5, 45 s)

Falter et al., 2012a	Simultaneity Judgments (0–100ms)	vis	16/16	24.7/26.7	94/88	FIQ 114/112 PIQ 114/114 VIQ 111/108	25 Autism; 75 AS	Simultaneity thresholds[2]
Falter et al., 2012b	Temporal Generalisation (300–1500ms)	vis aud aud-vis vis-aud	18/19	25.3/26.1	94/79	FIQ 112/113 PIQ 112/114 VIQ 110/110	22 Autism; 78 AS	– *d*'scores (ASD less sensitivity) – *c* scores (ASD more conservative responding)

Abbreviations: ACC = accuracy; ADS = absolute discrepancy scores; AS = Asperger Syndrome; aud = auditory; BP = Bisection Point; BPVS = British Picture Vocabulary Scale; Con = Control (neurotypical if not otherwise specified); CV = Coefficient of variation; DD = (individuals with) developmental disabilities; FA = functional age; FIQ = Full Scale Intelligence Quotient; LD = Learning Disability; MMN = Mismatch Negativity (i.e. an event-related potential associated with deviant stimulus processing in series of standard stimuli); MRT = mean reaction time; n.r. = not reported; n.s. = not significant; PDD = Pervasive Developmental Disorder; PDS = Proportional discrepancy scores (i.e. ADS/target duration); PIQ = Performance Intelligence Quotient; RPM = Raven's Standard Progressive Matrices; SD = standard deviation; SEN = special educational needs; SOA = Stimulus Onset Asynchrony; std. = standard; vis = visual; VIQ = Verbal Intelligence Quotient; WR = Weber Ratio (i.e. an index of sensitivity to durations).

focus is on pathophysiological differences in executive function (e.g., working memory, attention; see Russell, 1997) – these are abilities that are often related to some abstraction of time (event prediction, planning for events outside the present moment); and a third theory ascribing individuals with ASD reduced theory of mind skills (i.e., the ability to attribute mental states in order to understand behavior; see Baron-Cohen, Leslie, and Frith, 1985). It is beyond the scope of this chapter to discuss the different theories of ASD more fully, but see Belmonte et al. (2004).

Speculatively, a fourth neuropsychological theory of ASD could be formulated on the basis of anecdotal, clinical and behavioral observations of impairments in the *development* of timing and time perception in ASD (Allman, 2011; Boucher, 2001), although arguably, this is premature given the state of the field at this time. This is an attempt at relating autistic symptomology to posited timing differences, although the extent to which timing differences are found in other populations (i.e., ADHD) that present with very different symptoms is of interest (Falter and Noreika, 2011), and often speculations about which aspects of time processing are affected in ASD are untested. Boucher (2001) considered, "try to imagine periods of time longer than the lifetime of the universe...in fact, one cannot imagine a period of time longer than the lifetime of the universe except by thinking of a temporal succession of universes with cumulative lifetimes" (2001: 121). She suggests that there may be a close correspondence between the length (and complexity) of repeating behavioral units (e.g., stereotypies, rituals) and the ability to imagine extended time frames in ASD (mental concepts of time). As shorter and less complex stereotypies are usually observed with lower-functioning autistic individuals, and more complex, rigid routines are observed in those who are higher functioning. They may not be apparent at all in adults with high-functioning autism or Asperger syndrome (AS): it follows that a graded form of temporal insensitivity to duration might account for quantitative and qualitative differences in repetitive behaviors across the autistic spectrum (Boucher, 2001). A chapter by Allman and DeLeon (2009) extends this approach into formal models of interval timing, and provides a conceptual framework based upon indirect evidence linking posited differences in timing to diagnostic features of the disorder. For instance, repetitive, rhythmical behaviors may function to act as a form of behavioral clock, particularly if typical cognitive internal clock models are deficient (see Allman et al., 2014). "Stereotypies are typically produced in repeating cycles, and may be separated by (often short) intervals in time – continually measuring intervals in a repeating cycle requires less attentional resources (Lewis and Miall, 2003), and so repetitive motor behaviors may be a particularly effective time-parsing strategy for autistic individuals, and might function to concretize and reduce the stressor of an imposed disorientation in time. Peeters and Gillberg (1999: 87) report that "most people with autism feel lost in a sea of time…they will often try to develop routines and rituals by way of compensation. They want all activities to be undertaken in the same sequence every day…and if the sequence of activities changes on a certain day, then they have behavior problems" [which can include rhythmic lower-order motor behaviors, e.g., head-banging, self-injury]. To reiterate the main point here, an autistic impairment in the perception of duration may be compensated for by the production of repetitive motor behaviors, and an overreliance upon intact abilities, such as sequencing and order, and the stringing together of temporal units of perseveration or habits" (Allman and DeLeon, 2009: 70–1).

What these commentaries are attempting to do is associate various aspects of the autistic phenotype, in terms of deficits in language and communication, social interaction and structured imaginative play and behavior, to posited deficits in some underlying mechanism of 'keeping time'. But the nature of the interaction of timing and time processing to ASD and other developmental disabilities is unknown. "[Collectively] reports demonstrate complex associations between abnormally developing cognitive functions, and suggest that interval timing might play an important yet under-investigated role in developmental disorders by interacting with and modulating primary symptoms" (Falter and Noreika, 2011). Timing and time perception deficits likely play in a much larger symphony with other aspects of executive and cognitive functions in shaping the developing brain. The extent or "clout" that timing and time processing deficits play in characterizing not only the ASD phenotype but also other developmental and childhood disorders is currently uncertain.

There is neurophysiological evidence that the autistic brain has different absolute temporal windows for sensory integration and 'temporal binding' (Falter, Elliott, and Bailey, 2012; Falter et al., 2013; Foss-Feig et al., 2010; Rippon et al., 2007); and there are reports of abnormalities in the pre-attentive mismatch negativity (MMN) event-related potential to changes in duration (but not pitch) in speech sounds in individuals with ASD (Lepistö et al., 2005, 2006).

An MEG study reports enhanced temporal resolution for visual stimuli and reveals superior temporal event structure coding in individuals with ASD (Falter et al., 2013). These findings relate to short temporal intervals of ~57ms which are posited to correspond to units of subjective time (Brecher, 1932). An unpublished fMRI of supra-second duration visual ordinality comparison (by the first author) reveals that while unaffected children recruit cortico-cerebellar systems for timing relatively short durations (~2 s) and cortio-striatal systems for timing longer (~8 s) durations, children with ASD recruit cortico-striatal systems when timing short durations (but not long; the opposite pattern). A possible explanation for the pattern of neural activity observed is that children with ASD may overly recruit a "beat-based" system when timing a stimulus. This may have relevance to explanations of ASD which posit a behavioral clock time parsing mechanism in which repetitive and highly predictable events occur that help to ameliorate anecdotal and clinical reports of being "lost in a sea of time" (Allman and DeLeon, 2009; Boucher, 2001), and clinical evidence that external rhythmical/timing cues can improve the quality of social interactions in children with ASD (see Chapter 3 of this book).

For instance, Trevarthen and Daniel (2005) examined video recordings of monozygotic twin 11 month old infants, one of whom later went on to receive a diagnosis of ASD. They report differences in the affected sibling in respect to the ability for prospective awareness – being able to anticipate and expect future social and non-social events – the temporal patterning and form of parent-infant interactions differed between the twins, which the authors postulated to be due to "prospective control of movements and anticipations in awareness," related to changes in attention, motor tonus, emotion and initiative (2005: S25). Moreover, the discrimination of temporal synchrony between intermodal events (i.e., the sight and sound of the parent's speech) during parent/infant interactions "may be the first step in developing a capacity to discriminate more complex and specific forms of language" (Bebko et al., 2006: 96). These authors report that autistic children (aged 4–6 years) reveal atypical responding to multimodal temporal asynchrony with language-specific stimuli.

Higher visual temporal resolution in a perceptual simultaneity task was related to increased developmental communication difficulties in ASD (Falter, Elliott, and Bailey, 2012). At the same time, diagnostic communication scores were correlated with a more conservative response bias in a cross-modal interval timing task (Falter, Noreika, Wearden, and Bailey, 2013).

Additional indirect evidence for a timing mediated deficit on secondary symptoms is provided by Hill (2004) who examined aspects of executive function: planning, flexibility, and inhibition. Moreover, sleep difficulties and differences in circadian function and clock genes have also been noted in ASD (Richdale and Schreck, 2009; Nicholas et al., 2007; Wimpory, Nicholas, and Nash, 2002).

3 A Review of the Evidence

The nature of empirical studies of time processing in ASD is summarized in Table 2.1. These include studies examining millisecond and suprasecond timing, perceptual and performance tasks, using a variety of neurophysiological and psychological methods in a broad range of samples of individuals with

ASD. This chapter could have been presented in several different ways based upon the complex arrangement of this variation, and we have chosen to subdivide the review by duration range. Of course, within each duration range there are a variety of studies recruiting different durations and methods, and nature of autistic sample. We will summarize each study individually before presenting our own summary of the weight of the evidence and asking "so what?"

3.1 *Millisecond Range*

Mostofsky et al. (2000) examined procedural learning and duration discrimination in children (mean age 13 years) with high-functioning autism (IQ > 85). Their perceptual duration discrimination task involved the presentation of pairs of identical tones, 50ms in duration. At the beginning of a trial, the two tones were separated by 550ms; following this, a second pair of identical tones (50ms in duration) were separated by a variable interval, either shorter or longer than 550ms. Participants were required to make 'shorter' or 'longer' judgments on these empty intervals, which were entered into a computer by the experimenter. These authors report no significant difference between affected and non-affected individuals when discriminating between empty millisecond intervals, although the scores for those with ASD did tend to be higher (indicating poorer discrimination) for durations and lower (indicating better discrimination) for a pitch discrimination control task. This is in line with known sensory differences in ASD.

Another auditory perceptual task (Jones et al., 2009) used filled millisecond durations (tones) and found no group differences between duration (nor pitch or intensity) discrimination between adolescents with ASD (childhood autism and pervasive developmental disorder not otherwise specified) and non-affected controls. However, they did observe small subsets of the ASD sample that showed superior pitch discrimination, and inferior duration discrimination.

Gowen and Miall (2005) administered a battery of tasks designed to tap cerebellar function in adults with AS. Their timing tasks involved finger tapping, in a synchronization and continuation arrangement. The task included four identical tones, separated by identical intervals (that ranged from 400–800ms). The synchronization phase required button presses 'in time' with the third and forth tones, and in the continuation phase, the third and forth tones were omitted (button presses indicating the timing of the third and forth tones were still required). In the synchronization phase, they report that individuals with AS revealed greater absolute error, more variability and responded earlier than non-affected controls to the forth tone. In the continuation task, those with AS revealed the same pattern as in the synchronization task. The authors suggest that this may reflect "a fixed and nonadjustable mechanism of timing control" in affected individuals, and posit the collection of findings to differences in the integration of sensory signals with motor output in AS.

Foss-Feig et al. (2010) employed the flash-beep illusion, in which the presentation of a single light flash with multiple sounds results in the erroneous perception of multiple flashes. The breadth of the temporal window by which these multisensory stimuli are bound and produce the illusion was reported to be longer in those affected with ASD. In a related study, Kwakye et al. (2011) examined unisensory and multisensory temporal function on a temporal order judgment task by requiring children with ASD (8–17 years) to indicate which of two stimuli was presented first. Although there were no differences with respect to the visual modality, thresholds were higher for auditory stimuli, and multisensory processing was superior in children with ASD (see also Russo et al., 2012). However, in a separate study using visual stimuli in a perceptual simultaneity design, individuals with ASD judged two stimuli to be asynchronous at a smaller temporal difference between their onsets than non-affected controls, suggesting superior visual temporal resolution in ASD (Falter, Elliott, and Bailey, 2012). "Thus, increased temporal resolution might be a signature of the autistic cognitive profile differentiating it from other developmental and psychiatric disorders" (2012: 4). MEG signatures pointed towards enhanced access to early visual brain processes in ASD underlying superior visual temporal resolution (Falter et al., 2013). Speculatively, individuals with ASD might be parsing temporal events more than non-affected controls extending previous findings of increased spatial detail-focused perception, as proposed by Weak Central Coherence theory (Happe and Frith, 2006), to the temporal domain.

Hence, superior performance in aspects of time processing that are based on 'weak temporal coherence' seems to exist alongside impaired performance in other timing aspects, such as interval timing. Indeed, Falter et al. (2012) conducted a psychophysical study using a repeated standards version of a temporal generalization task with visual, auditory, and multimodal stimuli and found lower sensitivity to duration, particularly when the standard was an auditory stimulus. They employed two standard durations (600 and 1000ms) and found the scalar property to be preserved despite of impaired sensitivity to temporal intervals.

3.2 *Multisecond Range*

Szelag et al. (2004) reported that children with high-functioning autism (9–16 years) are severely impaired in their ability to reproduce both auditory and visual target durations (between 1–5 s). In this arrangement, a target stimulus duration was presented and after a 2 s interstimulus interval the stimulus would be presented again. Participants were required to press a button at the time of the judged target duration. Individuals with ASD tended to reproduce all durations as 3.0–3.5 s. A related study was conducted by Martin, Poirier, and Bowler (2010). They report an autistic tendency to make less accurate and more variable reproductions, and enhanced underestimation of durations over ~2.3 s. In a separate study, temporal reproduction of visual durations of 0.5–45 s was examined by Maister and Plaisted (2011) who employed an additional concurrent verbal task in an attempt to reduce the possibility of chronometric counting. These authors report difficulties by individuals with ASD with accurate reproductions of durations under 2 s and the longest duration (45 s), but not with durations around 4–30 s, in addition to increased overall variability in comparison to non-affected controls. The authors attributed the pattern of their findings to differences in attention (for short durations) and episodic memory processing (for the longest ones).

A temporal bisection perception procedure employing visual stimuli (Allman, DeLeon, and Wearden, 2011) reveals differences in time perception in children (aged 7–16 years) with ASD across two different duration ranges (1 to 4 s and 2 to 8 s). In the ASD sample, lower subjective points of equality were somewhat predictive of working memory and diagnostic language and communication impairment, and reduced temporal sensitivity for the longer set of durations was found. Parents of affected children also tended to rate their child's "sense of time" as poor. The scalar property was observed, although superimposition was generally not as well obtained for affected individuals. However in a separate bisection study (Gil et al., 2012) employing four pairs of duration ranges (two short: 0.5–1 s, 1.25–2.5 s; and two long: 3.12–6.25 s, 7.81–16.62 s) and auditory stimuli, no evidence of timing impairment in children of a similar age range (the majority of whom had AS) was found.

Wallace and Happe (2008) extended the range of durations tested with an ASD sample to 2–45 s on three different types of timing tasks: duration estimation (verbal numerical judgment), reproduction, and production, in which verbal "start" and "stop" cues were used to mimic more naturalistic arrangements. They report unimpaired time estimation and production, and more accurate reproduction (even to the longest duration of 45 s), with the possibility of increased task difficulty on shorter time intervals during reproduction.

3.3 *Mental Concepts of Time*

Boucher et al. (2007) report that children with ASD (7–16 years) are unable to i) think about past or future stages of current situation, ii) understand that things can change or evolve over time but are still the same thing, and iii) that successive events are part of a unitary process (see also Montangero, 1992). Zukauskas, Silton, and Assumpção et al. (2009) assessed temporality experience in individuals with AS qualitatively and found that past experiences dominated their sense of time while a “restricted mode of existence” (p. 104) was expressed with respect to present and future dimensions.

4 Summary

Surveying the spread of the extant findings just presented, it is clear there may be abnormal time processing in ASD – but evidence is relatively sparse, and there is much scientific diversity in the findings, including somewhat inconsistent empirical evidence that individuals with ASD display characteristic differences in aspects of time processing function. However a striking universal consistency amongst studies presented in this chapter is a tendency to observe increased variability in autistic time processing.

Here we present our main observations of the state-of-the findings as they relate to topics discussed in this chapter. These very general summary statements are not intended as definitive statements of timing ability in ASD, more a mere recapitulation of the ‘general ballpark’ of observed trends.

First, studies in the sub-second range have tended to find superior temporal discrimination differences in individuals with ASD. This may correspond to units of subjective time and sensory processing differences in ASD (Leekam et al., 2007).

Second, studies in the supra-second range have tended to find impairments in longer durations outside the bounds of the ‘psychological present’ (James, 1890), coupled with increased variability.

Third, there is initial evidence for problems with conceptual notions of time in ASD. These include empirical, clinical, and anecdotal reports of lack of a “time sense” (Boucher, 2001; Wing, 1996) to thinking forwards and backwards in time (Boucher, 2007).

The very fact that somewhat similar effects have been reported with a broad range of affected individuals suggests some fundamental significance or underlying tendency towards altered time processing. For instance, superior temporal visual acuity may be adaptive if other forms of time processing are impaired, or if other senses are differentially functional, as has been suggested in ASD. On the other hand, the apparent inconsistency between related studies that would otherwise be expected to provide a consistent pattern of results (e.g., Allman et al., 2011; Gil et al., 2012) is a challenge for the view that a core disability in ASD is to be found in time processing. Nevertheless, the study of time processing in ASD resembles that of other aspects of the autistic cognitive profile, such as visuo-spatial skills, which are also characterized by inconsistencies in findings (see recent meta-analysis by Muth, Hönekopp, and Falter, in press), but which are without doubt being considered a domain of altered processing in ASD.

5 The "So-what" for Autism Research

This chapter has outlined the findings from a diverse range of studies examining different aspects of time processing in ASD, in the hope that an understanding of the temporal scaffolding of the autistic mind might be elucidated. Given the discrepancy in stimulus duration, modality, and IQ between studies it is difficult to discern any clear picture at this time of the 'state' of autistic time processing. This will require considerably more research being done in this area, using a range of tasks and time scales in the same diagnostic samples, in addition to other developmental disorders.

It is not beyond the bounds of feasibility to see how a deficit in time processing might be relevant to the three dominant accounts of ASD (outlined earlier), and other new theories of ASD incorporate posited abnormalities in time processing in ASD (e.g., see Gepner and Feron, 2009; Welsh, Ahn, and Placantonakis, 2005). However, it is important to remember that the relationship between interval timing and higher mental representations of time is currently unspecified, and so any conceptualizations about time in the autistic mind should be made with caution. The association of interval timing abnormalities and other functional deficits is likely. "For instance, although a Theory of Mind deficit can hamper the understanding of social situations in its own right, an additional interval timing deficit could result in a lack of precise perception of temporal cues of eye gaze, and thereby increase misinterpretations of social situations. It has been proposed that different symptoms can be independent dimensions of impairment, which nevertheless interact with and modulate one another, leading to the characteristic phenomenology of an individual with a developmental disorder (Happe et al., 2006). In this line of thought, interval timing abnormalities might interact with primary dysfunctions" (Falter and Noreika, 2011: 1). It remains untested whether repetitive behaviors of the type outlined by Boucher (2001) and Allman (2011) are related to pathophysiological differences in timing and time perception. There is nevertheless much collective support for the notion that fundamental differences in time processing underlie aspects of the disorder (see also Allman and Meck, 2012).

Collectively the growing body of recent empirical findings reviewed in this chapter suggests there are quantifiable differences in functions of time processing in affected individuals, but this varies by duration range, possibly becoming superior for the millisecond range, probably tapping perception, and getting worse as duration increases into the bounds of secondary executive functions (working and episodic memory, sustained attention).

The extent of the influence of time processing differences to the manifestation of autistic qualities is unknown – although there may be links with communication (Allman et al., 2011; Falter, Elliott, and Bailey, 2012; Falter et al., 2013) – and is possibly mediated by its association with other cognitive abilities. For example, there are additional studies which we did not include into Table 2.1 but which are nonetheless worth mentioning: Zukauskas et al. (2009; AS only, no controls) using a semi-structured interview found individuals affected with ASD to have an abnormal perspective of themselves through time. Oram et al. (2005) studied rapid auditory processing – tasks that have been used to establish the 'temporal processing deficit' in dyslexia (which is somewhat different in nature than the timing mechanisms which are the focus of this chapter). Oram Cardy et al. suggested that impaired rapid temporal processing might rather be associated with language impairment than with ASD per se.

The extent to which any temporal deficit in ASD is fundamental to the disorder, and to other related disorders (ADHD, schizophrenia), and the role this may play in the ontology of the phenotype is clearly uncertain (Falter and Noreika, 2011). However, this is a fascinating and potentially fruitful question that would be informative to child development at a general level. The ultimate aim for future studies of ASD and time processing is to "fill the gaps" and try to sew the pieces together, and to examine the relevance of observed differences to clinical features of the disorder, and to temporal supports in their remediation. Additional studies in other childhood disorders are necessary to discern the selectivity of timing differences in autism and other developmental disabilities (see Falter and Noreika, 2011).

Perhaps the best lens by which to view time processing deficits in ASD is one that has more precise acuity (than is typical) with millisecond durations and increased variability (or blurriness) with longer durations, and with a 'big picture' view encompassing social and developmental aspects of timing; such as the apparent importance of timing to the development and maintenance of social interactions, and the perspective that different aspects of timing (sub-second, multi-second, and concepts of time) are component parts of a fundamental timing mechanism. Coupled with this is the view that interval timing is integral to other aspects of cognitive function (Lustig et al., 2005; Lustig and Meck, 2011).

As has been introduced in this chapter, there is an emerging conceptual "frame of reference" for understanding how a deficit in time processing might impact the autistic mind: speech and language difficulties; the importance of temporal information in social interactions; and the potential for a limited cognitive sense of time to influence higher-order cognitive functions (e.g., executive functions) and sensory perception. This is an exciting time for autism research in particular, and the study of time processing differences in other related populations in general.

References

Allman, M.J., B. Yin, and W.H. Meck. 2014. "Time in the Psychopathological Mind." In Dan Lloyd and Valtteri Arstila (Eds.), *Subjective time: An Anthology of the Philosophy, Psychology, and Neuroscience of Temporality*. Cambridge, MA: MIT Press.

Allman, M.J., and W.H. Meck. 2012. "Pathophysiological Distortions in Time Perception and Timed Performance" *Brain* 135: 656–77.

Allman, M.J., K.A. Pelphrey, and W.H. Meck. 2012. "Developmental Neuroscience of Time and Number: Implications for Autism and Other Neurodevelopmental Disabilities" *Frontiers in Integrative Neuroscience* 6: 7.

Allman, M.J., I.G. DeLeon, and J.H. Wearden. 2011. "A Psychophysical Assessment of Timing in Individuals with Autism" *American Journal of Intellectual and Developmental Disabilities* 116: 165–78.

Allman, M.J. 2011. "Deficits in Temporal Processing Associated with Autistic Disorder" *Frontiers in Integrative Neuroscience* 5: 2.

Allman, M.J., and I.G. DeLeon. 2009. "No Time Like the Present: Time Perception in Autism." In A.C. Giordano, and V.A. Lombardi (Eds.) *Causes and Risks for Autism* (pp. 65–76). New York: Nova Science Publishers.

Barkley, R.A. 1997. "Behavioral Inhibition, Sustained Attention, and Executive Functions: Constructing a Unifying Theory of ADHD" *Psychological Bulletin* 121(1): 65–94.

Baron-Cohen, S., A. Leslie, and U. Frith. 1985. "Does the Autistic Child Have a 'Theory of Mind'?" *Cognition* 21: 37–46.

Bebko, J.M. et al. 2006. "Discrimination of Temporal Synchrony in Intermodal Events by Children with Autism and Children with Developmental Disabilities without Autism" *Journal of Child Psychology and Psychiatry* 47(1): 88–98.

Belmonte, M.K. et al. 2004. "Autism as a Disorder of Neural Information Processing: Directions for Research and Targets for Therapy" *Molecular Psychiatry* 9: 646–63.

Boucher, J. 2001. "'Lost in a Sea of Time': Time-Parsing and Autism." In Hoerl C., McCormack T., (eds.) *Time and Memory*. Oxford: Clarendon. pp. 111–37.

Boucher, J. et al. 2007. "Temporal Cognition in Children in Autism Spectrum Disorders: Tests of Diachronic Thinking" *Journal of Autism and Developmental Disorders* 37: 1413–29.

Brecher, G.A. 1932. "Die Entstehung und biologische Bedeutung der subjektiven Zeiteinheit – des Momentes" *Zeitschrift für vergleichende Physiologie* 18: 204–43.

Falter, C.M. et al. 2013. "Enhanced Access to Early Visual Processing of Perceptual Simultaneity in Autism Spectrum Disorders" *Journal of Autism and Developmental Disorders* 43(8): 1857–66.

Falter, C.M., M.A. Elliott, and A.J. Bailey. 2012. "Enhanced Visual Temporal Resolution in Autism Spectrum Disorders" *PLoS One* 7(3): 1–6.

Falter, C.M., and V. Noreika. 2011. "Interval Timing Deficits and Abnormal Cognitive Development" *Frontiers in Integrative Neuroscience* 5: 26.

Foss-Feig, J.H. et al. 2010. "An Extended Multisensory Temporal Binding Window in Autism Spectrum Disorders" *Experimental Brain Research* 203: 381–9.

Frith, C. 2004. "Is Autism a Disconnection Disorder?" *The Lancet Neurology* 3: 577.

Gepner, B. and F. Feron. 2009. "Autism: A World Changing too Fast for a Mis-wired Brain? *Neuroscience and Biobehavioral Reviews* 33(8): 1227–42.

Gil, S. et al. 2012. "Children with Autism Have 'The Working Raw Material' for Time Perception" *PLoS One* 7(11).

Gowen, E., and R.C. Miall. 2005. "The Cerebellum and Motor Dysfunction in Neuropsychiatric Disorders" *The Cerebellum*, 6(3): 279–89.

Happe, F., and U. Frith. 2006. "The Weak Coherence Account: Detail-Focused Cognitive Style in Autism Spectrum Disorders" *Journal of Autism and Developmental Disorders* 36(1): 5–25.

Hill, E.L. 2004. "Evaluating the Theory of Executive Dysfunction in Autism" *Developmental Review* 24: 189–233.

James, W. 1890. *The Principles of Psychology*. Vol. 1 (pp. 605–42). New York: Holt.

Jones, C.R.G. et al. 2009. "Auditory Discrimination and Auditory Sensory Behaviours in Autism Spectrum Disorders" *Neuropsychologia* 47(13): 2850–8.

Kwakye, L.D. et al. 2011. "Altered Auditory and Multisensory Temporal Processing in Autism Spectrum Disorders" *Frontiers in Integrative Neuroscience* 4(129): 1–11.

Leekam, S.R. et al. 2007. "Describing the Sensory Abnormalities of Children and Adults with Autism" *Journal of Autism and Developmental Disorders* 37: 894–910.

Lepistö, T. et al. 2005. "The Discrimination of and Orienting to Speech and Non-Speech Sounds in Children with Autism" *Brain Research* 1066: 147–57.

Lepistö, T. et al. 2006. "Auditory Perception and Attention as Reflected by the Brain Event-Related Potentials in Children with Asperger Syndrome" *Clinical Neurophysiology* 117(10): 2161–71.

Lewis, P.A., and R.C. Miall. 2003. "Distinct Systems for Automatic and Cognitively Controlled Time Measurement: Evidence from Neuroimaging" *Current Opinion in Neurobiology* 13: 250–5.

Lustig, C., and W.H. Meck. 2011. "Modality Differences in Timing and Temporal Memory Throughout the Lifespan" *Brain and Cognition* 77(2): 298–303.

Lustig, C., M.S. Matell, and W.H. Meck. 2005. "Not "Just" a Coincidence: Frontal-Striatal Interactions in Working Memory and Interval Timing" *Memory* 13: 441–8.

Maister, L., and K.C. Plaisted-Grant. 2011. "Time Perception and Its Relationship to Memory in Autism Spectrum Disorders" *Developmental Science* 14(6): 1311–22.

Martin, J.S., M. Poirier, and D.M. Bowler. 2010. "Brief Report: Impaired Temporal Reproduction Performance in Adults with Autism Spectrum Disorder" *Journal of Autism and Developmental Disorders* 40(5): 640–6.

Montangero, J. 1992. The Development of a Diachronic Perspective in Children In F. Macar, V. Pouthas, and W.J. Friedman (Eds.) "*Time, Action and Cognition. Towards Bridging the Gap.*" (pp. 55–65). The Netherlands: Kluwer Academic Publishers.

Mostofsky, S.H. et al. 2000. "Evidence for a Deficit in Procedural Learning in Children and Adolescents with Autism: Implications for Cerebellar Contribution" *Journal of International Neuropsychological Society* 6(7): 752–9.

Muth, A., J. Hönekopp, and C.M. Falter. In press. "Visuo-Spatial Performance in Autism: A Meta-Analysis" *Journal of Autism and Developmental Disorders*.

Nicholas, B. et al. 2007. "Association of Per1 and Npas2 with Autistic Disorder: Support for the Clock Genes/Social Timing Hypothesis" *Molecular Psychiatry* 12(6): 581–92.

Noreika, V., C.M. Falter, and K. Rubia. 2013. "Timing Deficits in Attention-Deficit/ Hyperactivity Disorder (ADHD): Evidence from Neurocognitive and Neuroimaging Studies" *Neuropsychologia* 51: 235–66.

Oram Cardy, J.E. et al. 2005. "Magnetoencephalography Identifies Rapid Temporal Processing Deficit in Autism and Language Impairment" *Neuroreport* 16(4): 329–32.

Penney, T.B. et al. 2005. "Interval-Timing Deficits in Individuals at High Risk for Schizophrenia" *Brain and Cognition* 58: 109–18.

Peeters, T., and C. Gillberg. 1999. *Autism: Medical and Educational Aspects*. London: Whurr Publishers.

Richdale, A.L., and K.A. Schreck. 2009. "Sleep problems in Autism Spectrum Disorders: Prevalence, Nature, and Possible Aetiologies" *Sleep Medicine Reviews* 13(6): 403–11.

Rippon, G. et al. 2007. "Disordered Connectivity in the Autistic Brain: Challenges for the 'New Psychophysiology'" *International Journal of Psychophysiology* 63(2): 164–72.

Rubia, K. et al. 1999. "Synchronization, Anticipation and Consistency of Motor Timing in Dimensionally Defined Children with Attention Deficit Hyperactivity Disorder" *Perceptual and Motor Skills* 89(3): 1237–58.

Russell, J. 1997. *Autism as an Executive Disorder*. Oxford University Press, New York.

Russo, N. et al. 2012. "Parameters of Semantic Multisensory Integration Depend on Timing and Modality Order among People on the Autism Spectrum: Evidence from Event-Related Potentials" *Neuropsychologia* 50: 2131–41.

Szelag, E. et al. 2004. "Temporal Processing Deficits in High-Functioning Children with Autism" *British Journal of Psychology* 95(3): 269–82.

Trevarthen, C., and S. Daniel. 2005. "Disorganized Rhythm and Synchrony: Early Signs of Autism and Rett Syndrome" *Brain and Development* 27(1): 25–34.

Toplak, M.E., C. Dockstader, and R. Tannock. 2006. "Temporal Information Processing in ADHD: Findings to-Date and New Methods" *Journal of Neuroscience Methods* 15–29.

Wallace, G.L., and F. Happe. 2008. "Time Perception in Autism Spectrum Disorders" *Research in Autism Spectrum Disorders* 2(3): 447–55.

Welsh, J.P., E.S. Ahn and D.G. Placantonakis. 2005. "Is Autism Due to Brain Desynchronization?" *International Journal of Developmental Neuroscience* 23: 253–63.

Wing, L. 1996. *The Autistic Spectrum*. Constable, London.

Wimpory, D., B. Nicholas, and S. Nash. 2002. "Social Timing, Clock Genes and Autism: A New Hypothesis" *Journal of Intellectual Disability Research* 46(4): 352–8.

Yang, B. et al. 2007. "Time Perception Deficit in Children with ADHD" *Brain Research* 1170: 90–6.

Zukauskas, P.R., N. Silton, and F.B. Assumpcao. 2009. "Temporality and Asperger's Syndrome" *Journal of Phenomenological Psychology* 40(1): 85–106.

CHAPTER 3

A Social Timing Model of Autism, Informed by Typical Development

*Dawn Wimpory**

1 Introduction

The temporal structures of conversation are familiar examples of how timing is a critical element of human communication. Altering the temporal structures of conversation can have semantic implications and influence our sense of interrelatedness with the speaker. The anticipatory pause, for example, used on stage for comedic effect, uses timing to tease the audience, perhaps drawing out a double meaning that may have otherwise be missed. Similarly, in parent–infant play (in a game such as peek-a-boo) it is the temporal element that is the critical fun-factor in the game that delivers salience and amusement for the child.

Our media-filled social world also relies on timing in communication at both the macro- and micro-scale and deviation from the typical may result in failure of effective communication. Occasional audiovisual de-synchrony in broadcast and recorded media affords neurotypical individuals the experience of communication with unusual temporal constraints. Such de-synchronization increases cognitive workload and negatively impacts on feelings of interrelatedness.

Clinical experience of altered timing in the interactions of people with autism has led to hypotheses that focus on the role of timing in autism. To date, this field has been under-researched although there is general clinical consensus that forms of temporal impairment are common in autism. Whether, and to what degree, these timing deficits may be an intrinsic or causative element of the disorder remains to be determined. The present chapter will explore the notion of autism as a social timing disorder, first by considering timing phenomena in the context of typical child development and, subsequently, by reviewing what is known about altered timing in autism.

* Bangor University and Betsi Cadwaladr University Health Board, UK.

2 Social Timing in Typical Development

2.1 *Neonatal Preparedness for Engagement in Social Timing*

Feldman (2006, 2007b) outlines a biological propensity towards social rhythms. Allman, DeLeon, and Wearden (2011) observe how rhythmic changes in the seconds to minutes range often characterise biological systems and infant patterns of movement (for example, breathing, stereotypies, and babbling; Wolff, 1991). Neonatal activities, such as crying and sucking, typically comprise burst-pause rhythmic patterns (Wolff, 1991) that function and provide a structure for initial communication (Burke, 1977; Crook, 1979; Wolff, 1967). These are reflected in the burst-pause patterns of playful face-to-face interaction with young infants (Tronick, Als, and Brazelton, 1977) outlined in Section 2.4.2 later in this chapter.

2.2 *Typically Developing Infants' Perception of Duration, Rate, Rhythm*

At birth and even in the fetus, infants perceive time and have been shown to estimate durations of events (Decasper and Carstens, 1980; 1981; DeCasper and Fifer, 1980). Lewkowicz's (2000) review of infant perception of temporal information concluded that, within their first month, infants can discriminate unimodal auditory duration changes of 20ms as well as distinguish rhythm and rate. By 4–5 months infants can more finely discriminate duration, rate, and rhythm.

There is electroencephalography (EEG) evidence of the scalar property in 6 months olds tested during a task of interval timing and there are measurable developments in their ability to time interval durations over the subsequent few months (Allman, Pelphrey, and Meck, 2012, for a review; Brannon et al., 2004, 2008). Characteristically, interval timing shows scalar variability, i.e., there is a linear increase in the standard deviation of timing errors as the size of the test interval increases. Scalar timing is also identified later in this chapter as a feature of preverbal interaction in typical development. Developments in judgment of interval timing, from 6 to 10 months, may relate to the developments in temporal synchrony, over the second half of the first year, these are also outlined below in Section 2.4.2.

2.3 *Typically Developing Infants' Perception of Intra vs. Inter-sensory Synchronies (Re: Social and Non-social Stimuli)*

There are reports for perception of synchrony by 15–20 week olds and even newborns (Lewkowicz, 2000; Spelke and Cortelyou, 1981). Dodd (1979) presented 10- to 16-week-old infants with synchronous and asynchronous experiences of communication. Nursery rhymes were delivered through an auditory stimulus and lip movement display lasting approximately 4 minutes. The asynchronous display was anomalous by 400ms. Polygraphic measures showed that the infants showed preference for the synchronous presentation, suggesting affinity for the congruence of both auditory and visual information in speech.

Lewkowicz (2000) reports that infants gradually develop from a uni-sensory model (focusing on any one aspect of the stimulus either visual or auditory), progressing to a multi-sensory model between 4 and 8 months of age. At 4 months old they attend only to the featural (e.g., changes in visual characteristics like old/new face, gender, etc.) information whilst at 6 months they focus on asynchrony (e.g., mismatched auditory and visual information). But by 8 months old they achieve the ability to focus on each of these features independently and can process stimuli in a multi-model system.

Lewkowicz's (2000) review confirms that infant preference for synchrony extends to non-facial stimuli. Following familiarisation to synchronised presentations of auditory and visual stimuli, a sequence of asynchronous presentations was presented whereby the sound preceded or followed the visual stimulus. In adults, the threshold for asynchrony discrimination of sound preceding the visual stimulus was 65ms whereas it was significantly larger in infants (i.e., 350ms). When the sound followed the visual stimulus, adults detected asynchrony at a threshold of 112ms and infants at 450ms. A perceptual window of auditory-visual temporal synchrony and asynchrony was therefore postulated which becomes refined as the infant develops towards adulthood (Lewkowicz, 1996). Parameters of an inter-sensory temporal synchrony window for autism are considered later in this chapter (e.g., Section 3.3.2.3).

2.4 *Typically Developing Infants' Active Participation in Social Timing*

2.4.1 Intra-personal Timing in Typical Infants: Expressive Preverbal Babbling

L.A. Petitto studied hearing infants, in the second half of their first year, who were raised in the unusual circumstances of being exposed to signed, but not spoken, communication (Petitto et al., 2001, 2004). They produced "manual babbling" with rhythmic movements of the hands at a frequency (approximately 1 Hz) that contrasted with both their non-linguistic rhythmic hand activity and those of speech-reared infants (approximately 2.5 and 3 Hz, respectively). Petitto et al. (2004) concluded that these cross-modality findings indicate an innate sensitivity to specific rhythmic patterns at the core of human language and capacity to employ them (Petitto et al., 2000, 2001, 2004).

These findings complement those from Positron Emission Tomography (PET)/Magnetic Resonance Imaging (MRI) brain scanning research, showing that the brains of both deaf signers and hearing controls have tissues (e.g., the planum temporale) that are specifically sensitive to *rhythmic* patterns underpinning the structure of natural language (Petitto et al., 2000). Petitto et al. (2004) conclude that this rhythmic element is a fundamental abstract property of language that transcends the mode of communication (speech or sign-language). They further argue that this underpins the evolution, acquisition, and experience of language in humans (Petitto et al., 2004). Other aspects of intrapersonal timing in typical development, for example, prosody and rhythm/syllabicity in speech articulation, are covered through examination of control data that is outlined within Section 3.3.1 later in this chapter (Oller et al., 2010; Peppe et al., 2007, 2011).

2.4.2 Inter-personal Timing in Typical Development: Developments in Temporal Synchrony during Social Interaction

Feldman (2007b) informs the literature reviewed below and outlines three temporal parameters of interactional synchrony, functionally defined by their respective measurements. These are outlined as: (i) concurrent synchrony, where coherence is determined by the degree of matching between behaviour/states; (ii) sequential synchrony, where one partner's behavior/state leads another's, as determined by lead-lag analysis; and (iii) organised synchrony, whereby an ongoing patterned format is measured through time series analysis to determine time-lag associations between each partner's behavior streams once individual's periodicities have been removed.

Both (i) and (ii) form components of (iii) and a developmental perspective is employed below to outline developments in communicative synchrony over the first year of typical development. In short: the coherence of matching in mother–infant dyads (i, above) remains constant over this period (Feldman, 2005). Sequential synchrony (ii, above) shows an increasing decrease in the time-lag to synchrony from 3 to 9 months (Feldman, Greenbaum, and Yirmiya, 1999); and, the infant takes an increasingly responsive role in organized synchrony (iii, above) so that mutual synchrony is achieved by about 9 months (Stern, 1985).

Early observations indicated that newborns show 'entrainment' of limb movements to the syllabic rhythms of maternal speech in any language (Condon and Sander, 1974). Such observations have been supported through a musical type of temporal analysis of infant–parent interaction showing that newborn infants, through movement or vocalization, can synchronize exactly with particular salient moments in adult communication (Malloch, 1999). Even very low birthweight premature infants show such linked sequences between their movements and adult behavior (Eckerman et al., 1995).

Pioneering research by Stern (1977; Stern et al., 1977), using a solely durational approach, demonstrated that cross-modal timing structure operates in maternal communicative actions. Recognisable similarities are repeated fairly regularly across rhythmical action/vocal patterns, within burst-pause formats that fluctuate slightly around an average established tempo (Jaffe et al., 2001).

Sequential patterns between parent and infant communication frequently result in rhythmically chained interactive behaviours with stochastic organization (Gottman, 1981). Cycles of infant states are reflected in these stochastic processes (Cohn and Tronick, 1988). The latter are characterized by rhythmicity within a system of probabilities, whereby events can reliably predict subsequent ones, although the whole series itself lacks predetermined regularity (Di Paolo, 2001; Gottman, 1981).

Feldman's (2007b) review clarifies how the tendency of each communicative partner, to cycle independently between states, accounts for greater variance than does the cross-correlational early interaction data. It concludes that internal rhythmicity plays an important role in determining social interactions from infancy to adulthood, supported by data from both adult–adult (Warner, 1992) and adult–infant interactions (Feldman et al., 1997).

During early proto-conversations, turn taking in the infant–parent dyad is established at a slow beat of 1 per 900ms at 6 weeks old. The beat of shared infant–parent vocal play during animated games then accelerates to 1 beat per 700–500ms within the next month or two (Trevarthen and Aitken, 2001). Stern (1977; Stern et al., 1977) originally showed how the timing structure of maternal behaviour transcends modality. Jaffe et al. (2001) highlight how the variation of repeating patterns of rhythmical actions and vocalisations around an average tempo enables infant–parent dyads to co-create expectancies. This is recognised as a critical achievement in initial social relating, social cognition, and enculturation (Bruner, 1975; Haith, Hazan, and Goodman, 1988; Lewis, Goldberg, and Campbell, 1969).

Stern and Gibbon's (1979) solely durational approach, demonstrated a scalar timing process, whereby, within the durations of maternal repetitive kinesic and vocal behavior, the ratio of the mean to the standard deviation remains constant. They predicted that infants must also therefore use scalar timing processes to decode such temporal information. Zlochower and Cohn (1996) subsequently found that scalar timing is shown in the vocal timing of typical mother-4 month old infant dyads.

The structure of parent–infant interaction begins to be temporally well defined by 2–3 months with respect to matching, sequential linking, and time-series aspects. At this time, interactions comprise behaviour within "repetitive rhythmic cycles" of varied modalities such as indicators of arousal, posture, gaze, emotional expression, and hand movements including touch (Feldman,

2007b). From this age infants start to regulate parental emotions and parents reciprocate this (Trevarthen and Aitken, 2001).

Feldman (2007b) describes how the third month is characterised by the establishment of conventional synchrony, whereby separate "interactive behaviours" are expressed cross-modally in a pattern of continuous coordination. Shared gaze is the main modality of coordinated engagement at 3 months of age, characterizing this for a third to half of the time (Fogel, 1977; Harel et al., 2005; Messer and Vietze, 1984; Pawlby, 1977; Tronick, Als, and Brazelton, 1980).

Shared gaze affords a structure for coordinating the behavioural expressions of other modalities. For example, mutual vocalizations begin at this time, often accompanying shared gaze (Beebe and Gerstman, 1980). This 'coactive mode' is typical of 4-month olds' vocal and kinesic preverbal conversations (Jasnow and Feldstein, 1986). At this age, interactive turn taking is characterised by highly specific matching of such rhythm cycles (Beebe, Stern, and Jaffe, 1979) and computerised-acoustic analysis has shown infant vocalisations in time with maternal nursery rhyme chanting (Malloch, 1999; Malloch et al., 1997).

By 5 months, 1 second frame coded time-series analysis shows timed configurations of face-to-face engagement integrating shared gaze, co-vocalising and affectionate touching (Feldman, 2003). Overall, shared gaze decreases by approximately a third from 3 to 9 months, as typically developing infants develop their abilities in sharing their attention to objects (Landry, 1995). However, touch synchrony increases over this period (Granat, 2005, cited by Feldman, 2007b). For 3- to 4-month olds, synchronous interactions usually features infant-*leading* with mothers-*following*, at a time lag to synchrony of between 1.5 and 2 s (Feldman et al., 1996; Feldman, Greenbaum, and Yirmiya, 1999). Feldman (2007b) highlights how such parameters are very sensitive to biological risk in infancy.

Nine months olds show increasing readiness to be swayed by parental behaviour, thereby precipitating bi-directional mutual synchrony, where both partners respond to one another's rhythms (Feldman, 2005; Stern, 1985). However, the overall coherence of mother–infant synchrony remains stable over the first year (Feldman, 2005) as the infant time lag to *mutual* temporal synchrony parameters decreases from 3 to 9 months (Feldman et al., 1996; Feldman, Greenbaum, and Yirmiya, 1999). Nine months olds are also characterised by a decrease of the co-active mode (outlined above) whereupon the use of switching pauses (silences between turns) develops, so marking the boundaries of "conversational" turns, and enabling more conventional vocal, and ultimately verbal turn-taking (Jasnow and Feldstein, 1986).

There is matching of the switching pause durations of conversational partners within both mother–infant and adult dyads (Beebe et al., 1985, 1988;

Jasnow and Feldstein, 1986). The positive correlation that they show indicates that each partner similarly regulates the exchange: both pausing for a similar duration before taking their turn. Matching of preschoolers' switching pauses occurs before matching of the pauses within their turns, whilst child-to-child switching pauses begin to match from 5 to 7 years. Temporal synchrony is achieved in spoken conversations by 6.5 years, whereby the average pause durations within and between conversational turns are statistically similar (Welkowitz, Cariffe, and Feldstein, 1976).

Tronick and Cohn (1989; Tronick, 1989) observed that whilst most mother–infant interaction comprises mismatched states, most of these are repaired in the subsequent step(s). Therefore, preverbal interaction affords typically developing infants with sophisticated experience of an essentially negotiable process of co-regulation whereby mutual correction to attunement becomes a familiar part of ongoing relationships (Feldman, 2007b). This highlights qualities that remain missing from the interactions of people with autism throughout the lifespan, as outlined later in this chapter.

By the end of the first year, established nonverbal synchrony incorporates symbolic functioning, including gesture and spoken language (Bates, O'Connell, and Shore, 1987). Social use of language is acquired through such proto-conversations (Bates, Camaioni, and Volterra, 1975; Ninio and Snow, 1988, 1996). How frequently 4 month olds vocalise is predictive of Bayley developmental scores at 1 year whilst 3 month olds' responsivity of vocalising to their mothers (as opposed to strangers) predicts verbal cognitive scores at 3, 5, and 12 years (Roe, McClure, and Roe, 1982; Ruddy and Bornstein, 1982).

Tallal's (2004) review of communicative timing in language disordered and typically developing children observes that language development is predicted by early individual differences in Rapid Auditory Processing. Temporal/synchrony aspects of preverbal interaction are explored as developmental correlates of other aspects of typical development in the subsequent section of this chapter.

2.5 *Developmental and Physiological Correlates of Temporal/Synchrony Aspects of Preverbal Interaction in Typically Developing Infants*

The developmental correlates of preverbal interaction/temporal synchrony link physiological and behavioural phenomena; all have relevance to autism, especially those pertaining to attachment, symbolic play, Theory of Mind (ToM), and empathy. They vary from physiological factors that predict parameters of temporal synchrony months later to specific developmental abilities predicted years later by such temporal parameters.

For example, the neonatal sleep-wake cycles and cardiac vagal tone of 0- to 1-month-old infants, predict their mother–infant synchrony at 3 months

(Feldman, 2006). The level of temporal organisation of 3 months old infants' social attention predicts mother–infant synchrony at 9 months and 2 years (Feldman et al., 1996). Attunement of maternal and infant heart rates has been found during interactive synchrony (Moshe and Feldman, 2006, cited by Feldman, 2007b).

Coordination of bi-directional vocal timing measures (e.g., rhythmic coupling and bidirectional coordination) in the 1:1 interactions of 4 month olds with their mothers and with strangers predicts cognition at 1 year (Jaffe et al., 2001). More specifically, mid-range levels of vocal synchrony at 4 months predict security of infant attachment at 1 year (Jaffe et al., 2001). Mother–infant synchrony at 9 months also predicts cognition and self-control/regulatory skills at 2 years (after separating out possible effects of maternal style and infant IQ and temperament; Feldman, Greenbaum, and Yirmiya, 1999). This is built on earlier studies showing a linear relationship between face-to-face synchrony in infancy and better cognition levels in childhood (Feldman et al., 1996; Kirsh, Crnic, and Greenberg, 1995; Murray et al., 1996).

Infants' affect regulation (a nonrandom, stochastic-cyclic organization of affective states, coded from 250ms time frame analysis at 3 months) predicted all three areas of symbolic competence assessed at 2 years, including symbolic play and use of internal state terms, the latter are considered to be a precursor of ToM (Feldman and Greenbaum, 1997). Infant social gaze switching during infant–adult interactions have also been found to predict the ToM abilities at nearly 4 years (Charman et al., 2000).

Findings that face-to-face synchrony measures in infancy also predict better social-emotional adaptation (Feldman and Eidelman, 2004; Harrist et al., 1994) may be related to developments in both ToM and empathy. Face-to-face synchrony in the infant's first year has been found to predict empathy at 13 years (Feldman, 2005; Feldman, 2007a) and ToM at 5 years (reported for premature infants by Feldman, 2007b). Both interactive temporal synchrony and temporal processing in autism are considered in the following sections.

3 Social and Non-social Timing Anomalies in Autism

3.1 *Temporal Synchrony in Autistic Infancy and Childhood*

The subsequent sections on autism employ 'Autistic Disorder (AD)' to refer to the more severe manifestations of the condition and 'Autistic Spectrum Disorders (ASDs)' to refer to a range of more heterogenous related conditions pertaining to less severely affected individuals. Affected people of average or above average intelligence, without early language impairment, are referred to as having 'Asperger's Syndrome (AS)', 'High-Functioning Autism (HFA)' refers to those affected individuals who are also intellectually average, or able, but whose developmental history shows early language impairment. Where appropriate, there is a distinction made between high functioning AD and high functioning ASDs.

There are only a limited number of studies of synchrony in autism during infancy; two are twin studies. An early fraternal twin study unexpectedly observed a 4-month-old (later diagnosed) autistic infant who showed a lack of interactive turn taking in contrast to his typically developing (TD) sibling (Kubicek, 1980). This study employed continuous analysis of interactions to the accuracy of 1/24 s. The ASD infant's responses resembled those of TD infants interacting with their mothers over a time-delayed video-link whereupon the infants showed distress and avoidance, such as looking away (Murray and Trevarthen 1985; Nadel et al., 1999).

Hedenbro and Tjus (2007) studied a child, later diagnosed with both AD and 'mental retardation', from 3 to 48 months along with a cohort of TD children. Whilst a *temporal* synchrony approach was not employed, the study focused on dyadic and triadic synchrony between the child and her parent(s). The most obvious differences between the AD child and the TD cohort appeared when she was 9 months old. This is the age at which genuinely mutual temporal synchrony is typically established, as outlined in earlier sections of this chapter.

Temporal synchrony measures particularly focus on the flow of interaction within a dyad. Trevarthen and Daniel (2005) specifically applied these criteria to a pair of monozygotic (MZ) twins at 11 months. The study is unusual in that the video-recorded 1:1 adult:infant interactions were analysed to a resolution of 0.04 s. The twin later diagnosed with AD showed no well-timed co-regulation whilst the TD twin showed synchrony as well as cyclic reciprocation of expressions. Unfortunately, subsequent ASD studies (as reviewed in the paragraphs that follow) with larger infant numbers have not employed the fine resolution of Trevarthen and Daniel's temporal synchrony research.

Rozga et al. (2011) employed 1 and 5 s windows to analyse mother–infant free play and the effects of maternal still face procedure; however, this did not distinguish 6 months old who were later diagnosed with ASD from those who were later undiagnosed. Possibly higher temporal resolution of the infant–caregiver interactions and analysis over a longer duration might show differences in early social interaction of the infants with autism. Rozga et al. acknowledge that such a microanalytic approach to infant timing may have proved more productive. They cite Yirmiya et al. (2006) who, at a 1 s resolution, showed decreased synchrony in the infant–led interactions of unaffected siblings of children with autism.

In contrast to research employing individually based counts of social behaviour, synchrony-based videotape analysis has been able to distinguish (on the basis of social orientation) the first 6 months of infancies later diagnosed as AD (with Learning Disabilities, LD) from controls with TD or LD that remained undiagnosed (Saint-Georges et al., 2011).

A retrospective parental interview study found that interactive turn taking distinguished infants with autism (mostly AD, including some with developmental delay) from matched control infants with only developmental delay (Wimpory et al., 2000). It may be that the responsiveness or repertoire of infants and young children with autism makes temporal synchrony harder for parents to achieve with them. The prosody element of this is considered below in Section 3.3.1.

The chronographic presentation of verbal high-functioning adolescents and young adults with autism has been analysed during conversations by quantifying the structure of their vocalizations and silences with an interviewer in contrast to those with their parents (Feldstein et al., 1982). This revealed longer switching pauses for autistic participant–parent interaction as compared to parent–experimenter, and there were both longer pauses and switching pauses in the autistic participant–experimenter interaction. The findings demonstrated that autistic participants never achieved temporal synchrony in their interaction.

In typical conversations, congruent pattern matching occurs between participants so that the lengths of the pauses covary towards a match in mean pause duration between the conversational partners. TD children, even as young as 6.5 years, exhibit this pause duration synchrony in their conversations (Welkowitz, Cariffe, and Feldstein, 1976) and preverbal use of switching pauses typically develops from at around 9 months, as indicated earlier in this chapter (Jasnow and Feldstein, 1986). In temporal synchrony, the duration of each partner's switching pauses become increasingly statistically similar to the mean over time.

3.2 *Therapeutic Approaches to Synchrony in Autism Integration*

This section briefly considers attempts to facilitate synchrony in parental interactions with their children with autism. A randomised controlled trial of the Preschool Autism Communication Trial (PACT) evaluated the efficacy of specifically enhanced communicative synchrony in parental aspects of communication. This was found to correlate with significant improvement of child communication within the interactions of young autistic AD child–parent pairs (Green et al., 2010). Music Interaction Therapy (MIT) is an intervention aimed specifically at ameliorating *temporal* synchrony problems for young children with ASD during their familiar caregiver's spontaneous attempts to engage them in 1:1 interaction. MIT employs an accompanying musician to offer live timing support in an exaggerated and prolonged experience that aims to recapitulate, and capture the developmental potency of, preverbal interaction (Wimpory, Chadwick, and Nash, 1995).

More detailed temporal analysis of MIT is currently underway for at least sixteen recent cases (Wimpory, Nash, Muth, and Nicholas, in progress). MIT case studies of AD with LD have included findings showing that the child–parent dyad's accompaniment by live musical timing is associated with positive change in relation to previously established developmental trajectories. This has contrasted with, for example, developments recorded in child–parent interactive play during a period whilst the mother received social support, rather than MIT, from the Music Therapist (Wimpory, 1995; Wimpory and Nash, 1999a). Furthermore, communicative gains (in social acknowledgement, eye-contact, child-initiated interactive involvement) associated with seven months of MIT, significantly exceeded developmental trajectories established through 4 months baseline. These gains were maintained at 2 years follow-up (where MIT, for AD with LD, continued for just the initial 5 months; Wimpory, 1995; Wimpory, Chadwick, and Nash, 1995).

Other untrained related developments, such as teasing and symbolic play, were absent during the initial baseline but developed without training during the MIT phase and were maintained at 2 year follow-up where generally continuous child social attentiveness was recorded (Wimpory, 1995; Wimpory, Chadwick, and Nash, 1995). Where parents naturally achieve greater interactional synchrony with their AD child, with LD, this predicts child communication outcomes up to 16 years later (Siller and Sigman, 2002, 2008).

3.3 *Temporal Aspects of Social–Communication Skills in Autism*

3.3.1 Synchrony, Gesture, and Speech

Early research in autism highlighted how such preschool children show a delay in responses of a full second in some instances; this was noted as excessively asynchronous in comparison to the rapid responses of TD children (Condon, 1975; 1979). In more recent behavioural assessments of the perception of synchrony, 4–6 year old children with Autism (and TD and language-matched developmentally delayed controls) gave increased attention to temporally disturbed non-linguistic audio-visual stimuli (asynchronous by 3 s; Bebko et al., 2006). However, in contrast to both control groups, the children with autism were insensitive to disturbed temporal linguistic synchrony.

Deficiencies in gestural-speech synchrony have been found in the expressive communication of young people with HFA, AS, and ASD. Although the frequency of their gestures was comparable with those of typically developing adolescents during a standardised story-telling situation, the timing of their gestures (in relation to their speech) impacted negatively on the quality of their narrative. More severe autistic symptom scores correlated with worse quality ratings for story telling using a story telling element of the Autistic Diagnostic Observation Schedule (ADOS; De Marchena and Eigsti, 2010). Preschool children with AD and ASD are disinterested in the motherese that usually facilitates preverbal interaction (Klin, 1992; Kuhl et al., 2005, respectively). Indeed, children with autism HFA prefer synthesised voices that lack prosodic intonation (Burack et al., 2001). Prosody is the rhythm and intonation of speech that conveys nuances of emotion and meaning. Expressive and perceptive prosody difficulties characterise even HFA in both adults and children (Hesling et al., 2010; Peppe et al., 2007, 2011, respectively). Where TD, HFA, and AS groups are matched on expressive language, sub-group differences in expressive prosody emerge independent of language levels (Peppe et al., 2011).

Expressive and perceptive prosody difficulties are linked in autism (with respect to affect, emphasis, and rhythm; Hesling et al., 2010). Hesling et al. (2010) employed functional Magnetic Resonance Imaging (fMRI) to demonstrate that the neural network involved in prosody perception shows abnormal activation in the left Supra Marginal Gyrus (SMG) in HFA, correlating positively with intonation and emphasis. Such prosody tasks depend on both spectral and temporal information processing.

An acoustic analysis of children's voice spectra showed that rhythm/syllabicity was strongly correlated with, and predictive of, age in TD children. This association did not hold for the children with autism (Oller et al., 2010). Their abnormal temporal processing of the auditory stimuli in speech is considered below, along with other aspects of temporal processing findings that have been derived from a communicative context.

3.3.2 Temporal Processing in Autism

3.3.2.1 *Processing Speed Findings Derived from Social/Communication Stimuli*

An early study of left hemisphere recording sites showed that the Event-Related Potential (ERP) P300 response is reduced for phonetic stimuli in older children with autism with and without LD (Dawson et al., 1988). Children and adults with AD and HFA are often more able than TD controls to process simple auditory stimuli, like the (perfect) pitch of pure tones (Bonnel et al., 2003; Heaton et al., 2008; Lepistö et al., 2008) but less able to process auditory material that is more spectrally and temporally complex (O'Connor, 2012). It may be that the ERP findings of impaired semantic processing in AS and HFA are better interpreted through the enhanced visual processing effects outlined in the integrated timing section below (Russo et al., 2012). However, resultant functional timing difficulties could still adversely affect crucial timing aspects of early (and subsequent social) interaction outlined earlier in this chapter.

Visual ERP research in ASD shows delays in both perception and response systems with regard to facial stimuli. These are in addition to abnormalities in the temporal processing of auditory stimuli in speech. For example, there is delayed latency of the face-processing N170 component during passive observation of faces in HFA, AS (McPartland et al., 2004), and AD with LD (Webb et al., 2006). There are also ASD delays in the N300 component, which responds differentially to fearful faces. Dawson et al. (2004) found that these predict social orienting, joint attention skills, and time spent looking at a person expressing distress, in a sample of ASD and AD young children, mostly with LD.

Oberman, Winkielman, and Ramachandran (2009) found other evidence of temporal processing timing anomalies in high functioning AD and ASD in response to social stimuli, with no group differences either for intentional behavioural mimicry or for amplitude or selectivity of emotion-relevant electromyography (EMG) activity. However, there was delayed spontaneous mimicry in ASD across various stimulus presentation durations and expressions; this delay was longer (approximately 160ms) than the perceptual delays typically observed in ERP studies (approximately 20ms).

3.3.2.2 *Processing Speed: Findings Generally Derived from Non-social/ Communication Stimuli*

This section represents more recent developments from earlier research noting apparent temporal anomalies in rapid attention-switching in ASD older children and adults, for example, between spatial locations (Courchesne et al., 1994a,b, for AD/HFA; Haist et al., 2005, for AD/HFA and AS; Townsend et al., 2001, for AD/HFA; Townsend et al., 1999, for HFA). One physiological indication of a temporally anomalous alternative mode of processing is that peripheral visual stimuli produce a delayed frontal late positive ERP component in children with autism. This indicates a specific delay in spatial attention orientating, even when rapid attention-shifting demands are removed, and HFA subjects produce normal behavioural output (Townsend et al., 2001).

Magnetoencephalography (MEG) also shows altered temporal measures of processing in autism, indicating a disruption in encoding simple sensory information: an auditory evoked and characteristic neuromagnetic field response from the Superior Temporal Gyrus (STG) was delayed by 11ms in ASD (Roberts et al., 2010). Roberts et al. quantified a positive predictive value (PPV) of 86% for this electrophysiological measure for autism (with or without language impairment).

Allman (2011) highlights how social cognition is critically dependent on adaptive sensitivity to the duration of events and their intervening periods. Contradictory ASD findings concerning temporal processing of the shortest duration range may be due to methodological considerations, for example, tasks may or may not allow dependence on memory or counting strategies. Nevertheless they demonstrate at least a functional timing problem relevant to the speed of early and later social interaction. For example, difficulties have been found at the 0.4–0.8, 1–5.5, 1.7–4.1, and 0.5–2 s level (respectively, by Gowen and Miall, 2005, for older children and adults with AS; Szelag et al., 2004, for AD/HFA children; Martin, Poirier, and Bowler, 2010, for high functioning adults with ASD; and by Maister and Plaisted-Grant, 2011, for children with high functioning Autistic Spectrum Conditions). However, neither Mostofsky et al. (2000) found these at the millisecond level for HFA children nor did Martin, Poirier, and Bowler (2010) find them below 1.7 s for high functioning adults with ASD.

Allman, DeLeon, and Wearden (2011) studied perception of duration in high functioning children with ASD (mostly AD) using temporal bisection procedures (with 1–4 and 2–8 s standards). They found proportionally more responses that were too long, and an extinction effect in their precision to estimate durations longer than 5 s. Subsequently, Falter et al. (2012) tested visual, auditory, and cross-modal interval timing for 0.6 and 1 s standards in adolescents and adults with high-functioning ASD and TD controls. They found that the ASD group showed a clearer presentation of the scalar property (a characteristic of interval timing) but less ability to discriminate correctly whether the test stimuli were longer or shorter than the standards. The ASD group also showed a more conservative response bias.

Although Maister and Plaisted-Grant (2011, c.f. above) found no ASD deficit regarding reproductions of 4–30 s durations, they did find impairment of longer duration reproductions (45 s). Their paradigm removed opportunities for reliance on chronometric counting. Memory testing indicated possible episodic memory effects for the 45 s anomaly in the ASD group.

A publication by Gil et al. (2012) concluded that the substrates for time perception are intact in ASD. However, 75% of the 12 subjects in this study had AS (the remainder of this high-functioning sample had HFA or Pervasive Developmental Disorder-not otherwise specified: PDD-NOS). Gil et al.'s conclusions might therefore be more applicable to AS. The overall picture is not clear-cut, but even taking into account the only two other studies that did not find evidence of short interval timing deficit in older children with HFA (Mostofsky et al., 2000) or with AS or high functioning ASD (Wallace and Happe, 2008), it appears that there is, overall, empirical evidence for timing anomaly in ASD.

3.3.2.3 *Temporal Integration (Perceptual Simultaneity, Multisensory/ Temporal Binding, Local to Global/Integration over Time/Context)*

Difficulties/delays in integrating visual and auditory stimuli are noted in ASD and AS (Iarocci and McDonald, 2006; Kwakye et al., 2011). There are both increased errors in multi- and uni-sensory (e.g., auditory) temporal order judgement (Kwakye et al., 2011) in ASD whilst enhanced performance on perceptual simultaneity tasks indicates lower thresholds for vision in HFA and AS (Falter, Elliott, and Bailey, 2012). ERP evidence indicates that the parameters of semantic multisensory integration in ASD depend on both timing and modality order. In response to simultaneous presentations, congruent and incongruent waveforms diverge almost twice as fast in subjects with ASD (at around 0.15 s) than in TD individuals (Russo et al., 2012). Selective sensory temporal enhancement, of visual over auditory stimuli, correlates with sub-clinical ASD traits (Donohue, Darling, and Mitroff, 2012).

Russo et al. (2012) argue that enhanced processing speed in the visual domain may account for the wider temporal window reported in autism in which two cotemporaneous events (e.g., visual and auditory) are perceived to occur simultaneously (Foss-Feig et al., 2010 for high-functioning ASD and AD; Kwakye et al., 2011 for high functioning ASD and AS). Sound is processed faster that vision in the neurotypical brain and this differential may need to be maintained for neurotypical functioning (Der and Deary, 2006). The interdependency of the wider temporal binding window in ASD (Foss-Feig et al., 2010; Powers, Hillock, and Wallace, 2009) and the other difficulties, with multi-/uni-sensory temporal order and perceptual simultaneity, await clarification.

Disordered visual processing in ASD adults has been linked to oscillatory brain activity involved in neural binding (Grice et al., 2001). Gamma oscillation, a correlate of neural binding, is also anomalous in children with ASD (Brown et al., 2005). Furthermore, difficulties have been shown in ASD adults integrating local visual information into a global whole specifically over time (Nakano et al., 2010). Episodic memory, characterised by contextual specificity, pertains to the integration of information over time and depends on intact timing mechanisms to appropriately encode temporal and non-temporal contextual information. Episodic memory anomalies are found in individuals with autism (Millward et al., 2000) including HFA and AS (Boucher and Lewis, 1989; Bowler, Gardiner, and Grice, 2000; Salmond et al., 2005). Finally, problems with diachronic thinking, or thinking about time, are also characteristic of ASD including AS (Boucher et al., 2007).

3.4 *Motoric Timing Anomalies in Autism*

Whilst this chapter specifically focuses on temporal anomalies related to aspects of social communication in ASD, temporally measured motoric anomalies may also impact on these and/or inform understanding of relevant processes involved. For example, Gowen and Miall's (2005) assessments of adults with AS revealed motor coordination difficulties that are behavioural aspects of cerebellar function. The motor coordination temporal anomalies covered in this section range from, for example, posture to eye-blink. Their neurological implications and specific neurological findings for ASD are the focus of the subsequent section.

Temporally measured postural anomalies have been reported for HFA by Schmitz et al. (2003) and Minshew et al. (2004). With respect to classical eye blink conditioning, Sears, Finn, and Steinmetz (1994) found subjects with autism (including HFA) to be quicker at learning and un-learning the classically conditioned eye blink response. Furthermore, and in contrast to TD children, age did not modulate the process in autism, as the younger autistic children were as able as the older autistic children in acquisition and extinction.

Inui and Asama (2003) tested the bimanual finger tapping ability of children with LD and children with autism and LD. The autism group differed from both the TD controls and from the LD group. Those with autism produced faster rhythms in each of the tests at each of the required tempos (inter-tap intervals 200, 400, and 800ms), while the LD group showed only increased tapping speed in the 800ms test.

Altered motor behaviour in autism is not limited to the fine movements investigated in the above experiments. Vilensky, Damasio, and Maurer (1981) found anomalies in the gait of autistic individuals comparable with Parkinson's disease (PD). Rinehart et al. (2006) used gait analysis to conclude that both HFA and AS individuals are uncoordinated. Mutation of the circadian ubiquitin ligase *HERC2* is also causative of a phenotype of gait disturbance, nonsyndromic intellectual disability, and autism (Puffenberger et al., 2012).

3.5 *Timing-Related Brain Anatomy and Neurological Findings in Autism*

The temporally measured motor coordination anomalies outlined above implicate the cerebellum, cerebellar-hippocampal circuit, and basal-ganglia/frontostriatal region in autism (Rinehart et al., 2006; Sears et al., 1994; Vilensky et al., 1981) and cerebellar function and the basal-ganglia/frontostriatal region in AS (Gowen and Miall, 2005; Rinehart et al., 2006). Drawing overall conclusions from brain structure studies in autism is difficult due to inconsistency in: the degree of autism, participant's age, intelligence, and comorbidities across the different studies. Nevertheless, the cerebellum has long-remained a focus of interest with findings of cerebellar anomalies in autism (Bauman and Kemper, 1986) including HFA and AS (Scott et al., 2009). Cerebellar Purkinje neurons function in learning appropriate timing (Koekkoek et al., 2003; Kotani, Kawahara, and Kirino, 2003). It is therefore of interest that a reduction in Purkinje cell number is reported for autism (Kemper and Bauman, 2002; Ritvo et al., 1986). This may underlie other observations of cerebella dysfunction in the disorder (Courchesne, Townsend, and Saitoh, 1994).

Involvement of the cerebellum is reinforced by consistent genetic findings of association of the *EN2* gene with ASD (Benayed et al., 2005; Gharani et al., 2004). *EN2* regulates cerebellum development (Sillitoe et al., 2008), as does the clock gene, *RORA* (Boukhtouche et al., 2006) that shows altered methylation and expression patterns in individuals with autism (Nguyen et al., 2010).

Nuclear magnetic resonance imaging in autism (including high-functioning ASD) highlights regions of the temporal lobe (Ecker et al., 2010; Salmond et al., 2005). fMRI indicates altered intra-brain communication between neural networks (Williams and Minshew, 2007) that is supported by previous findings of a reduction in long-range axons in ASD shown through Diffusion Tensor Imaging (DTI; Barnea-Goraly et al., 2004). Atypical development of cerebellar systems and connection to the cortex are reported (Courchesne, 1997). These systems govern sensory and motor signal integration (Ivry et al., 2002). Altered connectivity that could impact temporal integration is also seen in anatomical differences in the corpus callosum, the brain's interhemispheric portal, in child and adult subjects with AD, some of whom also have LD (Piven et al., 1997).

3.6 *Physiological/Circadian-Related Aspects of Timing in Autism (Sleep Wake Cycle, Melatonin, Clock Genes, and Cardiac Vagal Tone)*

Sleep disturbance is associated with autism (AD, HFA, and AS, as well as low functioning autism: Elia et al., 2000; Limoges et al., 2005; Richdale and Prior, 1995). Measures of melatonin that regulate sleep timing, and its precursor, serotonin, are also disturbed in both adults and children (Cook and Leventhal, 1996; Nir et al., 1995; Richdale and Prior, 1992). A system of rhythmically expressed clock genes drives the circadian rhythm in melatonin production and time of day effects on timing performance are noted in humans (Aschoff, 1998).

Clock gene mutations modulate high frequency oscillators involved in fruit fly communication (Kyriacou and Hall, 1980) and pulsed hormone secretion in hamster (Loudon et al., 1994). Variation in clock genes in human populations are relevant to affect and cognitive processes (Lavebratt et al., 2010; Terracciano et al., 2010; Utge et al., 2010), and may be relevant to autism (Wimpory, Nicholas, and Nash, 2002); these points are outlined in more detail in Chapter 13 of this book. The association of clock gene variants, altered clock gene expression in twin studies and altered clock gene methylation patterns (case control) in autism support this hypothesis (Hu et al., 2009 for high functioning ASD; Melke et al., 2008 for ASD with and without LD; Nguyen et al., 2010 for high functioning ASD; Nicholas et al., 2007 for high functioning AD).

Sleep-wake cyclicity and cardiac vagal tone (circadian regulated) anomalies are associated with autism (Ming et al., 2005) and are important at a systemic level (Field and Diego, 2008). Measures of both sleep-wake cyclicity and cardiac vagal tone predict mother–infant synchrony in typical development (Feldman et al., 1996; Moore and Calkins, 2004), as outlined above.

4 A Timing-Based Developmental Model of Autism

4.1 *Summary: The Importance of Timing in Autism*

To summarise the earlier sections of this chapter, empirically measured temporal anomalies in autism (including HFA/AS) range from sleep architecture, circadian, and circadian-related hormonal anomalies, to brain oscillations involved in neural binding; temporal processing/interval timing; attention-switching motor coordination and communicative timing. Most importantly, for the model presented below, the temporal anomalies in autism also extend to the reciprocity/temporal synchrony skills required for adult and infant communication that are present from the early months in TD.

Appropriate temporal resolution plays a concurrent role in areas of social–communication relevant to ASD symptomatology, for example in interpersonal perception, communication of mood, empathy, understanding of intentions, and ToM abilities (Baldwin, 1993; Blakemore et al., 2003; Crown, 1982; Feldstein, 1982; Natale, 1976; Tomasello, 1999; Welkowitz and Feldstein, 1970). However, this chapter seeks to build beyond an understanding of the concurrent effects of timing in considering potential developmental effects.

4.2 *Timing Theories of Autism*

Various autism theories have proposed timing difficulties as being central to autism with hypotheses encompassing communicative, neurological and/or circadian aspects of timing (Allman, 2011; Boucher, 2000, 2001; Brock et al., 2002; Courchesne et al., 1994b; Grossberg and Seidman, 2006; Newson, 1984; Richdale and Prior, 1995; Segawa, 1985; Szelag et al., 2004; Welsh, Ahn, and Placantonakis, 2005; Wimpory, Nicholas, and Nash, 2002). For example, Boucher (2001) suggested a core deficit in timing due to systemic damage to an integrated system of neural and physiological oscillators.

Wimpory, Nicholas, and Nash (2002) proposed a causative, concurrent and developmental role for timing deficit in AD, suggesting that this deficit involved pathological variations in the structure/function of clock/clock-related genes.

However, it may be that *any* biological disruption of preverbal temporal synchrony, for example, through deaf-blindness (Fraigberg, 1977) may give rise to the more heterogenous presentations of ASD now more widely diagnosed.

This chapter conceives the timing deficit in autism *per se* as manifest through both temporally measured anomalies and apparently disparate symptomatology. It builds on the argument that there may be some temporal and/or clock gene dependency within what may otherwise appear to be disparate autism symptoms (Nicholas et al., 2007). Whilst the impact of temporal resolution on communication has been outlined above, cross-species research shows that functional variations in clock genes concomitantly influence circadian rhythms, emotional contextual memory (Garcia et al., 2000), and short period timing phenomena (Beaver and Giebultowicz, 2004; Kyriacou and Hall, 1980; Loudon et al., 1994) with possible relevance to: autistic communication; diachronic thinking (Boucher et al., 2007); and, perceptual learning (Plaisted, O'Riordan, and Baron-Cohen, 1998).

Perceptual learning generally depends upon sleep-mediated consolidation (Fenn, Nusbaum, and Margoliash, 2003; Karni et al., 1994; Maquet, 2001; Walker and Stickgold, 2004). Thus the perceptual learning deficit in autism may be exacerbated by circadian and sleep disturbance in autism detailed earlier. Clock gene variants can contribute to altered sleep architecture and circadian anomalies as well as other pleiotropic effects (these are concomitant effects of a given gene on more than one apparently unrelated trait). For example, knockout of the clock gene *Npas2* (in mice) reveals its impact upon sleep architecture, circadian rhythms, and emotional contextual memory (Garcia et al., 2000). Sequence variation within the human *NPAS2* clock gene is associated with AD (Nicholas et al., 2007).

Hierarchical accounts of the timing anomalies in autism may, of course, include those that are not dependent on clock genes. For example, as Russo et al. (2012) suggest, enhanced visual processing speed may account for the wider temporal binding window reported in autism (Foss-Feig et al., 2010; Kwakye et al., 2011). Allman (2011; Allman, DeLeon, and Wearden, 2011) builds on both Wimpory and Boucher's theories, to provide a temporal account of autism symptomatology regarding: executive functioning, memory, interactional synchrony, bonding, communication, weak central coherence, perception, restricted and repetitive behaviours, learning and memory problems, etc. (Boucher, 2001; Boucher et al., 2007; Wimpory, Nicholas, and Nash, 2002).

Allman's account assumes that duration sensitivity relates to the subjective sense of time and considers that autistic impairments in ToM, empathy, and imagination, might relate to deficits in the capacity for "mental time travel" (Allman, 2011; Suddendorf and Corballis, 1997). In accounting for these impairments, this chapter now draws on an understanding of the processes involved in typical development and the implications of timing anomalies for such processes in autism.

4.3 *How Typical Development Can Inform Our Understanding of Autism*

4.3.1 Developmental Correlates Relevant to Autism

Whilst there are some skills, relevant to deficits in autism, that may obviously be honed through early preverbal interaction (prosody, turn taking, facial expression, gesture, etc.), the apparent developmental role of early temporal synchrony in typical development may also enhance our understanding of important symbolic functioning deficits in autism. There are autistic deficits in each of the following areas, and each of these are shown to correlate developmentally with typical infant temporal synchrony: symbolic functioning/social pretence, social-emotional adaptation (Feldman and Eidelman, 2004; Harrist et al., 1994), empathy, ToM, and attachment (Jaffe et al., 2001).

4.3.2 Links from Preverbal Timing to Symbolic Functioning in Typical Development

Bergson (1911) originally argued that human experience and consciousness is essentially relational and inextricably linked with our evolved sense of time. The development of interpersonal synchrony, initially dependent on the infant's experience of his/her own rhythmic functions, and then on their relation to those of another, appears to support the development of procedural interactive pre-symbolic knowledge (Jaffe et al., 2001; Stern, 1974, 1985, 1995). More active turn taking, evolving the concept of self and other, may provide the basis for developing declarative knowledge (Feldman, 2007a and b). The appearance of symbolic functioning (at approximately 1 year) may further depend on the creation and acquisition of shared meaning that emerges from timing dependent parent–child interaction (Feldman, 2007b; Newson and Newson, 1975; Wimpory et al., 2000).

Symbols emerge within an interactive context during positive moments between caregiver and child whereby the synchronicity of experience (affective, communicative) is often crucial. Early joint action formats can typically be characterized as "always the same yet always different" (such as 'peek-a-boo' in different contexts, moods, speeds, etc.). Symbols may develop on the basis of playful repetitive encounters within such experiences. During synchronous preverbal interaction a supportive adult's communication may assist an infant in abstracting the shared characteristics of represented phenomenon into a single concept while overlooking minor differences, so enabling its symbolic encoding (Rogoff, 1990; Vygotsky, 1978). TD infants' earliest mental representations likely characterise their significant others, with whom they interact in a temporally specific manner.

Attachments are based around such experiences during a period when capacities for mental acts of substitution and referencing are only just becoming available to TD infants. The referential looking (from a person to an object and back to that person) is just one of the activities that characterises progression from primary to secondary inter-subjectivity (whereby external objects are incorporated into the interaction). Parental referencing of objects, and taking of disparate attitudinal perspectives towards them, is identified by Hobson (1993, 1994) as essential for the typical development of symbolic functioning in infants.

Hobson (1993, 1994) outlines how the disparate attitudes conveyed through parental roles, during the incorporation of joint attention to objects in preverbal interaction, could be pivotal in the development of pretence in TD. Wimpory (1995; Wimpory et al., 2000) has extended this to consider teasing in preverbal social interaction before the incorporation of physical objects into such engagement. Playful teasing by parent and infant is common in the preverbal social interaction of TD infants (Reddy, 1991, 2001, 2008). It is often facilitated through non-verbal play around issues of timing (for example, during the exaggerated speech and rotating hand movements of the action rhyme "s-l-o-w-l-y, s—l—o—w—l-y, *quickly!*"). As in this example, teasing involves playfully sabotaging another's expectations; these are often time dependent. Although this behaviour inevitably involves some mental representation of such parental expectations on the infant's part, it precedes other more formal manifestations of symbolic functioning (Hoicka and Gattis, 2008; Reddy, 1991, 2001, 2008; Trevarthen and Logotheti, 1987).

4.3.3 The Implications for Autism

Wimpory and colleagues have applied the above arguments to autism. In recognising timing as central to the development of social communication and symbolic functioning, they emphasize and report an early inability to partake in the proto-conversations that characterise TD in social communication and symbolic functioning (Nicholas et al., 2007; Wimpory et al., 2000; Wimpory, Hobson, and Nash, 2007; Wimpory, Nicholas, and Nash, 2002). Furthermore, parents of children with autism do not recall them engaging in teasing during infancy (Wimpory, 1995; Wimpory et al., 2000), and the teasing shown by older children with HFA and AS is much more limited (less playful and provocative) than that of matched controls (Heerey et al., 2005). Referential looking predicts ToM, and both these phenomena remain elusive in autism, along with broader aspects of symbolic functioning. Although autistic savants, with higher mathematical capacity, may be using symbols, their abilities do not extend to the full capacity to create symbols, particularly in an interactive context, as characterizes young TD children.

MIT offers opportunities to establish synchronous proto-conversational experience in autism (Wimpory et al., 1995, Wimpory and Nash, 1999b). MIT is recommended early and there is precedent for this conceptualization in the narrow time window during which chicks are able to learn their species-specific song from their parents. If this opportunity is missed, the young birds are not subsequently able to acquire the capacity for meaningful song (Bolhuis, Okanoya, and Scharff, 2010). Furthermore, song learning in birds is regulated by the circadian clock (Cassone and Westneat, 2012). In the fruit fly, *Drosophila*, the courtship song is an innate and genetically programmed behaviour regulated by the circadian clock gene, *period* (Kyriacou and Hall, 1980). Thus, there is cross-species evidence for co-regulation of the circadian clock and social, communicative capacity.

4.3.4 Conclusion

Earlier sections within this discussion draw on typical development to outline how timing anomalies in autism could account for: (i) temporally disabled relations, (ii) an impoverished sense of self and other, (iii) an impaired quality of attachments, and (iv) impaired symbolic functioning in terms of the creation of symbols and development of shared meaning. The focus on timing, as outlined above, has led to investigations identifying variations in clock genes associated with AD (*NPAS2* and *PER1*, Nicholas et al., 2007) as well as to therapeutic facilitation of temporal synchrony through MIT. MIT provides a unique opportunity to test the perspectives presented here; it is now the subject of a larger scale service-based project (Methley and Wimpory, 2011; Wimpory, Nash, Muth and Nicholas, in progress).

References

Allman, Melissa J. 2011. "Deficits in Temporal Processing Associated with Autistic Disorder" *Frontiers in Integrative Neuroscience* 5: 2.

Allman, Melissa J., Iser G. DeLeon, and John H. Wearden. 2011. "Psychophysical Assessment of Timing in Individuals with Autism" *American Journal on Intellectual and Developmental Disabilities* 116(2): 165–78.

Allman, Melissa J., K.A. Pelphrey, and W.H. Meck. 2012. "Developmental Neuroscience of Time and Number: Implications for Autism and Other Neurodevelopmental Disabilities" *Frontiers in Integrative Neuroscience* 6(7): 1–24.

Aschoff, J. 1998. "Human Perception of Short and Long Time Intervals: Its Correlation with Body Temperature and the Duration of Wake Time" *Journal of Biological Rhythms* 13(5): 437–42.

Baldwin, D.A. 1993. "Early Referential Understanding – Infants Ability to Recognize Referential Acts for What They are" *Developmental Psychology* 29(5): 832–43.

Barnea-Goraly, N. et al. 2004. "White Matter Structure in Autism: Preliminary Evidence from Diffusion Tensor Imaging" *Biological Psychiatry* 55: 54.

Bates, E., L. Camaioni, and V. Volterra. 1975. "The Acquisition of Performatives Prior to Speech" *Merrill Palmer Quarterly* 21: 205–26.

Bates, E., B. O'Connell, and C. Shore. 1987. "Language and Communication in Infancy," in *Handbook of Infant Development*, edited by J. Osofsky 149–203. New York, NY: Wiley.

Bauman, M.L., and T.L. Kemper. 1986. "Developmental Cerebellar Abnormalities – A Consistent Finding in Early Infantile-Autism" *Neurology* 36(4): 190.

Beaver, L.M., and J.M. Giebultowicz. 2004. "Regulation of Copulatioduration by Period and Timeless in *Drosophila Melanogaster*" *Current Biology* 14(16): 1492–7.

Bebko, J.M. et al. 2006. "Discrimination of Temporal Synchrony in Intermodal Events by Children with Autism and Children with Developmental Disabilities without Autism" *Journal of Child Psychology and Psychiatry* 47(1): 88–98.

Beebe, B., and L.J. Gerstman. 1980. "The Packaging of Maternal Stimulation in Relation to Infant Facial-Visual Engagement – A Case-Study at 4 Months" *Merrill-Palmer Quarterly-Journal of Developmental Psychology* 26(4): 321–39.

Beebe, B., D. Stern, and J. Jaffe. 1979. "The Kinesic Rhythm of Mother–Infant Interaction," in *Of Speech and Time: Temporal Speech Patterns in Interpersonal Contexts*, edited by W. Siegman and S. Feldstein, 23–34. Hillsdale, NJ: Lawrence Erlbaum Associates.

Beebe, B. et al. 1985. "Interpersonal Timing: The Application of an Adult Dialogue Model to Mother–Infant Vocal and Kinesic Interactions," in *Social Perception in Infants*, edited by T.M. Field and N.A. Fox 217–47. New York, NY: Ablex Publishing Corporation.

Beebe, B. et al. 1988. "Vocal Congruence in Mother Infant play" *Journal of Psycholinguistic Research* 17(3): 245–59.

Benayed, R. et al. 2005. "Support for the Homeobox Transcription Factor Gene ENGRAILED 2 as an Autism Spectrum Disorder Susceptibility Locus" *American Journal of Human Genetics* 77(5): 851–68.

Bergson, Henri, Creative Evolution (1911) tr. Arthur Mitchell, Henry Holt and Company.

Blakemore, S.J. et al. 2003. "The Detection of Contingency and Animacy from Simple Animations in the Human Brain" *Cerebral Cortex* 13(8): 837–44.

Bolhuis, Johan J., Kazuo Okanoya, and Constance Scharff. 2010. "Twitter Evolution: Converging Mechanisms in Birdsong and Human Speech" *Nature Reviews Neuroscience* 11(11): 747–59.

Bonnel, A. et al. 2003. "Enhanced Pitch Sensitivity in Individuals with Autism: A Signal Detection Analysis" *Journal of Cognitive Neuroscience* 15(2): 226–35.

Boucher, J. 2000. "Time Parsing, Normal Language Acquisition, and Language-Related Developmental Disorders," in *New direction in language development*, edited by M. Howard and S. Perkins, 13–23. USA: Springer.

Boucher, J. 2001. "Lost in a sea of time: Time Parsing and Autism," in *Time and memory*, edited by C. Hoerl and T. Cormack. Oxford, England: Clarendon Press.

Boucher, J., and Pons, F. (2003). Temporal Cognition and Mental Time Travel in Autism: Paper, Experimental Psychology Society, London, United Kingdom.

Boucher, J., and V. Lewis. 1989. "Memory Impairments and Communication in Relatively Able Autistic-Children" *Journal of Child Psychology and Psychiatry and Allied Disciplines* 30(1): 99–122.

Boucher, J. et al. 2006. "Temporal Cognition in Children with Autistic Spectrum Disorders: Tests of Diachronic Thinking" *Journal of Autism and Developmental Disorders* 37: 1413–29.

Boucher, J. et al. 2007. "Temporal Cognition in Children with Autistic Spectrum Disorders: Tests of Diachronic Thinking" *Journal of Autism and Developmental Disorders* 37(8): 1413–29.

Boukhtouche, F. et al. 2006. "Retinoid-Related Orphan Receptor Alpha Controls the Early Steps of Purkinje Cell Dendritic Differentiation" *Journal of Neuroscience* 26(5): 1531–8.

Bowler, D.M., J.M. Gardiner, and S.J. Grice. 2000. "Episodic Memory and Remembering in Adults with Asperger Syndrome" *Journal of Autism and Developmental Disorders* 30(4): 295–304.

Brannon, E.M. et al. 2004. "Timing in the Baby Brain" *Cognitive Brain Research* 21(2): 227–33.

Brannon, Elizabeth M. et al. 2008. "Electrophysiological Measures of Time Processing in Infant and Adult Brains: Weber's Law Holds" *Journal of Cognitive Neuroscience* 20(2): 193–203.

Brock, J. et al. 2002. "The Temporal Binding Deficit Hypothesis of Autism" *Development and Psychopathology* 14(2): 209–24.

Brown, C. et al. 2005. "Gamma Abnormalities during Perception of Illusory Figures in Autism" *Cortex* 41(3): 364–76.

Bruner, J.S. 1975. "The Ontogenesis of Speech Acts" *Journal of Child Language* 2: 1–19.

Burack, J.A. et al., editors, 2001. *Development and Autism: Messages from Developmental Psychopathology*. Mahwah, NJ: Lawrence Erlbaum Associates.

Burke, P.M. 1977. "Swallowing and Organization of Sucking in Human Newborn" *Child Development* 48(2): 523–31.

Cassone, Vincent M., and David F. Westneat. 2012. "The Bird of Time: Cognition and the Avian Biological Clock" *Frontiers in Molecular Neuroscience* 5: 32.

Charman, T. et al. 2000. "Testing Joint Attention, Imitation, and Play as Infancy Precursors to Language and Theory of Mind" *Cognitive Development* 15(4): 481–98.

Cohn, J.F., and E.Z. Tronick. 1988. "mother Infant Face-To-Face Interaction – Influence is Bidirectional and Unrelated to Periodic Cycles in Either Partners Behavior" *Developmental Psychology* 24(3): 386–92.

Condon, W.S. 1975. "Multiple Response to Sound in Dysfunctional Children" *Journal of Autism and Childhood Schizophrenia* 5(1): 37–56.

Condon, W.S., and L.W. Sander. 1974. "Synchrony Demonstrated between Movements of Neonate and Adult Speech" *Child Development* 45(2): 456–62.

Condon, W.S. (1979). "Neonatal Entrainment and Enculturation," in *Before Speech*, ed. M. Bullowa (Cambridge: Cambridge University Press), 131–148.

Cook, E.H., and B.L. Leventhal. 1996. "The Serotonin System in Autism" *Current Opinion in Pediatrics* 8(4): 348–54.

Courchesne, E. 1997. "Brainstem, Cerebellar and Limbic Neuroanatomical Abnormalities in Autism" *Current Opinion in Neurobiology* 7(2): 269–78.

Courchesne, E. et al. 1994a. "A New Finding: Impairment in Shifting Attention in Autistic and Cerebellar Patients," in *Atypical cognitive deficits in developmental disorder: Implications for brain function*, edited by Broman, S.H., Grafman, J., 101–37. Hillsdale, NJ: Lawrence Erlbaum Associates.

Courchesne, E. et al. 1994b. "Impairment in Shifting Attention in Autistic and Cerebellar Patients" *Behavioral Neuroscience* 108(5): 848–65.

Courchesne, E., J. Townsend, and O. Saitoh. 1994. "The Brain in Infantile-Autism – Posterior-Fossa Structures are Abnormal" *Neurology* 44(2): 214–23.

Crook, C.K. 1979. "The Organization and Control of Infant Sucking," in *Advances in Child Development and Behaviour*, edited by H.R. Reese and L.P. Lipsitt, 209–52. New York, NY: Academic Press.

Crown, C.L. 1982. "Impression Formation and the Chronography of Dyadic Interactions," in *Interaction rhythms: Periodicity in communicative behavior*, edited by M. Davis, 225–48. New York, NY: Human Sciences Press.

Dawson, G. et al. 1988. "Reduced P3 Amplitude of the Event-Related Brain Potential – Its Relationship to Language Ability in Autism" *Journal of Autism and Developmental Disorders* 18(4): 493–504.

Dawson, G. et al. 2004. "Early Social Attention Impairments in Autism: Social Orienting, Joint Attention, and Attention to Distress" *Developmental Psychology* 40(2): 271–83.

de Marchena, Ashley, and Inge-Marie Eigsti. 2010. "Conversational Gestures in Autism Spectrum Disorders: Asynchrony But Not Decreased Frequency" *Autism Research* 3(6): 311–22.

DeCasper, A.J., and W.P. Fifer. 1980. "Of Human Bonding – Newborns Prefer their Mothers, Voices" *Science* 208(4448): 1174–6.

DeCasper, A., and Carstens, A. (1980). Contingencies of Stimulation: Effects on Learning and Emotions in Neonates. *Infant Behav. Dev.* 4, 19–36.

DeCasper, A.J., and A.A. Carstens. 1981. "Contingencies of Stimulation – Effects on Learning and Emotion in Neonates" *Infant Behavior and Development* 4(1): 19–35.

Der, G., and Deary, I. (2006). Age and Sex Differences in Reaction Time in Adulthood: Results from the United Kingdom Health and Lifestyle Survey. *Psychology and Aging*, 21(1), 62–73. doi:10.1037/0882-7974.21.1.62.

Di Paolo, E.A. 2001. "Rhythmic and Non-Rhythmic Attractors in Asynchronous Random Boolean Networks" *Biosystems* 59(3): 185–95.

Dodd, B. 1979. "Lip Reading in Infants – Attention to Speech Presented in-Synchrony and Out-of-Synchrony" *Cognitive Psychology* 11(4): 478–84.

Donohue, Sarah E., Elise F. Darling, and Stephen R. Mitroff. 2012. "Links between Multisensory Processing and Autism" *Experimental Brain Research* 222(4): 377–87.

Ecker, Christine et al. 2010. "Describing the Brain in Autism in Five Dimensions-Magnetic Resonance Imaging-Assisted Diagnosis of Autism Spectrum Disorder Using a Multiparameter Classification Approach" *Journal of Neuroscience* 30(32): 10612–23.

Eckerman, C.O. et al. 1995. "The Development Prior to Term Age of Very Prematurely Born Newborns' Responsiveness in an En Face Exchange" *Infant Behavior and Development* 18: 283–97.

Elia, M. et al. 2000. "Sleep in Subjects with Autistic Disorder: A Neurophysiological and Psychological Study" *Brain and Development* 22(2): 88–92.

Falter, Christine M., Mark A. Elliott, and Anthony J. Bailey. 2012. "Enhanced Visual Temporal Resolution in Autism Spectrum Disorders" *PLoS One* 7(3): 32774.

Falter, Christine M. et al. 2012. "More Consistent, Yet Less Sensitive: Interval Timing in Autism Spectrum Disorders" *Quarterly Journal of Experimental Psychology* 65(11): 2093–107.

Feldman, R. 2003. "Infant–Mother and Infant–Father Synchrony: The Coregulation of Positive Arousal" *Infant Mental Health Journal* 24(1): 1–23.

Feldman, R. (2005). "Mother–Infant Synchrony and the Development of Moral Orientation in Childhood and Adolescence: Direct and Indirect Mechanisms of Developmental Continuity." Paper Presented in the Biennial Meeting of the Society for Research in Child Development, Atlanta, GA.

Feldman, R. 2006. "From Biological Rhythms to Social Rhythms: Physiological Precursors of Mother–Infant Synchrony" *Developmental Psychology* 42(1): 175–88.

Feldman, Ruth. 2007a. "Mother–Infant Synchrony and the Development of Moral Orientation in Childhood and Adolescence: Direct and Indirect Mechanisms of Developmental Continuity" *American Journal of Orthopsychiatry* 77(4): 582–97.

Feldman, Ruth. 2007b. "Parent–Infant Synchrony and the Construction of Shared Timing; Physiological Precursors, Developmental Outcomes, and Risk Conditions" *Journal of Child Psychology and Psychiatry* 48(3–4): 329–54.

Feldman, R., and A.I. Eidelman. 2004. "Parent–Infant Synchrony and the Social-Emotional Development of Triplets" *Developmental Psychology* 40(6): 1133–47.

Feldman, R. Greenbaum, C.W., Yirmiya, N., and Mayes, L.C. 1996. "Relations between Cyclicity and Regulation in Mother–Infant Interaction at 3 and 9 Months and Cognition at 2 Years" *Journal of Applied Developmental Psychology* 17(3): 347–65.

Feldman, R. Greenbaum, C.W., Mayes, L.C., and Erlich, H.S. 1997. "Change in Mother–Infant Interactive Behavior: Relations to Change in the Mother, the Infant, and the Social Context" *Infant Behavior and Development* 20(2): 151–63.

Feldman, R., and Greenbaum, C.W. (1997). "Affect Regulation and Synchrony in Mother–Infant Play as Precursors to the Development of Symbolic Competence" *Infant Mental Health Journal*, 18, 4–23.

Feldman, R., C.W. Greenbaum, and N. Yirmiya. 1999. "Mother–Infant Affect Synchrony as an Antecedent of the Emergence of Self-Control" *Developmental Psychology* 35(1): 223–31.

Feldstein, S. 1982. "Impression Formation in Dyads: The Temporal Dimension," in *Interaction rhythms: Periodicity in communicative behavior*, edited by M. Davis, 207–24. New York, NY: Human Sciences.

Feldstein, S. et al. 1982. "The Chronography of Interactions with Autistic Speakers – An Initial Report" *Journal of Communication Disorders* 15(6): 451–60.

Fenn, K.M., H.C. Nusbaum, and D. Margoliash. 2003. "Consolidation during Sleep of Perceptual Learning of Spoken Language" *Nature* 425(6958): 614–16.

Field, Tiffany, and Miguel Diego. 2008. "Vagal Activity, Early Growth and Emotional Development" *Infant Behavior and Development* 31(3): 361–73.

Fogel, A. 1977. "Temporal Organization in Mother–Infant Face-to-Face Interaction," in *Studies in mother–infant interaction*, edited by H.R. Schaffer. London, England: Academic.

Foss-Feig, Jennifer H. et al. 2010. "An Extended Multisensory Temporal Binding Window in Autism Spectrum Disorders" *Experimental Brain Research* 203(2): 381–9.

Fraigberg, S. 1977. *Insights from the blind*. London, England: Souvenir Press.

Garcia, J.A. et al. 2000. "Impaired Cued and Contextual Memory in NPAS2-Deficient Mice" *Science* 288(5474): 2226–30.

Gharani, N. et al. 2004. "Association of the Homeobox Transcription Factor, ENGRAILED 2, 3, with Autism Spectrum Disorder" *Molecular Psychiatry* 9(5): 474–84.

Gil, Sandrine et al. 2012. "Children with Autism Spectrum Disorders Have 'The Working Raw Material' for Time Perception" *PLoS One* 7(11): e49116.

Gottman, J.M. 1981. *Time-Series Analysis: A Comprehensive Introduction for Social Scientists*. Cambridge, England: Cambridge University Press.

Gowen, E., and R.C. Miall. 2005. "Behavioural Aspects of Cerebellar Function in Adults with Asperger Syndrome" *Cerebellum* 4(4): 279–89.

Granat, A. 2005. "Emotional Regulation Processes among Infants of Mothers with Depressive and Anxious Disorders: Affective-Behavioral Aspects" (PhD diss., Department of Psychology, Bar Ilan University).

Green, Jonathan et al. 2010. "Parent-Mediated Communication-Focused Treatment in Children with Autism (PACT): A Randomised Controlled Trial" *Lancet* 375(9732): 2152–60.

Grice, S.J. et al. 2001. "Disordered Visual Processing and Oscillatory Brain Activity in Autism and Williams Syndrome" *Neuroreport* 12(12): 2697–700.

Grossberg, Stephen, and Don Seidman. 2006. "Neural Dynamics of Autistic Behaviors: Cognitive, Emotional, and Timing Substrates" *Psychological Review* 113(3): 483–525.

Haist, F. et al. 2005. "The Functional Neuroanatomy of Spatial Attention in Autism Spectrum Disorder" *Developmental Neuropsychology* 27(3): 425–58.

Haith, M.M., C. Hazan, and G.S. Goodman. 1988. "Expectation and Anticipation of Dynamic Visual Events by 3.5-Month-Old Babies" *Child Development* 59(2): 467–79.

Harel, H. et al. 2005. "Maternal Representations and Mother–Infant Synchrony in Full-Term and Preterm Infants" (Paper Presented at the Biennial Meeting for the Society for Research in Child Development, Atlanta, Georgia).

Harrist, A.W. et al. 1994. "Dyadic Synchrony in Mother-Child Interaction – Relation with Childrens Subsequent Kindergarten Adjustment" *Family Relations* 43(4): 417–24.

Heaton, Pamela, Robert E. Davis, and Francesca G.E. Happe. 2008. "Research Note: Exceptional Absolute Pitch Perception for Spoken Words in an Able Adult with Autism" *Neuropsychologia* 46(7): 2095–8.

Hedenbro, M., and T. Tjus. 2007. "A Case Study of Parent–Child Interactions of a Child with Autistic Spectrum Disorder (3–48 Months) and Comparison with Typically-Developing Peers" *Child Language Teaching and Therapy* 23(2): 201–22.

Heerey, E.A. et al. 2005. "Understanding Teasing: Lessons from Children with Autism" *Journal of Abnormal Child Psychology* 33(1): 55–68.

Hesling, Isabelle et al. 2010. "The Integration of Prosodic Speech in High Functioning Autism: A Preliminary fMRI Study" *PLoS One* 5(7): e11571.

Hobson, R.P. 1994. "Perceiving Attitudes, Conceiving Minds," in *Origins of an understanding of mind*, edited by C.M. Lewis and P. Mitchell. Hillsdale, NJ: Lawrence Erlbaum Associates.

Hobson, R.P. 1993. *Autism and the Development of Mind*. Hillsdale, NJ: Lawrence Erlbaum Associates.

Hoicka, Elena, and Merideth Gattis. 2008. "Do the Wrong Thing: How Toddlers Tell a Joke From a Mistake" *Cognitive Development* 23(1): 180–90.

Hu, Valerie W. et al. 2009. "Gene Expression Profiling of Lymphoblasts from Autistic and Nonaffected Sib Pairs: Altered Pathways in Neuronal Development and Steroid Biosynthesis" *PLoS One* 4(6): e5775.

Iarocci, G., and J. McDonald. 2006. "Sensory Integration and the Perceptual Experience of Persons with Autism" *Journal of Autism and Developmental Disorders* 36(1): 77–90.

Inui, N., and K. Asama. 2003. "Timing of Bimanual Rhythmic Finger Tapping in Adolescents with Mental Retardation or Autism" *Journal of Human Movement Studies* 45(1): 59–80.

Ivry, R.B. et al. 2002. "The Cerebellum and Event Timing" *Cerebellum: Recent Developments in Cerebellar Research* 978(2002): 302–17.

Jaffe, J. et al. 2001. "Rhythms of Dialogue in Infancy: Coordinated Timing in Development" *Monographs of the Society for Research in Child Development* 66(2): 1–132.

Jasnow, M., and S. Feldstein. 1986. "Adult-Like Temporal Characteristics of Mother–Infant Vocal Interactions" *Child Development* 57(3): 754–61.

Karni, A. et al. 1994. "Dependence on Rem-Sleep of Overnight Improvement of a Perceptual Skill" *Science* 265(5172): 679–82.

Kemper, T.L., and M.L. Bauman. 2002. "Neuropathology of Infantile Autism" *Molecular Psychiatry* 7(2002): S12–3.

Kirsh, S.J., K.A. Crnic, and M.T. Greenberg. 1995. "Relations between Parent–Child Affect and Synchrony and Cognitive Outcome at 5 Years of Age" *Personal Relationships* 2(3): 187–98.

Klin, A. 1992. "Listening Preferences in Regard to Speech in 4 Children with Developmental-Disabilities" *Journal of Child Psychology and Psychiatry and Allied Disciplines* 33(4): 763–9.

Koekkoek, S.K.E. et al. 2003. "Cerebellar LTD and Learning-Dependent Timing of Conditioned Eyelid Responses" *Science* 301(5640): 1736–9.

Kotani, S., S. Kawahara, and Y. Kirino. 2003. "Purkinje Cell Activity during Learning a New Timing in Classical Eyeblink Conditioning" *Brain Research* 994(2): 193–202.

Kubicek, L.F. 1980. "Organisation in Two Mother–Infant Interactions Involving a Normal Infant and His Fraternal Twin Brother Who Was Later Diagnosed as Autistic," in *High-risk infants and children: Adult and peer interactions*, edited by Field, T.S. Goldberg, D. Stem, and A. Sostek. New York, NY: Academic.

Kuhl, P.K. et al. 2005. "Links between Social and Linguistic Processing of Speech in Preschool Children with Autism: Behavioral and Electrophysiological Measures" *Developmental Science* 8(1): F1–2.

Kwakye, Leslie D. et al. 2011. "Altered Auditory and Multisensory Temporal Processing in Autism Spectrum Disorders" *Frontiers in Integrative Neuroscience* 4: 129.

Kyriacou, C.P., and J.C. Hall. 1980. "Circadian-Rhythm Mutations in *Drosophila Melanogaster* Affect Short-Term Fluctuations in the Males Courtship Song." *Proceedings of the National Academy of Sciences of the United States of America-Biological Sciences* 77(11): 6729–33.

Landry, S.H. 1995. "The Development of Joint Attention in Premature Low Birth Weight Infants: Effects of Early Medical Complications and Maternal Attention-Directing Behaviors," in *Joint attention: Its origins and role in development*, edited by C. Moore and P.J. Dunham. Hillsdale, NJ: Lawrence Erlbaum Associates.

Lavebratt, Catharina et al. 2010. "CRY2 is Associated with Depression" *PLoS One* 5(2): e9407.

Lepistö, T. et al. 2008. "The Perception of Invariant Speech Features in Children with Autism" *Biological Psychology* 77(1): 25–31.

Lewis, M., S. Goldberg, and H. Campbell. 1969. "Perceptual-Cognitive Development in Infancy: A Generalized Expectancy Model as a Function of the Mother–Infant Interaction" *Merrill-Palmer Quarterly* 15: 81–100.

Lewkowicz, D.J. 1996. "Perception of Auditory-Visual Temporal Synchrony in Human Infants" *Journal of Experimental Psychology-Human Perception and Performance* 22(5): 1094–106.

Lewkowicz, D.J. 2000. "The Development of Intersensory Temporal Perception: An Epigenetic Systems/Limitations View" *Psychological Bulletin* 126(2): 281–308.

Limoges, E. et al. 2005. "Atypical Sleep Architecture and the Autism Phenotype" *Brain* 128: 1049–61.

Loudon, A.S.I. et al. 1994. "Ultradian Endocrine Rhythms are Altered by a Circadian Mutation in the Syrian-Hamster" *Endocrinology* 135(2): 712–18.

Maister, Lara, and Kate C. Plaisted-Grant. 2011. "Time Perception and Its Relationship to Memory in Autism Spectrum Conditions" *Developmental Science* 14(6): 1311–22.

Malloch, S. 1999. "Mothers and Infants and Communicative Musicality" *Musicae Scientiae* (Special Issue 1999–2000): 29–57.

Malloch, S. et al. 1997. "Measuring the Human Voice: Analysing Pitch, Timing, Loudness and Voice Quality in Mother/Infant Communication" *Proceedings of the Institute of Acoustics* 19: 195–500.

Maquet, P. 2001. "The Role of Sleep in Learning and Memory" *Science* 294(5544): 1048–52.

Martin, Jonathan S., Marie Poirier, and Dermot M. Bowler. 2010. "Brief Report: Impaired Temporal Reproduction Performance in Adults with Autism Spectrum Disorder" *Journal of Autism and Developmental Disorders* 40(5): 640–6.

McPartland, J. et al. 2004. "Event-Related Brain Potentials Reveal Anomalies in Temporal Processing of Faces in Autism Spectrum Disorder" *Journal of Child Psychology and Psychiatry* 45(7): 1235–45.

Melke, J. et al. 2008. "Abnormal Melatonin Synthesis in Autism Spectrum Disorders" *Molecular Psychiatry* 13(1): 90–8.

Messer, D.J., and P.M. Vietze. 1984. "Timing and Transitions in Mother Infant Gaze" *Infant Behavior and Development* 7(2): 167–81.

Methley, A., and Wimpory, D. 2011. "Music Interaction Therapy for Children with Autism" (ASD Training DVD Launched at ESRC-Funded Workshop on Early Intervention for Young Children with Autism, Bangor University).

Millward, C. et al. 2000. "Recall for Self and Other in Autism: Children's Memory for Events Experienced by Themselves and Their Peers" *Journal of Autism and Developmental Disorders* 30(1): 15–28.

Ming, X. et al. 2005. "Reduced Cardiac Parasympathetic Activity in Children with Autism" *Brain and Development* 27(7): 509–16.

Minshew, N.J. et al. 2004. "Underdevelopment of the Postural Control System in Autism" *Neurology* 63(11): 2056–61.

Moore, G.A., and S.D. Calkins. 2004. "Infants' Vagal Regulation in the Still-Face Paradigm is Related to Dyadic Coordination of Mother–Infant Interaction" *Developmental Psychology* 40(6): 1068–80.

Moshe, M., and Feldman, R. 2006. "Maternal and Infant Heart Rhythms and Mother–Infant Synchrony" (Paper Presented at the Biennial Conference of the World Association for Infant Mental Health, Paris, France).

Mostofsky, S.H. et al. 2000. "Evidence for a Deficit in Procedural Learning in Children and Adolescents with Autism: Implications for Cerebellar Contribution" *Journal of the International Neuropsychological Society* 6(7): 752–9.

Murray, L., and Trevarthen, C. 1985. "Emotional Regulation of Interactions between Two-Month Olds and Their Mothers," in *Social Perception in Infants*. Norwood, NJ: Alex.

Murray, L. et al. 1996. "The Impact of Postnatal Depression and Associated Adversity on Early Mother–Infant Interactions and Later Infant Outcome" *Child Development* 67(5): 2512–26.

Nadel, Jacqueline et al. 1999. "Expectancies for Social Contingency in 2-Month-Olds" *Developmental Science* 2(2): 164–73.

Nakano, Tamami et al. 2010. "Deficit in Visual Temporal Integration in Autism Spectrum Disorders" *Proceedings of the Royal Society B – Biological Sciences* 277(1684): 1027–30.

Natale, M. 1976. "Interviewee Matching of Interviewer Temporal Speech Behaviors as Related to Perceived Empathy, Warmth, and Genuineness" *Journal of Clinical Psychology* 32(2): 407–12.

Newson, E. 1984. "The Social Development of the Young Autistic Child" (Paper Presented at the National Autistic Society Conference, Bath, England).

Newson, J., and E. Newson. 1975. "Intersubjectivity and Transmission of Culture – Social Origins of Symbolic Functioning" *Bulletin of the British Psychological Society* 28: 437–46.

Nguyen, AnhThu et al. 2010. "Global Methylation Profiling of Lymphoblastoid Cell Lines Reveals Epigenetic Contributions to Autism Spectrum Disorders and a Novel Autism Candidate Gene, RORA, Whose Protein Product is Reduced in Autistic Brain" *FASEB Journal* 24(8): 3036–51.

Nicholas, B. et al. 2007. "Association of Per1 and Npas2 with Autistic Disorder: Support for the Clock Genes/Social Timing Hypothesis" *Molecular Psychiatry* 12(6): 581–92.

Ninio A., and C.E. Snow. 1996. *Pragmatic Development*. Boulder, Colorado: Westview Press.

Ninio A., and C.E. Snow. 1988. "Language Acquisition Through Language Use: The Functional Sources of Children's Early Utterances," in *Categories and Processes in Language Acquisition*, edited by Y. Levi, I. Schlesinger, and M. Braine, 11–30. Hillsdale, NJ: Lawrence Erlbaum Associates.

Nir, I. et al. 1995. "Brief Report: Circadian Melatonin, Thyroid-Stimulating Hormone, Prolactin, and Cortisol Levels in Serum of Young Adults with Autism" *Journal of Autism and Developmental Disorders* 25(6): 641–54.

Oberman, Lindsay M., Piotr Winkielman, and Vilayanur S. Ramachandran. 2009. "Slow Echo: Facial EMG Evidence for the Delay of Spontaneous, But Not Voluntary, Emotional Mimicry in Children with Autism Spectrum Disorders" *Developmental Science* 12(4): 510–20.

O'Connor, K. 2012. "Auditory Processing in Autism Spectrum Disorder: A Review" *Neuroscience and Biobehavioral Reviews* 36(2): 836–54.

Oller, D.K. et al. 2010. "Automated Vocal Analysis of Naturalistic Recordings from Children with Autism, Language Delay, and Typical Development" *Proceedings of the National Academy of Sciences of the United States of America* 107(30): 13354–9.

Pawlby, S.J. 1977. "Imitative Interaction," in *Studies in mother–infant interaction*, edited by H.R. Schaffer, 203–24. New York, NY: Academic Press.

Peppe, Susan et al. 2007. "Receptive and Expressive Prosodic Ability in Children with High-Functioning Autism" *Journal of Speech Language and Hearing Research* 50(4): 1015–28.

Peppe, Susan et al. 2011. "Expressive Prosody in Children with Autism Spectrum Conditions" *Journal of Neurolinguistics* 24(1): 41–53.

Petitto, L.A. et al. 2000. "Speech-Like Cerebral Activity in Profoundly Deaf People Processing Signed Languages: Implications for the Neural Basis of Human Language" *Proceedings of the National Academy of Sciences of the United States of America* 97(25): 13961–6.

Petitto, L.A. et al. 2001. "Language Rhythms in Baby Hand Movements" *Nature* 413(6851): 35–6.

Petitto, L.A. et al. 2004. "Baby Hands that Move to the Rhythm of Language: Hearing Babies Acquiring Sign Languages Babble Silently on the Hands" *Cognition* 93(1): 43–73.

Piven, J. et al. 1997. "An MRT Study of the Corpus Callosum in Autism" *American Journal of Psychiatry* 154(8): 1051–6.

Plaisted, K., M. O'Riordan, and S. Baron-Cohen. 1998. "Enhanced Discrimination of Novel, Highly Similar Stimuli by Adults with Autism during a Perceptual Learning Task" *Journal of Child Psychology and Psychiatry and Allied Disciplines* 39(5): 765–75.

PowersAlbert R. III, Andrea R. Hillock, and Mark T. Wallace. 2009. "Perceptual Training Narrows the Temporal Window of Multisensory Binding." *Journal of Neuroscience* 29(39): 12265–74.

Puffenberger, Erik G. et al. 2012. "A Homozygous Missense Mutation in HERC2 Associated with Global Developmental Delay and Autism Spectrum Disorder" *Human Mutation* 33(12): 1639–46.

Reddy, V. 1991. "Playing with Others' Expectations: Teasing and Mucking about in the First Year," in *Natural theories of mind*, edited by A. Whiten, 143–58. Oxford, England: Blackwell.

Reddy, V. 2001. "Infant Clowns: The Interpersonal Creation of Humor in Infancy" *Enfance* 3: 247–56.

Reddy, V. 2008. *How Infants Know Minds*. Cambridge, Massachusetts: Harvard University Press.

Richdale, A.L., and M.R. Prior. 1992. "Urinary Cortisol Circadian-Rhythm in a Group of High-Functioning Children with Autism" *Journal of Autism and Developmental Disorders* 22(3): 433–47.

Richdale, A.L., and M.R. Prior. 1995. "The Sleep-Wake Rhythm in Children with Autism" *European Child and Adolescent Psychiatry* 4(3): 175–86.

Rinehart, Nicole J. et al. 2006. "An Examination of Movement Kinematics in Young People with High-Functioning Autism and Asperger's Disorder: Further Evidence for a Motor Planning Deficit" *Journal of Autism and Developmental Disorders* 36(6): 757–67.

Ritvo, E.R. et al. 1986. "Lower Purkinje-Cell Counts in the Cerebella of 4 Autistic Subjects – Initial Findings of the Ucla-Nsac Autopsy Research Report" *American Journal of Psychiatry* 143(7): 862–6.

Roberts, Timothy P.L. et al. 2010. "MEG Detection of Delayed Auditory Evoked Responses in Autism Spectrum Disorders: Towards an Imaging Biomarker for Autism" *Autism Research* 3(1): 8–18.

Roe, K.V., A. Mcclure, and A. Roe. 1982. "Vocal Interaction at 3 Months and Cognitive Skills at 12 Years" *Developmental Psychology* 18(1): 15–16.

Rogoff, B. 1990. *Apprenticeship in Thinking: Cognitive Development in Social Context.* New York, NY: Oxford University Press.

Rozga, Agata et al. 2011. "Behavioral Profiles of Affected and Unaffected Siblings of Children with Autism: Contribution of Measures of Mother–Infant Interaction and Nonverbal Communication" *Journal of Autism and Developmental Disorders* 41(3): 287–301.

Ruddy, M.G., and M.H. Bornstein. 1982. "Cognitive Correlates of Infant Attention and Maternal Stimulation Over the 1st Year of Life" *Child Development* 53(1): 183–8.

Russo, N. et al. 2012. "Parameters of Semantic Multisensory Integration Depend on Timing and Modality Order among People on the Autism Spectrum: Evidence from Event-Related Potentials" *Neuropsychologia* 50(9): 2131–41.

Saint-Georges, Catherine et al. 2011. "Do Parents Recognize Autistic Deviant Behavior Long Before Diagnosis? Taking into Account Interaction Using Computational Methods" *PLoS One* 6(7): e22393.

Salmond, C.H. et al. 2005. "The Role of the Medial Temporal Lobe in Autistic Spectrum Disorders" *European Journal of Neuroscience* 22(3): 764–72.

Schmitz, C. et al. 2003. "Motor Control and Children with Autism: Deficit of Anticipatory Function?" *Neuroscience Letters* 348(1): 17–20.

Scott, Julia A. et al. 2009. "A Comprehensive Volumetric Analysis of the Cerebellum in Children and Adolescents with Autism Spectrum Disorder" *Autism Research* 2(5): 246–57.

Sears, L.L., P.R. Finn, and J.E. Steinmetz. 1994. "Abnormal Classical Eye-Blink Conditioning in Autism" *Journal of Autism and Developmental Disorders* 24(6): 737–51.

Segawa, M. 1985. "Circadian Rhythm in Early Infantile Autism" *Shinke Kenya No Shinpo* 29: 140–53.

Siller, M., and M. Sigman. 2002. "The Behaviors of Parents of Children with Autism Predict the Subsequent Development of Their Children's Communication." *Journal of Autism and Developmental Disorders* 32(2): 77–89.

Siller, Michael, and Marian Sigman. 2008. "Modeling Longitudinal Change in the Language Abilities of Children with Autism: Parent Behaviors and Child Characteristics as Predictors of Change" *Developmental Psychology* 44(6): 1691–704.

Sillitoe, Roy V. et al. 2008. "Engrailed Homeobox Genes Determine the Organization of Purkinje Cell Sagittal Stripe Gene Expression in the Adult Cerebellum" *Journal of Neuroscience* 28(47): 12150–62.

Spelke, E.S., and Cortelyou, A. 1981. "Perceptual Aspects of Social Knowing: Looking and Listening in Infancy," in *Infant social cognition: Empirical and theoretical considerations*, edited by M.E. Lamb and L.R. Sherrod. Hillsdale, NJ: Lawrence Erlbaum Associates.

Stern, D. 1974. "Mother and Infant at Play: The Dyadic Interaction Involving Facial, Vocal and Gaze Behaviors," in *The effect of the infant on its caregiver*, edited by M. Lewis and L. Rosenblum. New York, NY: Wiley.

Stern, D. 1977. *The First Relationship*. Cambridge, Massachusetts: Harvard University Press.

Stern, D.N. 1985. *The Interpersonal World of the Infant: A View from Psycho-Analysis and Developmental Psychology*. New York, NY: Basic Books.

Stern, D. 1995. *The Motherhood Constellation*. New York: Basic Books.

Stern, D., and J. Gibbon. 1979. "Temporal Expectancies of Social Behaviors in Mother–Infant Play," in *Origins of the Infant's Social Responsiveness*, edited by E. Thoman. Hillsdale, NJ: Lawrence Erlbaum Associates.

Stern, D. et al. 1977. "The Infant's Stimulus World during Social Interaction," in *Studies in mother–infant interaction*, edited by H.R. Schaffer. New York, NY: Academic.

Suddendorf, T., and M.C. Corballis. 1997. "Mental Time Travel and the Evolution of the Human Mind" *Genetic Social and General Psychology Monographs* 123(2): 133–67.

Szelag, E. et al. 2004. "Temporal Processing Deficits in High-Functioning Children with Autism" *British Journal of Psychology* 95: 269–82.

Tallal, P. 2004. "Opinion – Improving Language and Literacy is a Matter of Time" *Nature Reviews Neuroscience* 5(9): 721–8.

Terracciano, A. et al. 2010. "Genome-Wide Association Scan for Five Major Dimensions of Personality" *Molecular Psychiatry* 15(6): 647–56.

Tomasello, M. 1999. "Having Intentions, Understanding Intentions, and Understanding Communicative Intentions," in *Developing theories of intention: Social understanding and self control*, edited by P.D. Zelazo and J.W. Astington, 63–75. Mahwah, NJ: Lawrence Erlbaum Associates.

Townsend, J. et al. 1999. "Spatial Attention Deficits in Patients with Acquired or Developmental Cerebellar Abnormality" *Journal of Neuroscience* 19(13): 5632–43.

Townsend, J. et al. 2001. "Event-Related Brain Response Abnormalities in Autism: Evidence for Impaired Cerebello-Frontal Spatial Attention Networks" *Cognitive Brain Research* 11(1): 127–45.

Trevarthen, C., and K. Logotheti. 1987. "First Symbols and the Nature of Human Knowledge," in *Symbolism and Knowledge*, edited by J. Montangero, A. Tryphon and S. Dionnet. Cahier No. 8, Geneva: Jean Piaget Archives Foundation.

Trevarthen, C., and K.J. Aitken. 2001. "Infant Intersubjectivity: Research, Theory and Clinical Applications" *Journal of Child Psychology and Psychiatry* 42: 3–48.

Trevarthen, C., and S. Daniel. 2005. "Disorganized Rhythm and Synchrony: Early Signs of Autism and Rett Syndrome" *Brain and Development* 27: 25–34.

Tronick, E.Z. 1989. "Emotions and Emotional Communication in Infants" *American Psychologist* 44(2): 112–9.

Tronick, E.Z., and J.F. Cohn. 1989. "Infant–Mother Face-to-Face Interaction – Age and Gender Differences in Coordination and the Occurrence of Miscoordination" *Child Development* 60(1): 85–92.

Tronick, E.D., H. Als, and T.B. Brazelton. 1977. "Mutuality in Mother–Infant Interaction" *Journal of Communication* 27(2): 74–9.

Tronick, E., H. Als, and T.B. Brazelton. 1980. "Monadic Phases – Structural Descriptive Analysis of Infant–mother Face to Face Interaction" *Merrill-Palmer Quarterly-Journal of Developmental Psychology* 26(1): 3–24.

Utge, Siddheshwar J. et al. 2010. "Systematic Analysis of Circadian Genes in a Population-based Sample Reveals Association of TIMELESS with Depression and Sleep Disturbance" *PLoS One* 5(2): e9259.

Vilensky, J.A., A.R. Damasio, and R.G. Maurer. 1981. "Gait Disturbances in Patients with Autistic Behavior – A preliminary-study" *Archives of Neurology* 38(10): 646–9.

Vygotsky, L.S. 1978. *Mind in Society: The Development of Higher Mental Processes.* Cambridge, Massachusetts: Harvard University Press.

Walker, M.P., and R. Stickgold. 2004. "Sleep-Dependent Learning and Memory Consolidation" *Neuron* 44(1): 121–33.

Wallace, Gregory L., and Francesca Happe. 2008. "Time Perception in Autism Spectrum Disorders" *Research in Autism Spectrum Disorders* 2(3): 447–55.

Warner, R.M. 1992. "Sequential-Analysis of Social-Interaction – Assessing Internal Versus Social Determinants of Behavior" *Journal of Personality and Social Psychology* 63(1): 51–60.

Webb, Sara J. et al. 2006. "ERP Evidence of Atypical Face Processing in Young Children with Autism" *Journal of Autism and Developmental Disorders* 36(7): 881–90.

Welkowitz, J., and S. Feldstein. 1970. "Relation of Experimentally-Manipulated Interpersonal Perception and Psychological Differentiation to the Temporal Patterning of Conversation" *Proceedings of the 78th Annual Convention of the American Psychological Association* 5: 387–8.

Welkowitz, J., G. Cariffe, and S. Feldstein. 1976. "Conversational Congruence as a Criterion of Socialization in Children" *Child Development* 47(1): 269–72.

Welsh, J.P., E.S. Ahn, and D.G. Placantonakis. 2005. "Is Autism Due to Brain Desynchronization?" *International Journal of Developmental Neuroscience* 23(2–3): 253–63.

Williams, Diane L., and Nancy J. Minshew. 2007. "Understanding Autism and Related Disorders: What Has Imaging Taught us?" *Neuroimaging Clinics of North America* 17(4): 495.

Wimpory, D. 1995. "Social Engagement in Preschool Children with Autism" (PhD diss., University of Wales, Bangor, Gwynedd, UK).

Wimpory, D., and Nash, S. 1999a. "Musical Interaction Therapy for Children with Autism," in *Innovative Psychotherapy Techniques in Child and Adolescent Therapy*, 2nd Edition, edited by C. Schaeffer. New York, NY: Wiley.

Wimpory, D., and S. Nash. 1999b. "Musical Interaction Play Therapy" *Child Language Teaching and Therapy* 15: 17–28.

Wimpory, D., P. Chadwick, and S. Nash. 1995. "Musical Interaction Therapy for Children with Autism – An Evaluative Case-study with 2-year Follow-up – Brief report" *Journal of Autism and Developmental Disorders* 25(5): 541–52.

Wimpory, D.C. et al. 2000. "Are Infants with Autism Socially Engaged? A Study of Recent Retrospective Parental Reports" *Journal of Autism and Developmental Disorders* 30(6): 525–36.

Wimpory, D., B. Nicholas, and S. Nash. 2002. "Social Timing, Clock Genes and Autism: A New Hypothesis" *Journal of Intellectual Disability Research* 46: 352–8.

Wimpory, D., R.P. Hobson, and S. Nash. 2007. "What Facilitates Engagement in Preschool Children with Autism?" *Journal of Autism and Developmental Disorders* 37(3): 564–73.

Wimpory, D., S. Nash, A. Muth, and B. Nicholas. in progress. "Evaluating Temporal Aspects of Communication in Autistic Spectrum Disorders."

Wolff, P.H. 1967. "The Role of Biological Rhythms in Early Psychological Development" *Bulletin of the Menninger Clinic* 31: 197–218.

Wolff P.H. 1991. Endogenous Motor Rhythms in Young infants. In: Fagard J, Wolff P.H, editors. The Development of Timing Control and Temporal Organization in Coordinated Action. New York: Elsevier Science Publishers; 1991.

Yirmiya, N. et al. 2006. "The Development of Siblings of Children with Autism at 4 and 14 Months: Social Engagement, Communication, and Cognition" *Journal of Child Psychology and Psychiatry* 47(5): 511–23.

Zlochower, A.J., and J.F. Cohn. 1996. "Vocal Timing in Face-to-Face Interaction of Clinically Depressed and Nondepressed Mothers and Their 4-Month-Old Infants" *Infant Behavior and Development* 19(3): 371–4.

CHAPTER 4

Time Processing in Schizophrenia

Deana B. Davalos and Jamie Opper**

> Time is inside as well as outside of ourselves. Time is a perception. It is part of the outside world, but it is also a sensation immediately experienced in ourselves. We organize and crystalize the perception of time into the connotation of a continuous flowing time, which we measure by clocks, and we try to apply the same measures to the time experience in ourselves, to what we may call time sensation.... Time is an inherent part of the world of perception, outside and inside the body.
>
> (SCHILDER, 1936)

1 Introduction

Schilder (1936) described his conception of time in his paper, *Psychopathology of Time*, which addressed time perception and the idea that various types of psychopathology involved a disturbance in time perception. Schilder wrote the paper while at Bellevue Psychiatric Hospital and detailed a variety of disorders and the possible temporal distortion associated with each type of pathology. About schizophrenia, he quotes a patient who says, "I can't orient myself in the world – I am not clear anymore...I continue to live in eternity. There is no hour, no noon, no night.... Time does not move. I am wavering between past and future." Schilder's observations of timing dysfunction in his schizophrenic patients led him to ask, "...why does the schizophrenic give up his time experience? What does time mean for him?" The questions regarding time perception and the role timing plays in schizophrenia continue to perplex researchers today. Our understanding of time processing in schizophrenia has developed over the years from a once rather simplistic view of temporal dysfunction to an elaborate organization of temporal deficits that span from simple sensory measures of timing to higher order processes. And while it has been argued for decades that time processing is disrupted in schizophrenia, the breadth of the implications stemming from those temporal deficits have grown to include areas of clinical symptomatology, social and emotional processing, language, to higher order cognitive processes.

* Colorado State University, USA.

DOI 10.1163/9789004230699_005

Conceptually, our capacity to process time is viewed as an ability that plays a critical role in our perception of the world around us. Navon (1978) wrote that our perception of the world consists of a hierarchy of dimensions and that time is at the top of that hierarchy. The ability to process time has been associated with relatively basic tasks such as planning and sequencing and processing basic sensory input to higher order processes that are involved in athletic ability, driving, language, walking, and musical ability (Eagleman et al., 2005; Eagleman, 2009; Ferrandez et al., 2003; Macar et al., 2006; Mangels et al., 1998; Tracy et al., 1998). Given the widespread collection of behaviors, actions, and cognitive processes that appear to be influenced by timing, investigators have begun to speculate whether deficits in information processing and higher level cognitive processing that have been associated with dysexecutive syndrome, and schizophrenia specifically, may be in part due to temporal dysfunction (Macar and Vidal, 2009; Volz et al., 2001). And while time processing may not have a place in the scientific literature as extensive as other cognitive processes, such as attention, working memory, or inhibition, there are indicators that suggest that the magnitude of the importance of intact temporal processing is beginning to be understood. First, Head and colleagues (2008) have recently included time processing as one of the few possible "cognitive primitives" or what is described as basic neuropsychological processess that have broad influence on other cognitive functions, but cannot be separated into component processes themselves (Salthouse, 1985; Verhaeghen and Salthouse, 1997; Zacks and Hasher, 1994). While Head et al. have introduced the idea that time perception has a place in the hierarchy of cognitive processes alongside what have been viewed as fundamental cognitive capacities, such as inhibition and processing speed, there have been relatively few studies that can be used to support the argument. Cognitive primitives cannot, by definition, be separated into component processes and many of the current studies assessing time processing highlight the complexity of measuring temporal processing as most timing tasks generally involve some degree of attentional resources, decision making, and vigilance. Head's proposal, nonetheless, highlights the fact that timing is now considered as being a foundational cognitive process that may be involved in widespread dysfunction. The idea that temporal processing may lead to some type of dysfunction highlights the second indicator that stresses the magnitude of the importance of intact time processing.

The scientific literature is rich with studies assessing clinical populations and how time processing may be at the root of clinical symptomatology and deficits associated with various disorders. While it is beyond the scope of this chapter to address all of the types of psychopathology associated with temporal processing, included in the list of clinical disorders and/or clinical features are: aphasia, Alzheimer's disease, attention-deficit hyperactivity disorder, dyslexia, apraxia of speech, traumatic brain injury, autism-spectrum disorders, and schizophrenia (Anderson and Schmitter-Edgecombe, 2011; Barabasz, 1973; Bauermeister et al., 2005; Carroll, O'Donnell, Shekhar, and Hetrick, 2009; Condray, 2005; Efron, 1963; Kaminsky et al., 2002; Kraus, 2003; Merzenich et al., 1996; Rueda and Schmitter-Edgecombe, 2009; Wallace and Happe, 2008; Ziegler and von Cramon, 1986). These represent but a few of the many clinical disorders that are linked to time processing deficits. The breadth of disorders that link clinical symptomatology to temporal processing difficulties suggest a critical role between timing and clinical health, or more specifically clinical psychopathology. Regarding schizophrenia, the link between the disorder and timing dysfunction has been recognized for nearly a century. Minkowski (1931) argued that many clinical disorders could be linked to a disorder in space, time, or memory. Specifically, he described conditions such as dementia as being disturbed in terms of memory while the space and time "I-Here-Now" factors are intact. Schizophrenia, on the other hand, appears to involve intact memory, with disturbed "I-Here-Now" factors (Minkowski, 1928). Israeli (1932) eloquently reviews physician and researchers' perspectives on time processing in schizophrenia during the 1920's and 1930's and highlights the personal views of distorted time in patients with schizophrenia. Specifically, he discusses patient complaints presented by Fischer (1929, 1930a,b,c), which stress the "I-Here-Now" disturbance. Fischer reports that his patients complain of: feeling that time has come to a standstill which is compensated for by a dreamlike existence, demonic playing with time, praying for the destruction of time, and feelings of timelessness. Fischer goes on to speculate that there is no symptom of schizophrenia, which is not a product of a time-space disturbance.

The idea that most symptoms of schizophrenia, both cognitive and clinical, may be affected by a disturbance in timing has persisted over the years. Eagleman and Holcombe (2002) point to research, which argues that individuals with schizophrenia may believe that their own thoughts are caused by someone else or possess a delusion that unrelated events are attributable to their actions (Frith, Blakemore, and Wolpert, 2002). In Frith and colleagues' research, they speculate that these disturbances may be due to a lack of awareness of certain aspects of motor control. Eagleman and Holcombe have argued that rather than an emphasis placed on difficulties in the processing of motor control, these symptoms could instead be due to difficulties in temporal processing. Stetson et al. (2006) have shown that when healthy control subjects had a relatively simple sensory-motor task slightly adjusted in terms of the interval duration between when a button was pushed and a stimulus was presented, it induced a sense of not being in control of their physical response.

Specifically, participants were first asked to simply press a button, which instantly resulted in a nearby light to flash. The paradigm was then slightly changed to add a tenth of a second delay between the button press and the flash of light appearing. Following this condition with the slight delay, the experimenters removed the delay and returned to the initial presentation. The removal of the delay appeared to induce in the participants, a perception that the flash was occurring before the responder had pressed the key. This study, among others, has led to questions about whether individuals with schizophrenia may experience was has been termed credit misattribution, or difficulty recognizing actions as one's own. The thought is that a relatively brief temporal distortion may be associated with the delusion that someone other than the individual themselves is in charge of their behavior (Gandhi, Wassef, and Eagleman, 2007). These findings of slight temporal distortion leading to delusions or feelings of confusion in healthy young adults has been compelling evidence in support of an argument for the important role temporal dysfunction may play in the clinical pathology of schizophrenia.

While research continues to explore the role of temporal deficits in schizophrenia, our understanding of the neural underpinnings of temporal dysfunction remains elusive. Comprehending the neural mechanisms involved in timing is important for a number of reasons. Given the breadth of timing deficits observed in schizophrenia, one question that persists is whether there may be a specific type of neural dysfunction that affects a wide range of processes or whether there are multiple areas of dysfunction in the schizophrenic brain that each contribute a unique piece to the various types of timing dysfunction observed.

To understand the complexity of trying to unravel the neural underpinnings of temporal processing, one first has to appreciate the plethora of tasks that are subsumed under the heading of time processing. The types of task that are considered timing-related represent a diverse set of processes ranging from time perception of durations in milliseconds, time-based prospective memory, temporal reproduction, temporal order, to temporal estimation of hours. Many researchers have attempted to organize types of temporal tasks to better understand timing. This need to clarify temporal processes, specifically in the context of psychopathology, is clear in the literature as early as the 1960's. Lehmann (1967) presents one of the important distinctions in our understanding of time processing. He argues that there are two types of time, external time and internal time. External time is the "objective, universal, and absolute" time that is measured within a "conceptual framework." Internal time, on the other hand, is "subjective, individual, and relative" and he describes it as "the perception of enduring." It is this internal time that has been of most interest in the study of schizophrenia. Specifically, it is thought that disruptions in one's internal clock, even slight disruptions, appear to lead distortions in how individuals with schizophrenia conceive their environment and how they operate in their environment. And it is difficulties with the internal clock that appear to be at the root of a wide spectrum of clinical pathology in schizophrenia (e.g., hallucinations, delusions, psychomotor poverty, poverty of speech; Andreasen, 1999; Hyde, Ziegler, and Weinberger, 1993; McGlashan and Hoffman, 2000). In addition, temporal processing deficits have been argued by many to also affect *cognitive* functioning in schizophrenia.

For decades, researchers have posited that individuals with schizophrenia appear to process time differently than those who are not affected with the mental illness on a variety of behavioral and cognitive measures (Rabin, 1957). These deficits are noted on tasks assessing various types of time processing, including temporal perception, temporal estimation, and time reproduction (Davalos, Kisley, and Ross, 2002, 2003a; Rammsayer, 1990; Tysk, 1990). And these deficits of time processing seem to be widespread, spanning both auditory and visual temporal processing tasks (Schwartz, Mallott, and Winstead, 1988; Tysk, 1990). As will be addressed repeatedly in this Chapter, one of the primary limitations to understanding the neural underpinnings of *time processing per se* is the degree of nontemporal information that has historically been included in temporal processing tasks. Poynter and Homa (1983) point out that temporal perception tasks generally require the participant to be able to attend successfully to the nontemporal information that is often embedded in tasks. Given that most cognitive measures of timing employ attention and working memory, it is likely "temporal processing performance" in actuality reflects an amalgamation of multiple cognitive skills. The variations in attentional demands or recruitment of other cognitive resources have likely contributed to the variability in findings across behavioral studies and neurophysiology studies.

2 Electrophysiology

Mismatch negativity (MMN) has been utilized as a physiological measure of temporal processing. MMN is a brain response elicited to infrequent deviant stimuli in the context of recurring standard stimuli (Naatanen, 1992). One of the most important aspects of MMN is that it is generated in the absence of, or with minimal demands, on focused attention. Also, there is no motor response required nor is there a decision making element that is often involved in cognitive tasks and even other event related potentials (Sussman, Winkler, and Wang, 2003). MMN has been utilized in studies with a variety of clinical disorders, specifically those with suspected frontal deficits as it allows detection of deficits associated with frontal and temporal lobe dysfunction while minimizing confounding deficits associate with attention, vigilance, motivation, and decision making in these populations. MMN represents the neural process of detecting change in stimuli and generally exhibits maximal amplitude at frontal and central sites of the scalp. The main generators are thought to be located in the auditory and frontal cortices (Naatanen, 1992; Pulvermuller and Shtyrov, 2003). The frontal contribution to this waveform is thought to reflect an involuntary switch of attention to specific changes in the sensory environment (Deouell, Bentin, and Giard, 1998). MMN studies focusing on individuals with schizophrenia have included deviations in pitch, intensity, frequency, or, more specific to the current discussion, deviations in the duration of tones and/or interstimulus intervals (Baldeweg et al., 2002; Naatanen et al., 1993; Todd, Michie, and Jablensky, 2003). MMN paradigms utilizing duration intervals are thought to be valid representations of time processing as MMN amplitude is proportional to the relative discriminability of interval deviations, suggesting that the preattentive neural activity measured during MMN is that which is subsequently available for the conscious perception of time (Kisley et al., 2004). In schizophrenia, results consistently find that individuals with schizophrenia exhibit deficits in temporal processing measured via MMN (Baldeweg et al., 2002; Davalos, Kisley, and Ross, 2003b; Light and Braff, 2005). Light and Braff (2005a) have emphasized the practical applications of these deficits by linking these time-based MMN deficits to functional status (i.e., GAF Scale ratings and level of independence in community living situation). In their studies, an MMN utilizing a subtle temporal deviant (90% standard, 50 msec, and 10% deviant, 100 msec, duration stimuli) was assessed in patients with schizophrenia. MMN deficits in this population were accounted for up to 42% of the variance in functional status. As the authors note, the correlation between time-based MMN deficits and impaired functioning in their study was observed as being much higher than the correlation between traditional neurocognitive measures and functional status and outcome (Green, 1996; Green et al., 2000; Palmer et al., 2002). In a separate study, the authors extend their findings by proposing that time-based MMN may also serve as a useful method to track progressive changes in neural substrate dysfunction and functional impairments over the course of schizophrenia (Light and Braff, 2005b). Light and Braff bolster this argument by highlighting past research indicating that while clinically unaffected relatives of individuals with schizophrenia and those with chronic schizophrenia exhibit mismatch negativity deficits, first-episode patients exhibit intact mismatch negativity (Light and Braff, 2005b, Michie et al., 2002; Salisbury et al., 2002a; Umbricht et al., 2003). These findings support the utility of using temporal MMN indices as a tool for tracking brain-based changes associated with the progression of schizophrenia and possibly the role that MMN may play in serving as a biomarker for schizophrenia. While Light and Braff have highlighted the potential clinical utility of MMN, the authors point out that further research is needed to fully appreciate the potential strengths and limitations of time-based MMN in diagnoses and understanding of progression of schizophrenia. They argue that greater clarification is needed to understand the longitudinal course of time-based MMN deficits and the role that temporal processing measured via MMN may play in understanding neurodegeneration in schizophrenia.

It should be noted that if the reader questions whether it is the general electrophysiological index, MMN, independent of type of deviant, that appears to be the basis for the relationship between neurocognitive functional, functional status, and tracking progression of the disease in schizophrenia, the literature suggests otherwise. Specifically, past studies suggest that there is something unique about temporal deviants that are not present with other types of deviants. MMN response to duration deviants has been compared to frequency and intensity in the past to assess test-retest reliability and duration only was most replicable amplitude and latency among the deviances assessed. The authors argue then that duration-dependent MMN may be the best measure to utilize in the study of cognitive brain functioning (Tervaniemi et al., 1999). In general, these findings, along with others, illustrates the role that temporal processing measured via MMN plays in detecting vulnerability for schizophrenia, indirectly assessing the state of NMDA receptor functioning, and measuring the integrity of frontotemporal brain systems.

It should be noted, however, that recent electrophysiological research suggests that the timing deficits observed in schizophrenia are not generalized across all timing paradigms nor are they apparent across all areas of the brain. Rather, poor temporal processing was only noted for the conditions in which the difference between the standard interstimulus interval and the deviant interstimulus interval was more subtle (e.g., 15% different rather than 50% different; Davalos, Kisley, and Freedman, 2005). What was notable during the study was that there was a dissociation between behavioral performance and neurophysiological indices of temporal processing in the schizophrenia group. While their neurophysiological response was comparable to controls on the "easy" deviant, their behavioral performances were impaired compared to controls on both the "easy" and "difficult" deviants. The finding suggesting differences between conditions (easy vs. difficult) and across method of assessment (event-related potentials versus behavioral) suggests that the role of attention, motivation, and other mediating cognitive factors cannot be minimized when examining behavioral indices of time processing in populations that may be more affected by impairments in these cognitive areas. Salisbury et al. have argued in the past that the use of traditional behavioral measures of cognition paired with electrophysiological measures of brain activity allow for a unique multifaceted examination of behavior and the underlying processes (Salisbury et al., 2002b). And while ERP measures allow us to begin to understand that there *are* neurophysiological differences in the brains of patients with schizophrenia and healthy controls when it comes to temporal processing, it has not told us about the 'where' and 'how' of these processes.

3 Functional Imaging/Localization

The links between neurophysiology and behavioral indices of temporal processing highlight that there are clearly shared neural networks between both types of measures of time processing as well as distinct processes involved. This has led to questions about the neural etiology of time processing. Is there truly an "internal clock" that may be associated with all temporal information (Mangels et al., 1998)? And is this impaired internal clock the common link that is shared across an array of clinical disorders, such as PD, amnesia, cerebellar, and basal ganglia insult, in addition to frontal syndromes (Artieda et al., 1992; Casini and Ivry, 1999; Harrington and Haaland, 1998; Ivry and Keele, 1989; Nichelli et al., 1995; Rammsayer, 1993). Or is it more likely that there are specific types of temporal processing that rely on different neural networks?

For years it has been argued that the cerebellum is the key structure associated with short-duration processing while longer duration processing may be dependent on prefrontal cortical functioning (Clarke et al., 1996; Heatherington et al., 2000; Mangels, Ivry, and Shimizu, 1998). While the cerebellar argument appears to be relatively well established in terms of the role in automatic short-duration processing, the processing of longer durations appears more complex, with the relative roles of the basal ganglia and frontal cortex still contested (Hazeltine, Helmuth, and Ivry, 1997; Ivry, 1996). Specifying the neural underpinnings to processing longer durations is inherently more complicated because of the possible recruitment of other cognitive processes. Mangels et al. (1998) point out that longer durations (e.g., 3 or 4 s) utilized in interval comparison tasks may require working memory to keep those interval durations available for comparison. These types of temporal judgments may be particularly sensitive to frontal lobe functioning. Nichelli et al. (1995) supports this argument, noting that timing beyond 2 or 3 s may exceed more automatic "motor routines" and may require sustained attention and/or strategy use. Whether one argues for the necessity of sustained attention or working memory, it is likely that, unlike short intervals, longer intervals recruit the prefrontal cortex. Given the well-established prefrontal dysfunction associated with schizophrenia, this potential confound has made measuring the "true" neural correlates of temporal process difficult (Goldman-Rakic and Selemon, 1997; Goldman-Rakic, 1999; Mull and Seyal, 2001; Perry et al., 2001; Wall and Messier, 2001). Volz and colleagues began to explore the neuroanatomy of time processing by assessing patterns of brain activation during a functional magnetic resonance image (fMRI) study during which patients with schizophrenia were assessed on time estimation and pitch discrimination. Their findings, using psychophysical threshold levels, indicated that the groups differed significantly only in time estimation. In addition, their comparison of pitch versus time discrimination indicated a specific dysfunction of the right putamen, right medial/superior prefrontal cortex, and right anterior thalamus to underlie that was associated with timing dysfunction in schizophrenia. The authors speculated that, based on their findings, timing deficits in schizophrenia might be due to impairment in both the basal timing mechanisms of the basal ganglia and/or thalamus paired with impaired attentional or working memory resources that are facilitated by the prefrontal cortices. An additional finding, that is important when studying individuals with schizophrenia, is that their findings were not simply a function of the patients performing worse than the controls, as difficulty was adjusted to an individual performance level. The authors, like many to follow, argue that timing in the brain is associated with a disturbed fronto-thalamo-striatal circuit.

Davalos, Rojas, and Tregellas (2011) followed up on Volz and colleagues work with some minor changes, including varied levels of difficulty assessed between patients and controls, a larger cohort of patients and healthy comparison subjects, and the use of a 3T scanner in contrast to the 1.5T scanner used in the earlier study. The study not only sought to elaborate on the work of Volz and colleagues, but also to provide a follow up to the previous electrophysiological study (Davalos, Kisley, and Freedman, 2005), which indicated a dissociation between behavioral indices of time processing and neurophysiological responses to temporal deviants. Similar to the previous ERP study, participants were assessed on an "easy" and a "difficult" condition. In the "easy" condition, participants heard two tones, the first 200 msec in duration, the second either shorter in duration (70ms, 100ms) or longer (300ms, 330ms) than the first tone. The paradigm utilized an interval, separated by 500ms. In the "difficult" condition, the duration of the second tone was more similar to the standard tone, either shorter (160ms, 170ms) or longer (230ms, 240ms; Figure 4.1).

Finally, in the base condition, both tones were 200ms. On both the "easy" and "difficult" conditions, the control group performed significantly better than the schizophrenia group. Regarding hemodynamic responses, group differences

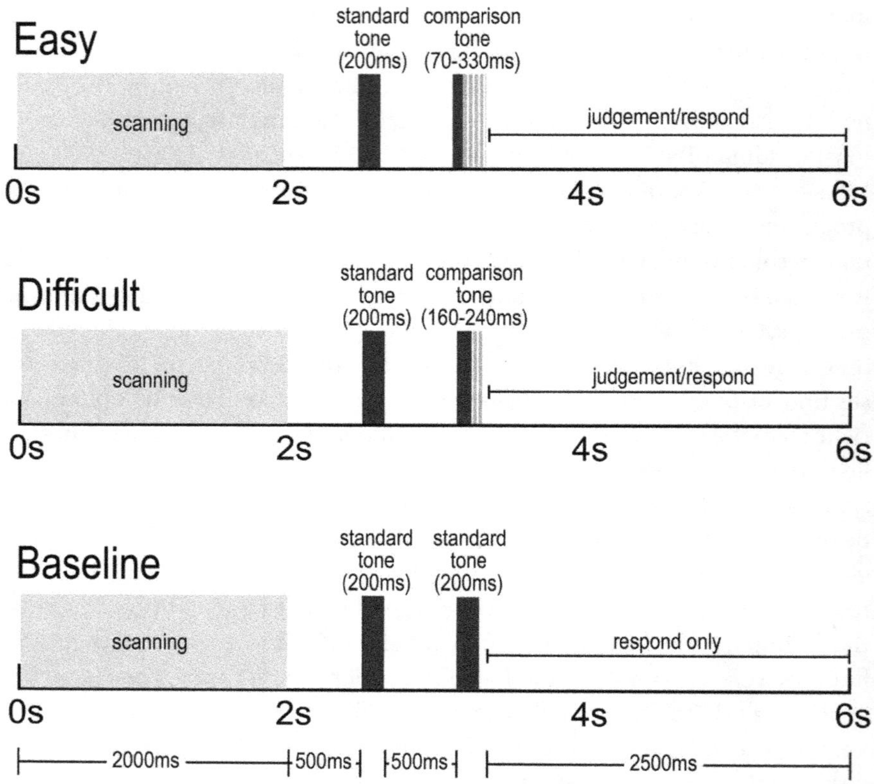

FIGURE 4.1 *Schematic representation of experimental design. During the "easy" condition, the 200ms standard tone was followed by a comparison tone of 70, 100 300 or 330ms (±50 or 65%). During the "difficult" condition, the comparison tone was 160, 170, 230 or 240ms (±15, 20%). During "baseline," subjects pressed a button, but did not judge, following two 200ms tones. Reprinted from Schizophrenia Research, Vol. 127, D.B. Davalos, D.C. Rojas, and J.R. Tregellas, Temporal processing in schizophrenia: Effects of task-difficulty on behavioral discrimination and neuronal responses, pp. 123–130.*

(control vs. schizophrenia) were assessed for three contrasts. In the easy discrimination condition, individuals with schizophrenia exhibited reduced activation in the supplementary motor area (SMA), insular/opercular cortex, and dorsolateral prefrontal cortex (DLPFC; Table 4.1, Figure 4.2).

In the "difficult" condition, the individuals with schizophrenia showed a more robust reduction in response in the SMA, insular/opercular cortex, and DLPFC. Also, reduced activation was observed in the striatum and thalamus (Figure 4.3). Regarding the difficult–easy contrast, findings indicate that the schizophrenia group exhibited reduced activation in the insular/opercular

TABLE 4.1 *MNI coordinates and statistics for brain regions with greater activation in controls compared to individuals with schizophrenia. Reprinted from Schizophrenia Research, Vol. 127, D.B. Davalos, D.C. Rojas, and J.R. Tregellas, Temporal processing in schizophrenia: Effects of task-difficulty on behavioral discrimination and neuronal responses, pp. 123–130. Copyright 2011.*

	x	y	z	t value	P value
Easy					
Operculum/Insula (L)	−36	24	−3	1.79	0.042
Pre-SMA (R)	12	21	51	1.69	0.051
DLPFC (L)	−48	9	36	1.88	0.035
Difficult					
Operculum/Insula (R)	51	24	0	2.06	0.03
	30	24	−6	2.25	0.016
Operculum/Insula (L)	36	27	−9	2.69	0.006
Pre-SMA (R)	12	24	51	2.8	0.004
Putamen (R)	18	12	6	1.85	0.37
Putamen (L)	−24	−3	3	2.24	0.016
DLPFC (R)	45	12	18	1.95	0.03
DLPFC (L)	−45	12	36	1.77	0.043
Difficult-Easy					
Operculum/Insula (R)	48	27	3	2.07	0.023
Putamen (R)	21	18	3	2.46	>0.001
Putamen (L)	-24	18	−6	1.67	0.052

cortex and striatum (Figure 4.4) and striatum ROIs. The findings described a network of brain regions with reduced temporal processing-related responses in the participants diagnosed with schizophrenia. Included in this network were the SMA, DLPFC, striatum, thalamus, and insula/operculum. Also observed were differing levels of activation, or to be precise, more pronounced differences between patients with schizophrenia and controls in striatum and insula/operculum function under conditions of high task difficulty.

What appears to be clear from neuroimaging data is that timing in the schizophrenic brain is complex. While there are similarities to healthy controls, there are clearly regions that appear to "work harder" to process temporal information. There are also regions that seem to underperform when engaged to make temporal distinctions. For example, the (pre) SMA appears to be underactive in schizophrenia, independent of task difficulty. The SMA is thought to

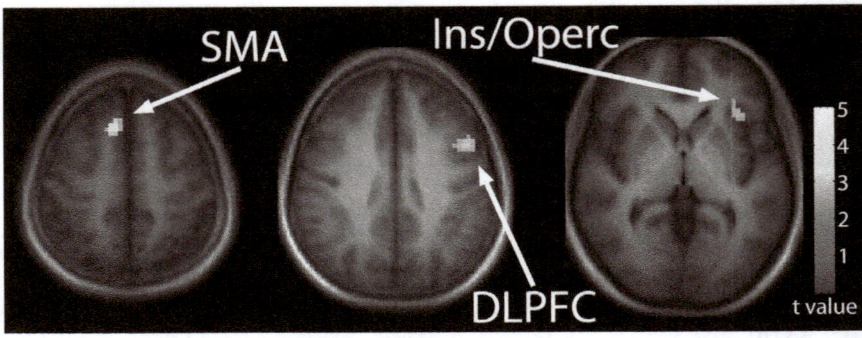

FIGURE 4.2 *Temporal processing during "easy" temporal processing. Reduced activation in the schizophrenia group relative to controls was observed in the supplementary motor area* (SMA)*, insula/ opercular cortex, and the dorsolateral prefrontal cortex* (DLPFC)*. Statistical parametric maps thresholded at p <0.01, overlaid onto the average T1-weighted anatomy of all subjects. Reprinted from Schizophrenia Research, Vol. 127, D.B. Davalos, D.C. Rojas, and J.R. Tregellas, Temporal processing in schizophrenia: Effects of task-difficulty on behavioral discrimination and neuronal responses, pp. 123–130.*

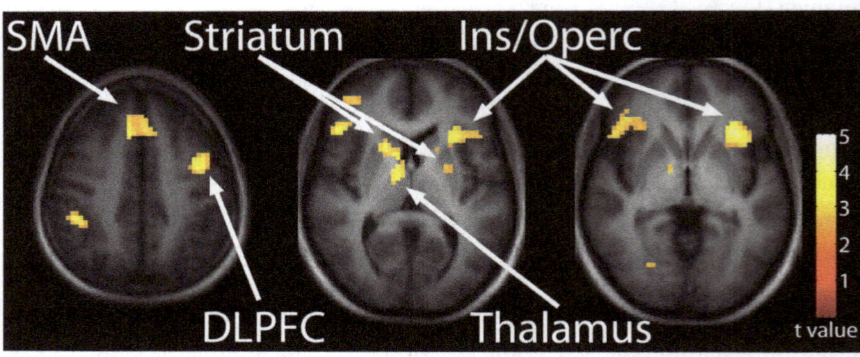

FIGURE 4.3 *Temporal processing during "difficult" temporal processing. Reduced activation in the schizophrenia group relative to controls was observed in the* SMA*, insula/opercular cortex, and the* DLPFC*, as well as the striatum and thalamus. Statistical parametric maps thresholded at pb0.01, overlaid onto the average T1-weighted anatomy of all subjects. Reprinted from Schizophrenia Research, Vol. 127, D.B. Davalos, D.C. Rojas, and J.R. Tregellas, Temporal processing in schizophrenia: Effects of task-difficulty on behavioral discrimination and neuronal responses, pp. 123–130.*

be a key structure during temporal processing (Ferrandez et al., 2003; Macar et al., 2002; Rao et al., 2001; Tregellas et al., 2006). Specifically, the SMA has been associated with the 'pulse accumulation' process (Gibbon, Church, and Meck, 1984; Macar et al., 2004) and timing comparisons (Coull et al., 2004). Another

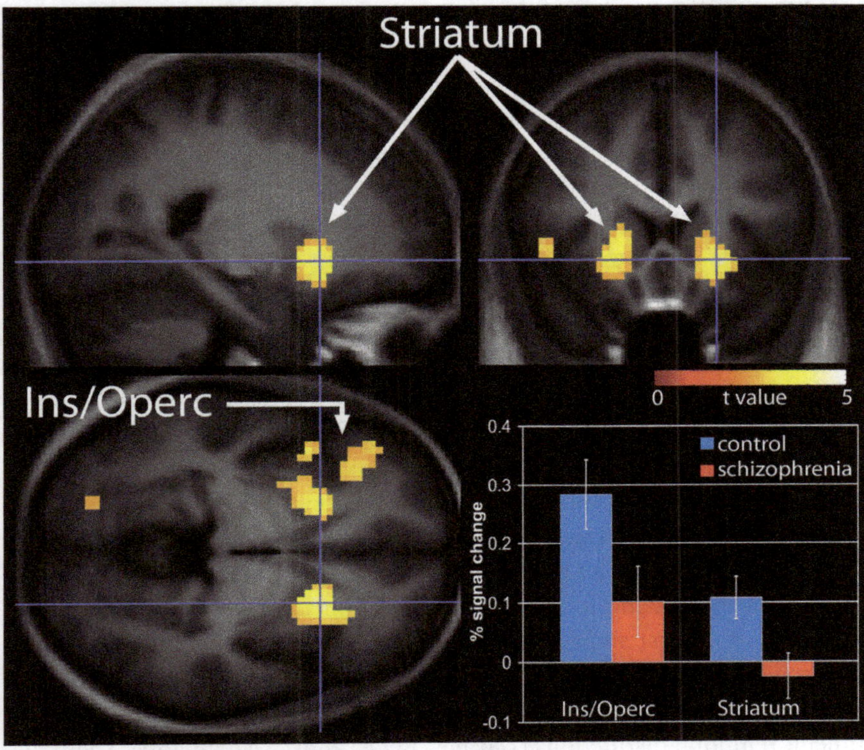

FIGURE 4.4 *"Difficult" compared to "easy" temporal processing. Reduced activation in the schizophrenia group relative to controls was observed in the insula/opercular cortex and the striatum. Statistical parametric maps thresholded at pb0.01, overlaid onto the average T1-weighted anatomy of all subjects. Reprinted from Schizophrenia Research, Vol. 127, D.B. Davalos, D.C. Rojas, and J.R. Tregellas, Temporal processing in schizophrenia: Effects of task-difficulty on behavioral discrimination and neuronal responses, pp. 123–130.* COPYRIGHT 2011, WITH PERMISSION FROM ELSEVIER.

region that appears to be underactive in schizophrenia is the insular/opercular cortices. The insular/opercular cortex has been identified in a number of timing studies and is thought to be involved in a variety of aspects of timing. These timing processes include attention to time components (Coull et al., 2000, 2004), encoding time-based sequences (Schubotz et al., 2000), and perceiving interval duration information (Hinton et al., 2004; Maquet et al., 1996; Ferrandez et al., 2003; Lewis and Miall, 2003). The DLPFC is a region of the brain that is linked to various types of cognitive deficits in schizophrenia, including higher level cognitive functioning (Glahn et al., 2005; Wilmsmeier et al., 2010). Some have speculated that the DLPFC differences observed in schizophrenia versus control studies represents a difference that emerges primarily as a function of task-difficulty (Manoach et al., 1999). Davalos, Rojas, and Tregellas (2011), however, found reduced activity in patients with schizophrenia regardless of task difficulty. The involvement of the DLPFC in time processing has been documented previously across a variety of timing tasks and for both healthy controls and clinical populations. DLPFC is thought to be involved in duration perception (Rao et al., 2001; Lewis and Miall, 2003; Smith et al., 2003), interval time estimation (Macar et al., 2002; Basso et al., 2003), and motor timing (Rubia et al., 1998; Jancke et al., 2000). As noted previously, there continues to be a debate regarding the role of the DLPFC in time processing. The argument persists that DLPFC may be secondarily involved in time processing, when it's primary role is associated with task difficulty or keeping information online successfully for temporal discrimination and comparison (Abi-Dargham, 2004). However, relatively recent data suggests that the DLPFC is also involved in time processing paradigms that minimize working memory components, suggesting that the reduced response in the region in schizophrenia reflects a timing-specific deficit (Davalos, Rojas, and Tregellas, 2011; Smith et al., 2003; Tregellas et al., 2006). The regions discussed so far all seem to be involved in general time processing, independent of task difficulty. All appear to be underactivated, or at least show a reduced response to timing stimuli in patients with schizophrenia compared to healthy controls. In contrast, the thalamus has been shown to be significantly diminished in schizophrenia, but only for difficult temporal distinctions (Davalos, Rojas, and Tregellas, 2011). The thalamus is one of the key structures previously identified in timing literature, both in healthy controls, and more specifically in schizophrenia. It is one of the integral structures in the fronto-striatal network that is involved in complex temporal processing tasks (e.g., perceptual timing; Menon et al., 2000; Rubia and Smith, 2004; Volz et el., 2001). Rao and colleagues' (2001) work in which thalamic responses were observed during timing, but not pitch perception, has added weight to the argument that, like the DLPFC, the thalamus plays a specific role in time processing.

Another region of the brain that has emerged as being critical in timing, and possibly associated with load-dependent temporal processing for individuals with schizophrenia is the striatum. The striatum, or more specifically the putamen, was identified as being "most impaired" in terms of the spatial extent of response differences between patients with schizophrenia and healthy controls (Davalos, Rojas, and Tregellas, 2011). The striatum has been shown to play a central role in temporal processing in animal model studies and studies assessing healthy controls compared to populations with disorders affecting the striatum (e.g., PD; Coull et al., 2004; Ferrandez et al., 2003; Harrington et al., 1998; Maricq et al., 1981; Meck, 1996; Schubotz et al., 2000). Regarding schizophrenia, in particular, the striatum was also identified as being a critical structure involved in temporal processing dysfunction in Volz and colleagues' (2001) work. As was described with the thalamus, timing-related responses in the striatum were reduced in the schizophrenia group only in the "difficult" condition. These findings suggests that there may be at least a couple of distinct neural networks involved in temporal processing in schizophrenia and that timing judgments that are relatively easy may result from a different system that does not involve the striatum and thalamus. It is clear that there may be core areas of the brain that are involved in timing, independent of task difficulty, while there may be other areas of the brain that are recruited for temporal processing on an as-needed basis.

In summary, it appears that the neuroanatomical basis for temporal processing deficits in schizophrenia involves a widespread network of brain regions, including the SMA, PFC, striatum, thalamus, and insula/operculum. As is consistent with behavioral and electrophysiological data, there appears to be dissociation in behavioral responses, ERP responses and activation observed in neuroimaging that is dependent on task difficulty. Specifically, difficult temporal processing paradigms appear to involve the recruitment of other cognitive processes (e.g., decision making) and supplemental brain regions (e.g., striatum and insula/operculum).

4 Summary

Eagleman (2008) questioned whether subjective time should be viewed as a unitary phenomenon, or whether there are separate neural mechanisms that typically work together, but can be dissociated under certain circumstances? Research to date suggests that there is a general temporal dysfunction in schizophrenia that appears across a variety of measures, which all loosely assess timing. What is also clear from research is that understanding and rectifying the etiology of that dysfunction is complex. We know that individuals with schizophrenia do not have *only* temporal processing difficulties. Schizophrenia is characterized by a variety of cognitive difficulties that may contribute to time processing deficits or simply confound our ability to truly understand timing in the schizophrenic brain. What appears fairly safe to conclude is that individuals with schizophrenia exhibit timing deficits, regardless of the method used to assess timing (e.g., behaviorally, neurophysiologically). These deficits appear to be hard-wired in some ways and appear to affect many aspects of the lives of individuals with schizophrenia. These deficits are likely exacerbated when they are combined with other demands that are deficient in this population (e.g., working memory, decision making) and appear to be worsened, as temporal judgments are made more difficult. It seems clear that temporal dysfunction is embedded in the psychopathology of schizophrenia and that the construct of timing needs to move up on the hierarchy of schizophrenia dysfunction. Rather than generally taking a backseat to executive dysfunction, disinhibition, and poor working memory, as it has in the past, time processing dysfunction in schizophrenia needs to be appreciated as a core deficit that may be the key to improving cognition and clinical presentation in this population.

References

Abi-Dargham, Anissa. 2004. "Do We Still Believe in the Dopamine Hypothesis? New Data Bring New Evidence" *International Journal of Neuropsychopharmacology* 7(1): S1-5.

Andreasen, Nancy C. 1999. "A Unitary Model of Schizophrenia. Bleuler's 'Fragmented Phrene' as Schizencephaly" *Archives of General Psychiatry* 56: 781–7.

Anderson, Jonathon W., and Maureen Schmitter-Edgecombe. 2011. "Recovery of Time Estimation Following Moderate to Severe Traumatic Brain Injury" *Neuropsychology* 25: 36–44.

Artieda, Julio et al. 1992. "Temporal Discrimination is Abnormal in Parkinson's Disease" *Brain* 115: 199–210.

Baldeweg, Torsten et al. 2002. "Impairment in Frontal But Not Temporal Components of Mismatch Negativity in Schizophrenia" *International Journal of Psychophysiology* 43: 111–22.

Barabasz, Arreed F. 1973. "Temporal Orientation: A Review of the Literature" *Child Study Journal Monograph* (1–3): 43–9.

Basso, Gianpaolo et al. 2003. "Distributed Neural Systems for Temporal Production: A Functional MRI Study" *Brain Research Bulletin* 59: 405–11.

Bauermeister, Jose J. et al. 2005. "Time Estimation and Performance on Reproduction tasks in Subtypes of Children with Attention Deficit Hyperactivity Disorder" *Journal of Clinical and Child Adolescent.Psychology* 34: 151–62.

Carroll, Christine A. et al. 2009. "Timing Dysfunctions in Schizophrenia Span from Millisecond to Several-Second Durations" *Brain and Cognition* 70: 181–90.

Casini, Laurence and Richard Ivry. 1999. "Effects of Divided Attention on Time Perception in Patients with Lesions of the Cerebellum or Frontal Lobe" *Neuropsychology* 13(1): 10–21.

Clarke, Sean et al. 1996. "Exploring the Domain of the Cerebellar Timing System" In Maria A. Pastor and Julio Artieda (Eds.), *Time, Internal Clocks and Movement. Advances in Psychology*, Vol. 115, pp. 257–80. Amsterdam, Netherlands: North-Holland/Elsevier.

Condray, Ruth. 2005. "Language Disorder in Schizophrenia as a Developmental Learning Disorder" *Schizophrenia Research* 73: 5–20.

Coull, Jennifer T. et al. 2000. "Orienting Attention in Time: Behavioural and Neuroanatomical Distinction between Exogenous and Endogenous Shifts" *Neuropsychologia* 38: 808–19.

Coull, Jennifer T. et al. 2004. "Functional Anatomy of the Attentional Modulation of Time Estimation" *Science* 303: 1506–8.

Davalos, Deana B., Michael A. Kisley, and Robert Freedman. 2005. "Behavioral and Electrophysiological Indices of Temporal Processing Dysfunction in Schizophrenia" *Journal of Neuropsychiatry and Clinical Neurosciences* 17: 517–25.

Davalos, Deana B., Michael A. Kisley, and Randall G. Ross. 2002. "Deficits in Auditory and Visual Temporal Perception in Schizophrenia" *Cognitive Neuropsychiatry* 7: 273–82.

Davalos, Deana B., Michael A. Kisley, and Randall G. Ross. 2003a "Effects of Interstimulus Interval Manipulation on Auditory Temporal Processing in Schizophrenia" *Brain and Cognition* 52: 295–301.

Davalos, Deana B., Michael A. Kisley, and Randall G. Ross. 2003b. "Mismatch Negativity in Detection of Interval Duration Deviation in Schizophrenia" *Neuroreport* 14: 1283–6.

Davalos, Deana B., Donald C. Rojas, and Jason R. Tregellas. 2011. "Temporal Processing in Schizophrenia: Effects of Task-Difficulty on Behavioral Discrimination and Neuronal Responses" *Schizophrenia Research* 127: 123–30.

Deouell, Leon Y., Schlomo Bentin, and Marie-Hélène Giard. 1998. "Mismatch Negativity in Dichotic Listening: Evidence for Interhemispheric Differences and Multiple Generators" *Psychophysiology* 35: 355–65.

Eagleman David M. 2008. "Human time perception and its illusions." *Current Opinion in Neurobiology* 18(2): 131–6.

Eagleman, David M. 2009. "Temporality, Empirical Approaches" In: *The Oxford Companion to Consciousness.* Patrick Wilken, Tim Bayne, and Axel Cleeremans (Eds). Oxford University Press: Oxford.

Eagleman, David M., and Alex O. Holcombe. 2002. "Causality and the Perception of Time" *Trends in Cognitive Sciences* 6: 323–5.

Eagleman, David M. et al. 2005. "Time and the Brain: How Subjective Time Relates to Neural Time" *Journal of Neuroscience* 25: 10369–71.

Efron, Robert. 1963. "Temporal Perception, Aphasia, and Deja Vu" *Brain* 86: 403–24.

Ferrandez, Anne Marie et al. 2003. "Basal Ganglia and Supplementary Motor Area Subtend Duration Perception: An fMRI Study" *Neuroimage* 19: 1532–44.

Fischer, F. 1929. "Zeitstruktur und Schizophrenie" *Zeitschrift für die Gesamte Neurologie und Psychiatrie* 121: 544–74.

Fischer, F. 1930a. "Raum-Zeit-Struktur und Denkstorung in der Schizophrenic" *Zeitschrift jur die gesamte Neurologie und Psychiatrie* 124: 241–56.

Fischer, F. 1930b. "Die Zeitstorung als Schizophreniesymptom" Zentbl. f. d. ges. *Neur. u. Psyckiat.* 56: 455–6.

Fischer, F. 1930c. "Weitere Mitteilung uber das schizophrene Zeiterleben (zugleich ein Beitrag zum Verlaufsproblem)" *Neur. u. Psychiat* 57: 563–4.

Frith, Chris D., Sarah-Jayne Blakemore, and Daniel M. Wolpert. 2000. "Explaining the Symptoms of Schizophrenia: Abnormalities in the Awareness of Action" *Brain Research Reviews* 31: 357–63.

Gandhi S.K. et al. 2007. "Timing Judgments in Schizophrenia" (Paper Presented at the Society for Neuroscience Annual Meeting; San Diego, CA. 2007).

Gibbon, John, Russell M. Church, and Warren H. Meck. 1984. "Scalar Timing in Memory" *Annals of the New York Academy of Science.* 423: 52–77.

Glahn, David C. et al. 2005. "Beyond Hypofrontality: A Quantitative Meta-Analysis of Functional Neuroimaging Studies of Working Memory in Schizophrenia" *Human Brain Mapping* 25: 60–9.

Goldman-Rakic, Patricia S., and Lynn D. Selemon. 1997. "Functional and Anatomical Aspects of Prefrontal Pathology in Schizophrenia" *Schizophrenia Bulletin* 23: 437–58.

Goldman-Rakic, Patricia S. 1999. "The Physiological Approach: Functional Architecture of Working Memory and Disordered Cognition in Schizophrenia" *Biological Psychiatry* 46: 650–61.

Green, Michael F. 1996. "What are the Functional Consequences of Neurocognitive Deficits in Schizophrenia?" *American Journal of Psychiatry* 15: 321–30.

Green, Michael F. et al. 2000. "Neurocognitive Deficits and Functional Outcome in Schizophrenia: Are We Measuring the 'Right Stuff'?" *Schizophrenia Bulletin* 26: 119–36.

Harrington, Deborah L., and Kathleen Y. Haaland. 1998. "Temporal Processing in the Basal Ganglia" *Neuropsychology* 12: 3–12.

Hazeltine, Eliot, Laura Helmuth, and Richard B. Ivry. 1997. "Neural Mechanisms of Timing" *Trends in Cognitive Science* 1: 163–9.

Head, Denise et al. 2008. "Neuroanatomical and Cognitive Mediators of Age-Related Differences in Episodic Memory" *Neuropsychology* 22: 491–507.

Heatherington, Ross, Maureen Dennis, and B. Spiegler. 2000. "Perception and Estimation of Time in Long-Term Survivors of Childhood Posterior Fossa Tumors" *Journal of the International Neuropsychological Society* 6: 682–92.

Hinton, Sean C. et al. 2004. "Neural Systems Supporting Timing and Chronometric Counting: An fMRI Study" *Cognitive Brain Research* 21: 183–92.

Hyde, Thomas M., James C. Ziegler, and Daniel R. Weinberger. 1993. "Psychiatric Disturbances in Metachromatic Leukodystrophy. Insights into the Neurobiology of Psychosis" *Archives of Neurology* 49: 401–6.

Israeli, N. 1932. "The Psychopathology of Time" *Psychological Review* 39(5): 486–91.

Ivry, Richard B. 1996. "The Representation of Temporal Information in Perception and Motor Control" *Current Opinions in Neurobiology* 6: 851–7.

Ivry, Richard B., and Steven W. Keele.1989. "Timing Functions of the Cerebellum" *Journal of Cognitive Neuroscience* 1: 136–51.

Jancke, Lutz et al. 2000. "Cortical Activations during Paced Finger-Tapping Applying Visual and Auditory Pacing Stimuli" *Cognitive Brain Research* 10: 51–66.

Kaminsky, Matti, Zohar Eviatar, and Joel Norman. 2002. "The Timing Deficit Hypothesis of Dyslexia and Its Implications for Hebrew Reading" *Brain and Cognition* 48: 394–8.

Kisley, Michael A. et al. 2004. "Small Changes in Temporal Deviance Modulate Mismatch Negativity Amplitude" *Neuroscience Letters* 358: 197–200.

Kraus, Nina. 2003. "Auditory Pathway Encoding and Neural Plasticity in Children with Learning Problems" *Audiology and Neurotology* 6: 221–7.

Lehmann, Heinz E. 1967. "Time and Psychopathology" *Annals of the New York Academy of Sciences* 138: 798–821.

Lewis, Penny A., and R. Chris Miall. 2003. "Brain Activation Patterns During Measurement of Sub- and *Supra*-Second Intervals" *Neuropsychologia* 41: 1583–92.

Light, Gregory A., and David L. Braff. 2005a. "Mismatch Negativity Deficits are Associated with Poor Functioning in Schizophrenia Patients" *Archives of General Psychiatry* 62: 127–36.

Light, Gregory A., and David L. Braff. 2005b. "Stability of Mismatch Negativity Deficits and Their Relationship to Functional Impairments in Chronic Schizophrenia" *American Journal of Psychiatry* 162: 1741–3.

Macar, Francois, and Franck Vidal. 2009. "Timing Processes: An Outline of Behavioral and Neural Indices Not Systematically Considered in Timing Models" *Canadian Journal of Experimental Psychology* 63: 227–39.

Macar, Francois, Jennifer Coull, and Franck Vidal. 2006. "The Supplementary Area in Motor and Perceptual Time Processing: fMRI Studies" *Cognitive Processing* 7: 89–94.

Macar, Francois et al. 2002. "Activation of the Supplementary Motor Area and of Attentional Networks during Temporal Processing" *Experimental Brain Research* 142: 475–85.

Macar, Francois et al. 2004. "Timing Functions of the Supplementary Motor Area: An Event-Related fMRI Study" *Cognitive Brain Research* 21: 206–15.

Maquet, Pierre et al. 1996. "Brain Activation Induced by Estimation of Duration: A PET Study" *Neuroimage* 3: 119–26.

Mangels, Jennifer A., Richard Ivry, and Naomi Shimizu. 1998. "Dissociable Contributions of the Prefrontal and Neocerebellar Cortex to Time Perception" *Cognitive Brain Research* 7: 15–39.

Manoach, Dara S. et al. 1999. "Schizophrenic Subjects Activate Dorsolateral Prefrontal Cortex during a Working Memory Task, as Measured by fMRI" *Biological Psychiatry* 45: 1128–37.

Maricq, Andres V., S. Roberts, and Russell M. Church. 1981. "Methamphetamine and Time Estimation" *Journal of Experimental Psychology; Animal Behavior* 7: 18–30.
McGlashan, Thomas H., and Ralph E. Hoffman. 2000. "Schizophrenia as a Disorder of Developmentally Reduced Synaptic Connectivity." *Archives of General Psychiatry* 57: 637–48.
Meck, Warren H. 1996. "Neuropharmacology of Timing and Time Perception" *Cognitive Brain Research* 3: 227–42.
Menon, Vinod et al. 2000. "Basal Ganglia Involvement in Memory-Guided Movement Sequencing" *Neuroreport* 11: 3641–5.
Merzenich Michael M. et al. 1996. "Temporal Processing Deficits of Language-Learning Impaired Children Ameliorated by Training" *Science* 271: 77–81.
Michie Patricia T. et al. 2002. "Duration Mismatch Negativity in Biological Relatives of Patients with Schizophrenia Spectrum Disorders" *Biological Psychiatry* 52: 749–58.
Minkowski, E. 1928. "Quelques remarques sur la psychopathologie de la demence senile" *Journal de Psychologie* 25: 79–90.
Minkowski, E. 1931. "Das Zeit- und das Raumproblem in der Psychopathologie" *Wiener klinische Wochenschrift,* 44: 380–4.
Mull, B.R., and Masud Seyal. 2001. "Transcranial Magnetic Stimulation of Left Prefrontal Cortex Impairs Working Memory" *Clinical Neurophysiology* 112: 1672–5.
Naatanen, Risto. 1992. *Attention and Brain Function*. Hillsdale, NJ, Erlbaum.
Naatanen, Risto et al. 1993. "Event-Related Potentials Reveal a Memory Trace for Temporal Features" *Neuroreport* 5: 310–2.
Navon, David. 1978. "On a Conceptual Hierarchy of Time, Space, and Other Dimensions" *Cognition* 6: 223–8.
Nichelli, Paolo et al. 1995. "Duration Processing after Frontal Lobe Lesions" In Jordan Grafman, Keith J. Holyoak, and F. Boller (Eds.), *Annals of the New York Academy Sciences* (Vol. 769, pp. 183–90). New York: New York Academy of Sciences.
Palmer, Barton W. et al. 2002. "Heterogeneity in Functional Status Among Older Outpatients with Schizophrenia: Employment History, Living Situation, and Driving" *Schizophrenia Research* 55: 205–15.
Perry, William et al. 2001. "Working Memory in Schizophrenia: Transient "Online" Storage Versus Executive Functioning" *Schizophrenia Bulletin* 27(1): 157–76.
Poynter, W. Douglas, and Donald Homa. 1983. "Duration Judgment and the Experience of Change" *Perception and Psychophysics* 33: 549–60.
Pulvermuller, Friedemann, and Yury Shtyrov. 2003. "Automatic Processing of Grammar in the Human Brain as Revealed by the Mismatch Negativity" *Neuroimage* 20: 159–72.
Rabin, A.I. 1957. "Time Estimation of Schizophrenics and Non-Psychotics" *Journal of Clinical Psychology* 13: 88–90.
Rammsayer, Thomas. 1990. "Temporal Discrimination in Schizophrenic and Affective Disorders: Evidence for a Dopamine-Dependent Internal Clock" *International Journal of Neuroscience* 53: 111–20.

Rammsayer, Thomas. 1993. "On Dopaminergic Modulation of Temporal Information Processing" *Biological Psychiatry* 36: 209–22.

Rao, Stephen M., Andrew R. Mayer, and Deborah L. Harrington. 2001. "The Evolution of Brain Activation during Temporal Processing" *Nature Neuroscience* 4: 317–23.

Rubia, Katya, and Anna Smith. 2004. "The Neural Correlates of Cognitive Time Management: A Review" *Acta Neurobiologiae Experimentalis* 64: 329–40.

Rubia, Katya et al. 1998. "Prefrontal Involvement in "Temporal Bridging" and Timing Movement" *Neuropsychologia* 36: 1283–93.

Rueda, Alicia D., and Maureen Schmitter-Edgecombe. 2009. "Time Estimation Abilities in Mild Cognitive Impairment and Alzheimer's Disease" *Neuropsychology* 23: 178–88.

Salisbury, Dean F. et al. 2002a. "Mismatch Negativity in Chronic Schizophrenia and First-Episode Schizophrenia" *Archives of General Psychiatry* 59: 686–94.

Salisbury, Dean F. et al. 2002b. "Semantic Bias, Homograph Comprehension, and Event-Related Potentials in Schizophrenia" *Clinical Neurophysiology* 113: 383–95.

Salthouse, Timothy 1985. "A. Speed of Behavior and Its Implications for Cognition" In James E. Birren and K. Warner Schaie (Eds.), *Handbook of the psychology of aging* (pp. 400–26). New York: Van Nostrand Reinhold.

Schilder, Paul. 1936. "Psychopathology of Time" *The Journal of Nervous and Mental Disorders* 83: 530–46.

Schubotz, Ricarda I., Angela D. Friederici, and D. Yves von Cramon. 2000. "Time Perception and Motor Timing: A Common Cortical and Subcortical Basis Revealed by fMRI" *Neuroimage* 11: 1–12.

Schwartz, Barry D., David B. Mallott, and Daniel K. Winstead. 1988. "Preattentive Deficit in Temporal Processing by Chronic Schizophrenics" *Biological Psychiatry* 23: 664–9.

Smith, Anna et al. 2003. "A Right Hemispheric Frontocerebellar Network for Time Discrimination of Several Hundreds of Milliseconds" *Neuroimage* 20: 344–50.

Stetson, Chess et al. 2006. "Motor-Sensory Recalibration Leads to an Illusory Reversal of Action and Sensation" *Neuron* 51: 651–9.

Sussman, Elyse, István Winkler, and Wenjung Wang. 2003. "MMN and Attention: Competition for Deviance Detection" *Psychophysiology* 40: 430–35.

Tervaniemi, Mari et al. 1999. "Test-Retest Reliability of Mismatch Negativity for Duration, Frequency and Intensity Changes" *Clinical Neurophysiology* 110(8): 1388–93.

Todd, Juanita, Patricia T. Michie, and Assen V. Jablensky. 2003. "Association between Reduced Duration Mismatch Negativity (MMN) and Raised Discrimination Thresholds in Schizophrenia" *Clinical Neurophysiology* 114: 2061–70.

Tracy, Joseph I. et al. 1998. "Information-Processing Characteristics of Explicit Time Estimation by Patients with Schizophrenia and Normal Controls" *Perceptual and Motor Skills* 86: 515–26.

Tregellas, Jason R., Deana B. Davalos, and Donald C. Rojas. 2006. "Effect of Task Difficulty on the Functional Neuroanatomy of Temporal Processing" *Neuroimage* 32: 307–15.

Tysk, Lennart. 1990. "Estimation of Time by Patients with Positive and Negative Schizophrenia." *Perceptual and Motor Skills* 71: 826.

Umbricht, Daniel et al. 2003. "How Specific are Deficits in Mismatch Negativity Generation to Schizophrenia?" *Biological Psychiatry* 53: 1120–31.

Verhaeghen, Paul, and Timothy A. Salthouse. 1997. "Meta-Analyses of Age-Cognition Relations in Adulthood: Estimates of Linear and Nonlinear Age Effects and Structural Models" *Psychological Bulletin* 122: 231–49.

Volz, Hans-Peter et al. 2001. "Time Estimation in Schizophrenia: An fMRI Study at Adjusted Levels of Difficulty" *NeuroReport* 12: 313–6.

Wall, Philip M., and Claude Messier. 2001. "The Hippocampal Formation–Orbitomedial Prefrontal Cortex Circuit in the Attentional Control of Active Memory" *Behavioural Brain Research* 127: 99–117.

Wallace, Gregory L., and Francesca Happe. 2008. "Time Perception in Autism Spectrum Disorders" *Research in Autism Spectrum Disorders* (2): 447–55.

Wilmsmeier, Andreas et al. 2010. "Neural Correlates of Set-Shifting: Decomposing Executive Functions in Schizophrenia" *Journal of Psychiatry and Neuroscience* 35(5): 321–9.

Zacks, Rose T., and Lynn Hasher. 1994. "Directed Ignoring: Inhibitory Regulation of Working Memory" In Dale Dagenbach and Thomas H. Carr (Eds.), *Inhibitory Processes in Attention, Memory, and Language.* (pp. 241–64). San Diego, CA: Academic Press.

Ziegler, Wolfram, and D. Yves Von Cramon. 1986. "Disturbed Coarticulation in Apraxia of Speech: Acoustic Evidence" *Brain and Language* 29: 34–47.

CHAPTER 5

Sense of Time Continuity: Possible Mechanisms of Disruption in Schizophrenia

Anne Giersch, Laurence Lalanne*, Mitsouko van Assche*, Patrick E. Poncelet*, and Mark A. Elliott**†*

1 Introduction

Several psychiatrists have described clinical symptoms that suggest a disturbed sense of time continuity (i.e., a fragmentation of time) in patients with schizophrenia (Fuchs, 2007; Vogeley and Kupke, 2007). All aspects of mental life involve the succession of events in time, and according to several authors (Andreasen, 1999; Minkowski, 1933), time disturbance might play a central role in the pathophysiology of disorders such as schizophrenia. Approaching the phenomenological experience of time in patients with schizophrenia is complex, particularly given patients' difficulty in reporting their inner mental life. Patients nonetheless report disturbances that can be taken as hints as to how they experience their disorder. For example, Chapman (1966) describes several patients' reports as follows:

> Things go too quick for my mind. [...] It's as if you were seeing one picture one minute and another picture the next.
>
> You are dying from moment to moment and living from moment to moment and you're different each time.
>
> When I start walking I get a fast series of pictures in front of me. Everything seems to change and revolves around me. Something goes wrong with my eyes and I've got to stop and to stand still.

What are the mechanisms underlying the sense of time continuity in healthy adults and what type of impairment leads to the disturbances experienced by patients? On a more practical level, how can we address questions of the

* INSERM U1114; Dept of Psychiatry; University Hospital of Strasbourg; Fédération de Médecine Translationnelle de Strasbourg (FMTS); 1, pl de l'Hôpital; 67000 Strasbourg.

** Department of Psychology; National University of Ireland, Galway, University Road, Galway, Ireland.

† Department of Human Sciences, Faculty of Design, Kyushu University, Fukuoka, Japan.

DOI 10.1163/9789004230699_006

experience of time continuity using an experimental approach and objectify clinical descriptions? These are questions that are fundamental to the work described in the following paragraphs. Mechanisms involved in the sense of time continuity may operate at various time scales (reviewed in Wittmann, 2011). In our work, we have chosen to focus on very short time scales below 100ms, with the aim of assessing the elementary mechanisms that subtend processing of successive stimulus events in patients. A focus on short time scales was motivated by several results from the literature: elementary timing mechanisms have been related directly with neuronal synchronization (reviews in Elliott, 2005; Elliott et al., 2006; van Wassenhove, 2009), which seems appropriate given that schizophrenia is very precisely characterized by abnormalities in synchronization at frequencies in the EEG gamma band (i.e., at around 40 Hz), and, thus, correspond to intervals of around 25ms (reviewed in Uhlhaas and Singer, 2010). In addition, impairments are also observed at a behavioural level that may be explained by elementary time disorders, such as in motor actions (Carroll et al., 2009; Delevoye-Turrell et al., 2003): for example, when participants lift an object or hit a target with an object, the peak grip force is usually synchronized with the time of impact or maximal load. In patients, however, grip force is delayed by around 30 to 100ms (Delevoye-Turrell et al., 2003). This means a difficulty at precisely synchronizing grip force in time. Another example of abnormalities at short time scales can be found in visual perception tasks: it has been proposed that patients have difficulties to efficiently allocate attention in time (i.e., to focus attention at precise moments in time), when two stimuli follow each other at delays of between 50 to 250ms Stimulus Onset Asynchrony (SOA; Granholm et al., 2009; Lalanne et al., 2012a). Impairments in timing at the shortest time scales have all the more to be considered in schizophrenia that these time scales might be involved in the sense of time continuity.

In fact, different time scales must be involved in the sense of time continuity: first, there is a lower margin in our ability to distinguish separate events in time. Even though two spatially separate stimuli are presented at different times for very brief intervals they may appear to be presented simultaneously. Exner (1875) suggested minimum time differences in temporal order discrimination for intervals of up to 17 and 44ms. Considering invariances, a common measure that extends across sensory modalities seems to be the minimum time required for temporal order discrimination following the successive presentation of more than two stimuli. For tactile and visual stimuli, and irrespective to the precise structure of the visual stimuli concerned, simultaneity thresholds have been determined with remarkably little variation: Brecher (1932) showed what he referred to as units of 'subjective time' corresponded to average periods of 55.3ms for tactile stimulation and periods of 56.9ms for visual stimulation, with standard deviations of no greater than 1.4ms.

This observation relates to the concept of a window of time, or perceptual moments, within which all events are processed together and perceived as occurring simultaneously (Brecher, 1932; Elliott et al., 2006, 2007; Lalanne, 1876; van Wassenhove, 2009; von Baer, 1864; Wittmann, 2011). Although referred to in terms of perception, time windows of this order of magnitude would represent an elementary quantum without experienced duration. Only larger time windows of up to 2 to 3 s would be associated with the experience of duration, and these would represent the second time scale involved in the sense of time continuity. At these larger time scales, information would not be temporally fused and as such perceived in terms of a perceptual unity but would be separate events grouped together within the same moment of experience. A popular example is the experience of present time arising when listening to a melody: when listening to the present tone, the previous tone is still in mind, while the coming tone is usually anticipated. Because past, present, and future tones are all momentarily present in mind, all of them are part of the subjective present. The past tone is nonetheless known as being past, and is thus both past (in objective time) and present (in subjective present time). Similar reasoning holds for the future tone, and this leads to the concept of specious present (James, 1896). Husserl (1893–1917) has proposed that the integration of past, present, and future represents a key mechanism in our sense of time continuity. It is not known to which amount and how the shorter 30ms time windows are integrated within the experienced 3-s moments (Pöppel, 1997, 2009; Szelag et al., 2004). This requires an understanding of how windows overlap and integrate with one another (Dainton, 2010; Elliott, 2005). It might be proposed that temporal windows and their overlap are brought about by neuronal synchronization (Varela, 1999). Because neuronal synchronization of action potentials requires time, even the processing of the simplest event, like the display of a square on a computer screen, is time-consuming, and is coded within a temporal window. Different events would then overlap in time even if their onset were shifted in time. The result would be a sense of continuity rather than the perception of discrete moments.

There are, thus, several candidate mechanisms bringing about a sense of time continuity, but a number of questions require clarification: are events judged to be simultaneous really processed as co-temporal? Is the overlap of events enough to bring about continuity of time, or is it required to integrate events with one another? At which time scale do these mechanisms occur? We first explored the length of the elementary time window in patients with schizophrenia. We found these elementary time windows to be altered in patients, but these disturbances led us to wonder about the processing of events within time windows. To answer this question, we devised a new method of analyzing participants' responses. The results question the usual assumption that events are treated as co-temporal within temporal windows, and lead us to reconsider the mechanisms underlying the sense of time continuity. We summarize our findings in the following.[1]

2 Measuring the Windows of Visual Simultaneity

Visual stimuli separated in time by an SOA of less than 20 to 30ms are judged as being simultaneous. Our aim was to compare the associated time window in patients with schizophrenia against that found in healthy participants. The paradigm used to determine this window is relatively simple: two visual stimuli (two bars or two squares, for instance) are shown on a computer screen. They appear either simultaneously or with a short asynchrony and participants decide whether the two stimuli are simultaneous or asynchronous. In our experiments, participants gave their response by pressing on a left response key in case of simultaneity or a right response key in case of an asynchrony. To date, four studies have shown that patients need larger asynchronies than healthy participants to report two stimuli as appearing at different times (Foucher et al., 2007; Giersch et al., 2009; Lalanne et al., 2012b; Schmidt et al., 2011). Several possible confounds have been eliminated, i.e. the role of eye movements, inter-hemispheric transfer, modality specificity, bias, and access to consciousness. Hereafter, we review shortly how these possibilities have been discarded.

It has been shown that the impairment persists if patients are required to look at a central fixation point until the stimuli appear (Lalanne et al., 2012b) indicating that the impairment is not related to abnormal eye movements (Holzman, 2000; Phillips and Davis, 1994; Trillenberg et al., 2004). In addition, the deficit is similar if stimuli are displayed within the same hemifield or across hemifields (Lalanne et al., 2012b), thus eliminating a possible role for impaired

1 Our groups of patients with schizophrenia (around 20 in each study) were mildly symptomatic outpatients, usually treated with antipsychotics, but without benzodiazepines. Their performance was compared with those of control participants individually matched on sex, age, and education level (Giersch et al., 2009; Lalanne et al., 2012a,b,c). Until now, no result allows us to firmly exclude an effect of antipsychotic medication, although the doses of antipsychotic medication do not correlate with performance. It is to be noted that clinical descriptions of time continuity disruptions predate the establishment of antipsychotic medication. Besides, the mechanisms of the sense of time continuity and its impairments need to be unraveled even in treated patients.

interhemispheric connections (Knöchel et al., 2012; Mohr et al., 2008; Schwartz et al., 1984; but see David, 1993). A difficulty at discriminating simultaneous from asynchronous events is observed also in auditory (Foucher et al., 2007) and cross-modal conditions (audio-visual; Martin et al., 2013). Finally, a possible decisional bias has been carefully discarded (Giersch et al., 2009; Schmidt et al., 2011). A decisional bias is independent of perceptual ability and may occur during response selection: in this case it might be that patients process asynchronies as do the healthy participants but need larger asynchronies to select an 'asynchronous' response. Classical methods based upon signal detection theory (Green and Swets, 1974) showed no difference between patients and healthy participants regarding bias (Giersch et al., 2009; Schmidt et al., 2011).

We used an additional method, based upon priming, which bypasses the problem of an explicit judgment on the part of the patients. The possibility of impairment due to the explicit nature of the response is distinct from a decisional bias. A deficit at making explicit judgment is related to the mechanisms allowing information to become conscious (Del Cul et al., 2006), whereas decisional biases are related to variations in the use of information to give a response. Each time a patient is explicitly asked to make a judgment, and show abnormal performance, it can be asked whether the impairment is related to the difficulty the patients experience in formulating an explicit judgment. Patients have often been shown to be thus impaired although implicit information processing remains unimpaired (Del Cul et al., 2006). In the present context, a seeming difficulty to detect an asynchrony between two items might be due to a non-specific difficulty at making a subjective judgment, rather than impairment in processing the timing of events. In our task, we used a procedure developed by Elliott et al. (2007) and took an implicit measure of the effects of an asynchrony on a subsequent explicit judgment of simultaneity/asynchrony. In this paradigm, two priming bars were displayed on a computer screen, either simultaneously or asynchronously while a series of 6 distracter bars were rapidly switched on and off around the priming bars, thus making the temporal relation between the priming bars impossible to accurately report. After the distracters switched off the priming bars remained on screen and after a short interval increased in luminance separately and with a variable SOA (which included a simultaneous increase; Figure 5.1). Participants reported whether this luminance increase was simultaneous or asynchronous across the two bars.

Elliott et al. (2007) showed that in healthy volunteers, the masked asynchronous bars biased participants towards reporting an asynchrony in the subsequent increase in bar luminance. We showed, in addition, that this bias increases with the asynchrony of the priming bars: the larger the prime's

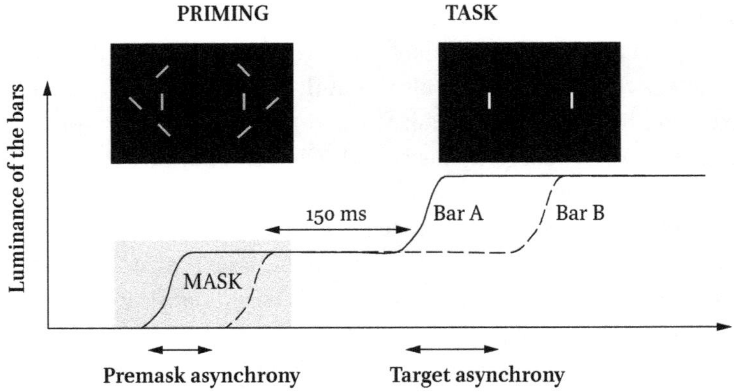

FIGURE 5.1 *Illustration of the paradigm used to check for a non-specific effect of subjective judgments. The curves represent the increase in luminance of the two target bars, A and B. The first increase in luminance is used as a prime and masked by the distracters ('priming' figure). The prime is asynchronous when the two bars increase their luminance asynchronously. The participant's task is to decide whether the second increase in luminance is simultaneous or asynchronous. With the permission of schizophrenia bulletin.*

asynchrony, the larger the bias (Giersch et al., 2009). This was observed even though the prime's asynchrony was individually adapted so that it was below threshold in all cases, indicating that the effect is due to the implicit priming of the asynchrony. We reasoned that if the difficulties of the patients are due to impaired explicit judgments per se, there should be a dissociation between implicit and explicit processing of the asynchrony: explicit judgments of asynchrony would be impaired while their implicit influence would remain unaffected. If this were the case, we expected priming to be identical between groups for equivalent sub-threshold asynchronies. This is not what was observed, however: similar priming effects were observed in the two groups, but for larger asynchronies in patients than in healthy participants. This suggests that the enlarged temporal window observed in patients reveals a true difficulty at discriminating events over time, and is not explicable in terms of a non-specific difficulty in formulating the required judgment.

Inasmuch as the subjective experience of the present is built upon elementary time windows, it can be expected that a lengthened interval of subjective simultaneity distort the sense of time continuity. Before drawing this conclusion however, we questioned whether this lengthened interval entailed the temporo-perceptual fusion of all the events within the associated temporal window, i.e., whether it meant that events are treated as co-temporal even at an implicit level. It was indeed surprising to observe the amplitude of the deficit in some patients with schizophrenia, who were unable to detect an asynchrony of more than 100ms. Did this really mean that patients fuse events in time over periods longer than 100ms? The next step was, thus, to check the integrity of implicit information processing within the window of simultaneity.

3 Implicit Processing of Information within the Temporal Window

In order to evaluate the implicit processing of information within the temporal window, we devised an original method of analysis, by examining the Simon effect. The Simon effect refers to the fact that participants are faster and more accurate when a visual stimulus is presented within the same perceptual hemifield as their responding hand (Hommel, 2011a). The precise mechanisms of the Simon effect are a matter of discussion (i.e., Hommel, 2011a,b; van der Lubbe and Abrahamse, 2011), but the important point is that the effect is independent of the explicit instructions given to the participant. In our paradigm, it allowed us to check whether or not the stimuli are fused in time. During our task, two stimuli are displayed on the screen, one to the left and one to the right and participants give their response – 'simultaneity' or 'asynchrony' by pressing the left or right response key, respectively. When the two stimuli are displayed simultaneously, a Simon effect cannot occur: the participants cannot be biased to respond on any particular side since the displayed information is equivalent to both sides. However, when the stimuli are asynchronous there is an asymmetry and under these conditions we observed a Simon effect (Lalanne et al., 2012b,c). Healthy participants were systematically biased to answer to the side of the second stimulus independent of its right or left location (Lalanne et al., 2012c). Given that physical information is identical on both sides, it is only the temporal difference between the two stimuli that can lead to such a Simon effect. The important point is that this Simon effect was observed even at short asynchronies when participants reported the majority of stimulus presentations to be simultaneous (Lalanne et al., 2012b,c). This result suggests that the asynchrony is processed even though participants are unable to report it. Interestingly, a Simon effect was also observed in patients with schizophrenia. At large asynchronies patients were biased to the side of the second stimulus to the same extent as the healthy participants (Lalanne, 2012b). However, at small asynchronies, patients were biased to the side of the first stimulus (Lalanne et al., 2012b,c). This effect has been observed in three different studies, with three different groups of 18 to 20 patients with schizophrenia. It has been observed for asynchronies of 8 to 17ms, i.e., at delays that lead to rates of 'simultaneous' responses identical to the rates observed for perfect synchrony (Lalanne, 2012c). As with healthy participants, this means that the asynchronies are processed in patients with schizophrenia even for delays of below 20ms and are not fused in time. Had the stimuli been fused in time, then the two stimuli would have been processed as if identical. In that case, the Simon effect could not have occurred, i.e., the subjects could not have been biased to either stimulus. The Simon effect shows that patient's process very short asynchronies even though they need much larger asynchronies than healthy participants to explicitly report them. These data show a clear dissociation between the implicit response – indicated by the Simon effect, and the subjective experience of temporal relations between stimuli – indicated by the explicit judgments of those relations. A similar dissociation has also been observed with multisensory stimuli (Martin et al., 2013). Our results however do not indicate a preserved implicit processing. On the contrary, the results in patients with schizophrenia differ qualitatively from those in healthy participants: at short asynchronies, the responses of healthy participants are biased to the side of the second stimulus whereas the responses of patients are biased to the side of the first stimulus. Hence, consistent with our previous studies (Giersch et al., 2009), it seems that both the implicit and explicit processing of asynchronies is affected in patients with schizophrenia.

The critical question at this stage is the meaning of the bias to the side of the first or second stimulus. The bias to the side of the second stimulus initially seems surprising, given the effect of prior entry. This effect has been demonstrated using similar tasks, involving simultaneity/asynchrony or temporal order judgments on successive or simultaneous stimuli (review in Spence and Parise, 2010). It has been frequently shown that cueing the first stimulus in a sequence of two facilitates subsequent judgments of order or simultaneity/asynchrony. A cue, for example, an indicator or a flash, encourages the deployment of attention to the cued location. This is believed to speed up processing of the first stimulus, thus facilitating the detection of an asynchrony or of a succession between first and second stimuli (Spence and Parise, 2010). This shows that focusing attention on the first stimulus in a sequence is important for the processing of sequential order. By contrast, we found that the response of healthy participants was biased towards the second stimulus. The major differences between the two observations is, first, that our observations take place when subjects are not aware of any asynchrony, and second, that prior entry takes place before critical stimuli are presented. In contrast, the Simon effect is recorded at the time of the participant's response, i.e., after the display of the stimuli. From these observations it is tempting to conclude that healthy participants move from the first to the second stimulus during the sequence, as if biased towards the last occurring event. Conversely, patients would be unable to do so and would remain stuck on the first stimulus.

A number of difficulties and questions still remain with this interpretation: first, it might be possible that deployment of visuo-spatial attentional mechanisms rather than time disorders account for the difficulties of patients with schizophrenia. Our results suggest a form of move between the first and second stimulus. Given that the two stimuli differ not only on their time onset but also on their spatial location, it might be asked whether patients have a difficulty to move from the first to the second stimulus as the result of spatial or visual organization impairments. In fact, patients with schizophrenia are known to be impaired at grouping items when those items are spatially separate (review in Silverstein and Keane, 2011), and this might have made it difficult for them to sequentially process from the first to the second stimulus. To check for this possibility, we used the fact that patients' difficulties with perceptual grouping are alleviated when clear grouping cues relate the stimuli (Giersch and Rhein, 2008; van Assche and Giersch, 2011).We slightly adapted our paradigm and added line-segments between the stimuli in order to facilitate grouping (Figure 5.2).

We also manipulated the spatial predictability of the second stimulus to further facilitate the move towards this stimulus in patients: in one experiment there was only two possible locations for the stimuli and the second stimulus location was always predictable, whereas in a second experiment, 4 locations were used, and the location of the second stimulus was uncertain (Figure 5.3).

Since perceptual grouping promotes the perception of synchrony in time (Nicol and Shore, 2007), it was expected that all Simon effects would be reduced by making right and left stimuli more similar in their temporal properties. As expected, healthy participants' bias to the side of the second stimulus was reduced when stimuli were connected (Lalanne et al., 2012c). On the other hand, the Simon effect to the side of the second stimulus was increased when the location of the second stimulus was predictable. The results in healthy participants, therefore, confirmed that our experimental manipulations had the expected effect. However, in patients there was a very clear bias to the side of the first stimulus when the first and second stimulus were related with a line-segment, and when the second stimulus location was predictable in 100% of the cases (Figure 5.4).

This suggests that in patients the abnormal bias to the side of the first stimulus persists when grouping and spatial difficulties are alleviated by experimental manipulations. Overall, these results suggested difficulties related to time rather than to spatial and visual organization impairment.

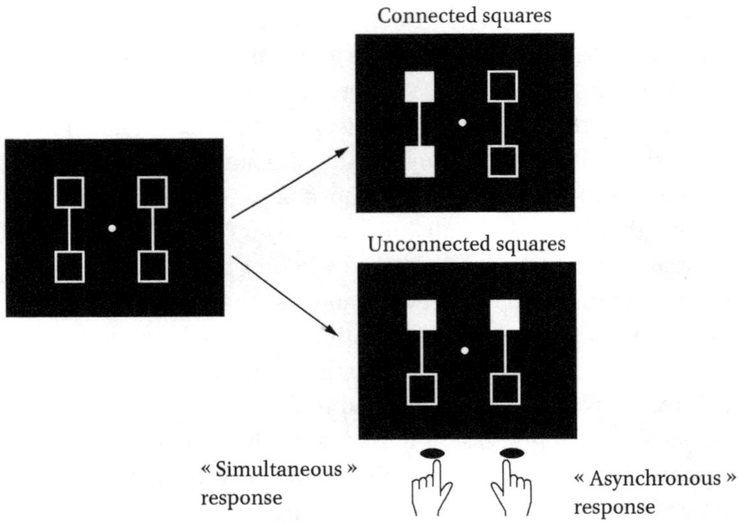

FIGURE 5.2 *Illustration of the procedure designed to compare simultaneity/asynchrony discrimination for connected and unconnected squares and squares displayed in the same hemifield vs. in different hemifields. Two squares are filled in, in gray, either simultaneously or asynchronously. These two squares are either connected or not. Participants are instructed to hit the right key when they think the squares are filled in asynchronously and the left key when they think the filling occurs simultaneously. In the examples, filled squares are connected and located within the same hemifield in the right top figure and unconnected and in different hemifields in the right lower figures. The intra- vs. across-hemifield manipulation has been conducted without connecters in a first experiment (Lalanne et al., 2012b) and with connecters in a second, distinct experiment (Lalanne et al., 2012c). The examples correspond to the second experiment. In this experiment, the location of the connecters (vertical vs. horizontal) and the location of the targets (within the same vs. different hemifields) yielded four main possibilities (targets connected within the same or different hemifields and targets unconnected within the same or different hemifields), which were equally represented and displayed in random order.*

A difficulty remained, however, in the interpretation of the Simon effect, because we do not know whether participants are coding a succession between the first and the second stimulus when the asynchronies are very short. This point is critical when considering the sense of time continuity. As emphasized above, it has been proposed that the sense of time continuity arises from the integration of past, present, and future moments within the subjective present. If such succession is coded within the most elementary time windows, integration would take place on a shorter time scale than previously believed. It would then have to be explained how events are coded one relative to another on time scales shorter than 20ms. Automatically ordering events in time might

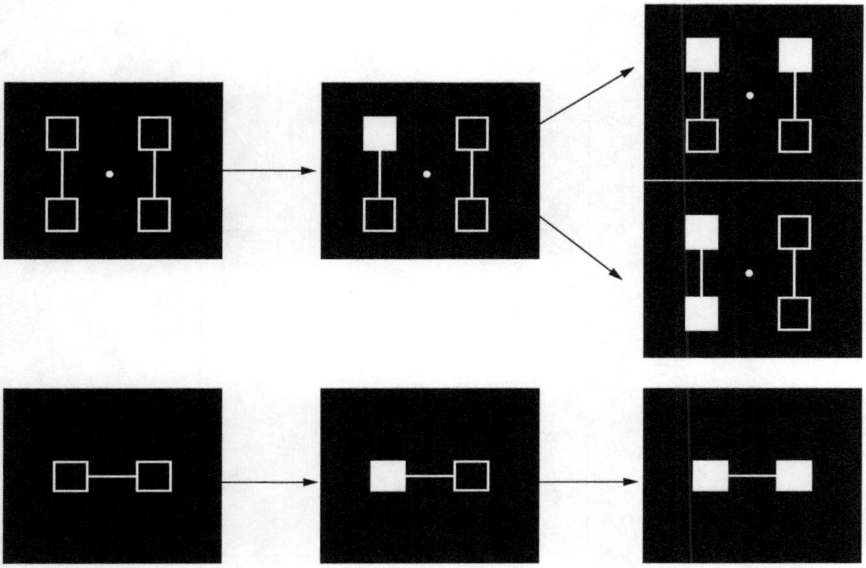

FIGURE 5.3 *Illustration of the successive events in the simultaneity/asynchrony discrimination task in case of an asynchrony (from left to right). When 4 locations are used (upper row) and the first square is filled in (middle figure in the upper row), there are two possible locations for the second one (figure on the right), and there is, thus, an uncertainty regarding the location of the second stimulus. The spatial location of the second stimulus is always predictable when only two locations are used (lower row). These two experiments have been conducted in two different groups of participants (Lalanne et al., 2012c).*

indeed be costly. In our experiments, stimuli were sometimes displayed in two different hemifields and, thus, processed in different cerebral hemispheres (cf. Figure 5.2). Comparing their onsets is bound to involve long-distance connectivity (at least to transfer information regarding the stimulus onset) and some specialized comparison mechanisms. If this is to be generalized in the natural environment, it would mean permanent comparisons between unrelated stimuli. The usefulness of such computation might be questioned. Beside, the computation cost would increase exponentially in a crowded environment. It might, thus, be proposed that even if events delayed by short asynchronies are not processed as being co-temporal, their succession is not coded automatically. This possibility is supported by our most recent results in healthy participants (Poncelet et al., in preparation). We used a priming paradigm during which two square frames (two empty squares) are displayed, one to the right and one to the left of fixation (Figure 5.5). These two priming frames are displayed either simultaneously or asynchronously, but the asynchrony is only

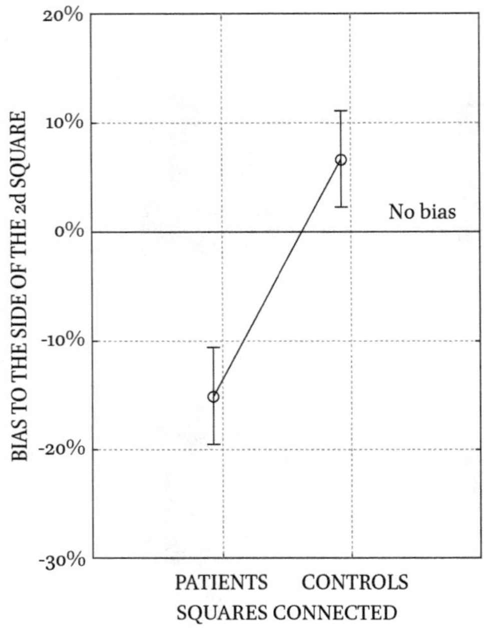

FIGURE 5.4 *Amplitude of the bias (in %) to the side of the 1st or 2nd stimulus. A negative bias corresponds to a bias to the side of the first stimulus (in patients), whereas a positive bias corresponds to a bias to the side of the second stimulus (in healthy participants).*

17ms and is not consciously detected by the participants. After a subsequent delay of between 20 to 100ms, one of the priming frames is filled in and represents the target. Participants have to decide about the right or left location of the filling in and give their response by pressing a right or left response key on the side of the target. At delays of less than 50ms between priming frame presentations and target filling in participants are faster when filling in is in the location of the first frame, consistent with the prior entry effect. At frame-target delays of 100ms, the participants are faster when the filling in is in the location of the second rather than the first frame, consistent with the Simon effect observed in our previous studies. The RT difference is of about 20ms. This result has been found to be significant in three different groups of 12 healthy participants. RTs in the neutral condition are intermediate between the two asynchronous conditions. These results are consistent with the idea of some kind of move between the stimuli delayed by short asynchronies.

We then adapted our paradigm to answer the critical question whether participants also coded the succession between frames. Accordingly, instead of filling in only one frame, two frames were filled in with an onset asynchrony of

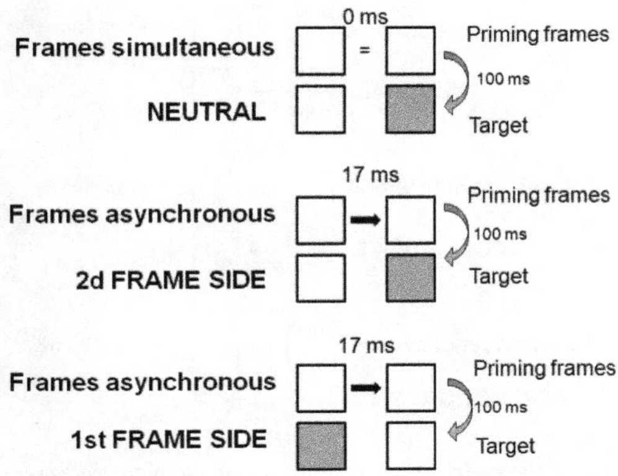

FIGURE 5.5 *Illustration of the experimental conditions used in our location priming paradigm. Priming frames were either simultaneous or asynchronous with a SOA of 17ms between frames. One of the frames was then filled in and participants had to press a response key to the side of the target. The results correspond to a SOA of 100ms between prime frames and the target.*

100ms, providing a perfectly detectable target order. The order of target filling in was in the same or in the opposite direction compared to priming frame order. This permits a check on whether participants automatically coded the order of the priming frames at an asynchrony of 17ms. It was expected that if participants coded the priming-frames succession, then this should facilitate perception of the order of target filling in when it matched the order of the frames. However, this is not what was observed: no succession priming was found while instead a facilitation was observed when prime and filling in order were in opposite directions.[2]

The results of the second priming study can be understood easily by taking into account the prior entry effect (Figure 5.6). The previous studies have shown that participants put more emphasis on the location of the second priming frame. This emphasis facilitated the processing of the subsequently

2 The results of the priming studies allow us to eliminate several alternative accounts for the bias to the side of the second stimulus. The first study shows a bias even though the task does not require any judgment of asynchrony or order. This indicates that the bias to the side of the second stimulus does not depend on a particular task-set activation. Additionally, the second study shows that apparent motion does not play a role. A processing of apparent motion between the prime frames should have led to the coding of a succession between the frames and should have enabled a priming of the stimulus order. This was not the case (see also Lalanne et al., 2012c, for a discussion on this point).

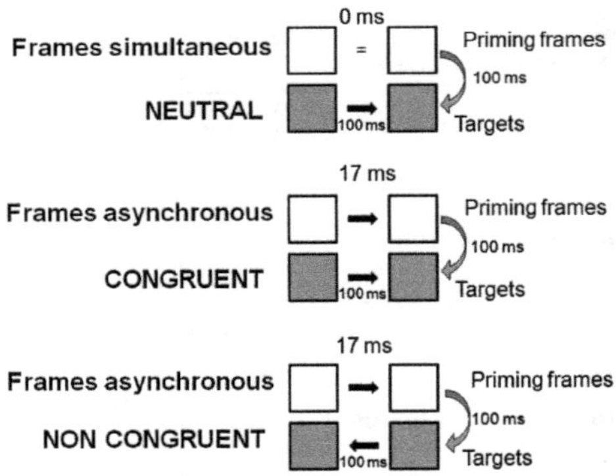

FIGURE 5.6 *Illustration of the experimental conditions used in our succession priming paradigm. Priming frames were either simultaneous or asynchronous with a SOA of 17ms between frames. The two frames were then filled in with an asynchrony of 100ms, with these filled in squares representing the targets. Participants had to press a response key to the side of the 2nd target. The results correspond to a SOA of 100ms between prime frames and the targets.*

filled in frame in the same location. Like for the prior entry effect, this led to a facilitated response when the first filled in target was in the location of the second priming frame (i.e., when the frame and filling in order was in opposite directions). The RT difference is again of about 20ms and has also been reproduced in several groups of 12 healthy participants.

These results, thus, confirm an automatic bias towards the last event in control subjects. They allow us to eliminate alternative explanations of the impact of the short asynchronies, in terms of a coding of motion, direction or task instructions. This has also implications regarding the results observed in patients, inasmuch it suggests the impairments observed in patients are, like for controls, independent from a difficulty in coding motion or in following instructions. The results suggest that the impairment in patients is rather related to a difficulty to follow events over time. Once we have reached this conclusion, however, we can ask about the significance of this impairment. For example, it might be asked how patients are able to follow a stream of incoming information over time in every day life. The impact on duration perception should also be investigated. Inasmuch duration perception relies on the accumulation of information as time moves forward; an inability to follow information over time can be expected to disturb perception duration. However, duration perception concerns time scales that are much longer than those studied here, and theoretical models relating these different time scales are missing (van Wassenhove, 2009; Grondin, 2010). More generally, the consequences of an elementary impairment at moving attention over time, is necessarily speculative at this stage. Yet some possibilities seem likely in light of the known impairments described in patients with schizophrenia, and one of them is briefly evoked in the next section.

4 Possible Impact of Elementary Timing Impairments on Visual Perception

In every day life, the outer world is usually dynamic rather than static and events succeed each other rapidly. Wind can make tree leaves move, alternately uncovering or occluding visual information. We usually experience no difficulty in distinguishing the tree from the objects located behind it. Yet moving objects like the moving leaves create much ambiguity in the organization of information, and some processing is required in order to attribute elements of information to the right objects. In dynamic environments, this requires discrimination in both space and time: an item has to be precisely focused in time in order to avoid confusion with distracter information, especially if it is displayed in the same location but at a different time. As a matter of fact, the use of masking experiments has shown that patients with schizophrenia have difficulties at distinguishing target information when it is closely preceded or followed by a distracter (Butler et al., 2003; Green and Walker, 1986; Saccuzzo and Braff, 1981). Explanations for this impairment are diverse (Breitmeyer, 1984; Butler et al., 2003; Green et al., 1994; Green et al., 2011; Schuck and Lee, 1989), but the impairment certainly confirms that patients have a difficulty with stimuli shown in close succession. Besides, based on masking experiments, it has been proposed that patients with schizophrenia have a difficulty in the temporal precision of target-directed attention (Granholm et al., 2009; Lalanne et al., 2012a). All in all, temporal difficulties mean heightened difficulties at resolving ambiguities arising in case of dynamic visual information.

5 Relating the Simon Effect with the Sense of Time Continuity

In summary, it seems that items are distinguished in time at shorter asynchronies than previously believed. However, this is not associated with succession processing, which instead appears to be processed only at larger intervals between events, perhaps associated with the explicit processing of succession. Dissociation between the processing of asynchronies and the sense of succession is frequently reported (reviewed in Wittmann, 2011). What is more surprising is the bias to the side of the second stimulus without any coding of succession, although this might be explained relatively easily. Even if the lack of succession coding makes it impossible to tag stimuli as 'first' and 'second', the second one might nevertheless be identified as the latest occurring event. In fact, the bias to the side of the second stimulus in a succession of two suggests the existence of a mechanism putting a priority on the last occurring information. This might be analogous to what happens when attention is driven reflexively by an external event. It is known that attention deploys to novel information, meaning that associated brain mechanisms are designed to continually check for novel information, not only in space but also in time. Feedback loops described in visual perception, predictive coding or forward models might provide neural bases for these mechanisms (Elliott and Müller, 2000; Friston, 2008; Miall et al., 1993; Wolpert et al., 1998). Although the results suggest some kind of prioritization and possibly an involvement of attention, the processes making this prioritization possible are not available to consciousness. Participants do not perceive very short asynchronies, and they put nonetheless more weight on the second and last stimulus. Even if this means the involvement of attention, it cannot be induced by some cognitive expectation. In our priming paradigm, participants have to detect a target and the task does not incite them to expect successive stimuli. This means that some subconscious signal is powerful enough to induce a priming effect and suggests that at least part of the mechanisms making it possible for the last stimulus to be prioritized are automatic in nature.

What is the relationship between the ability to distinguish stimuli at very short asynchronies and the sense of time continuity? In so far as the succession of stimuli is coded only at longer delays, the integration of past, present, and future moments proposed as a mechanism of time continuity should emerge only at large delays. However, what we observe at the shortest delays is an automatic priority for the latest events, possibly sub-tended by mechanisms allowing us to permanently look forward. It might be asked whether this bias to anticipate future events also participates in the sense of continuity, thereby providing an elementary basis for the expectation of what is coming next. Husserl described the concept of a protention, which allows us to anticipate the future during the present time. Although what Husserl describes might be more easily related to conscious phenomena taking place at larger time scales, the bias towards the latest stimulus might nonetheless be considered as an elementary mechanism subserving protention.

Related to that, it can be asked whether patients' disrupted sense of time continuity is related to their difficulty at assigning priority to the latest occurring events. It certainly reminds of the observation of Minkowski (1933), who described a loss of life force in patients with schizophrenia.

More concretely, it might be asked how patients perceive a rapid succession. If they cannot move easily from the first to the second event, do they miss information in everyday life? How is this related with the way they perceive duration (Allman and Meck, 2012; Davalos et al., 2005; Elvevåg et al., 2004; Volz et al., 2001) or time actions (Carroll et al., 2009; Delevoye-Turrell et al., 2012; Turgeon et al., 2012)?

6 Conclusion

Although many questions remain, our results show that information is not fused in time at very short asynchronies, either in healthy participants or in patients. Our results also indicate that healthy participants move in time very rapidly between succeeding events and that this capability is impaired in patients. These observations may mean new additional mechanisms underlying the sense of time continuity. It remains to be investigated whether and how these mechanisms are really involved in the sense of time continuity, and how the impairments of these mechanisms impact cognitive functions and clinical symptoms in schizophrenia.

References

Allman, Melissa J., and Warren H. Meck. 2012. "Pathophysiological Distorsions in Time Perception and Timed Performance" *Brain* 135: 656–77.

Andreasen, Nancy C. 1999. "A Unitary Model of Schizophrenia: Bleuler's 'Fragmented Phrene' as Schizencephaly" *Archives of General Psychiatry* 56: 781–7.

Brecher, Gerhard A. 1932. "Die Entstehung und biologische Bedeutung der subjektiven Zeiteinheit-des Momentes" *Zeitschrift für Vergleichende Physiologie* 18: 204–43.

Breitmeyer, Bruno G. 1984. *Visual Masking: An Integrative Approach* New York: Oxford University Press.

Butler, Pamela D. et al. 2003. "Visual Backward-Masking Deficits in Schizophrenia: Relationship to Visual Pathway Function and Symptomatology" *Schizophrenia Research* 1: 199–209.

Carroll, Christine A. et al. 2009. "Timing Dysfunctions in Schizophrenia" *Brain Cognition* 71: 345–53.

Chapman, James. 1966. "The Early Symptoms of Schizophrenia" *British Journal of Psychiatry* 112: 225–51.

Dainton, Barry. 2010. "Temporal Consciousness." In *The Stanford Encyclopedia of Philosophy*, edited by Edward N. Zalta. Available at <http://plato.stanford.edu/archives/fall2010/entries/consciousness-temporal>.

Davalos, Deana B., Michael A. Kisley, and Robert Freedman. 2005. "Behavioral and Electrophysiological Indices of Temporal Processing Dysfunction in Schizophrenia" *The Journal of Neuropsychiatry and Clinical Neurosciences* 17: 517–25.

David, Anthony S. 1993. "Callosal Transfer in Schizophrenia: Too Much or Too Little?" *Journal of Abnormal Psychology* 102: 573–9.

Del Cul, Antoine, Stanislas Dehaene, and Marion Leboyer. 2006. "Preserved Subliminal Processing and Impaired Conscious Access in Schizophrenia" *Archives in General Psychiatry* 63: 1313–23.

Delevoye-Turrell, Yvonne, Hélène Wilquin, and Anne Giersch. 2012. "A Ticking Clock for the Production of Sequential Actions: Where Does the Problem Lie in Schizophrenia?" *Schizophrenia Research* 135: 51–4.

Delevoye-Turrell, Yvonne, Anne Giersch, and Jean-Marie Danion. 2003. "Abnormal Sequencing of Motor Actions in Patients with Schizophrenia: Evidence from Grip Force Adjustments during Object Manipulation" *American Journal of Psychiatry* 160: 134–41.

Elliott, Mark A. 2005. "Temporal Tolerance Circumscribed" *Visual Cognition* 12: 655–61.

Elliott, Mark A., and Hermann J. Müller. 2000. "Evidence for a 40-Hz Oscillatory Short-Term Visual Memory Revealed by Human Reaction Time Measurements" *Journal of Experimental Psychology: Learning, Memory and Cognition* 26: 703–18.

Elliott, Mark A., Zhuanghua Shi, and Sean D. Kelly. 2006. "A Moment to Reflect Upon Perceptual Synchrony." *Journal of Cognitive Neuroscience* 18: 1663–5.

Elliott, Mark A., Zhuanghua Shi, and Fatma Sürer. 2007. "The Effects of Subthreshold Synchrony on the Perception of Simultaneity" *Psychological Research* 71: 687–93.

Elvevåg, Brita et al. 2004. "Identification of Tone Duration, Line Length, and Letter Position: An Experimental Approach to Timing and Working Memory Deficits in Schizophrenia" *Journal of Abnormal Psychology* 113: 509–21.

Exner, Sigmund. 1875. "Experimentelle Untersuchungen der einfachsten psychischen Processe" *Pflügers Archives* 11: 403–32.

Foucher, Jack R. et al. 2007. "Low Time Resolution in Schizophrenia Lengthened Windows of Simultaneity for Visual, Auditory and Bimodal Stimuli" *Schizophrenia Research* 97: 118–27.

Friston, Karl. 2008. "Hierarchical Models in the Brain" *PLOS Computational Biology*, 4: e1000209.

Fuchs, Thomas. 2007. "The Temporal Structure of Intentionality and Its Disturbance in Schizophrenia" *Psychopathology* 40: 229–35.

Giersch, Anne et al. 2009. "Extended Visual Simultaneity Thresholds in Patients with Schizophrenia" *Schizophrenia Bulletin* 35: 816–25.

Giersch, Anne, and Virginie Rhein. 2008. "Lack of Flexibility in Visual Grouping in Patients with Schizophrenia" *Journal of Abnormal Psychology* 117: 132–42.

Granholm, Eric, Scott C. Fish, and Steven P. Verney. 2009. "Pupillometric Measures of Attentional Allocation to Target and Mask Processing on the Backward Masking Task in Schizophrenia" *Psychophysiology* 46: 510–20.

Green, David M., and John A. Swets. 1974. *Signal Detection Theory and Psychophysics* Melbourne, Florida: Robert E. Krieger Publishing Co.

Green, Michael F. and Elaine Walker. 1986. "Symptom Correlates of Vulnerability Backward Masking in Schizophrenia" *American Journal of Psychiatry* 143: 181–6.

Green, Michael F., Keith H. Nuechterlein, and Jim Mintz. 1994. "Backward Masking in Schizophrenia and Mania. II. Specifying the Visual Channels" *Archives of General Psychiatry* 51: 945–51.

Green, Michael F. et al. 2011. "Visual Masking in Schizophrenia: Overview and Theoretical Implications" *Schizophrenia Bulletin* 37: 700–8.

Grondin, Simon. 2010. "Timing and Time Perception: A Review of Recent Behavioral and Neuroscience Findings and Theoretical Directions" *Attention Perception and Psychophysics* 272: 561–82.

Holzman, Philip S. 2000. "Eye Movements and the Search for the Essence of Schizophrenia" *Brain Research: Brain Research Reviews* 31: 350–6.

Hommel, Bernhard. 2011a. "The Simon Effect as a Tool and Heuristic" *Acta Psychologica* 136: 189–202.

Hommel, Bernhard. 2011b. "Attention and Spatial Stimulus Coding in the Simon Task: A Rejoinder to Van Der Lubbe and Abrahamse (2010)" *Acta Psychologica* 136: 265–68.

Husserl, Edmund. 1893–1917. *On the Phenomenology of the Consciousness of Internal Time (1893–1917)* Brough JB, trans-ed. Dordrecht, Boston: Kluwer Academic Publishers, 1991.

James, William. 1896. "The Perception of Time" In *The Principles of Psychology, Chapter XV*, London: McMillan.

Knöchel, Christian et al. 2012. "Interhemispheric Hypoconnectivity in Schizophrenia: Fiber Integrity and Volume Differences of the Corpus Callosum in Patients and Unaffected Relatives" *Neuroimage* 59: 926–34.

Lalanne, Léon. 1876. "Sur la durée de la sensation tactile" *Note des Comptes rendus de l'Académie des Sciences de Paris*, 1314–6.

Lalanne, Laurence et al. 2012a. "Attention and Masking in Schizophrenia" *Biological Psychiatry* 71: 162–8.

Lalanne, Laurence, Mitsouko Van Assche, and Anne Giersch. 2012b. "When Predictive Mechanisms Go Wrong: Disordered Visual Synchrony Thresholds in Schizophrenia" *Schizophrenia Bulletin* 38: 506–13.

Lalanne, Laurence et al. 2012c. "Looking Forward: An Impaired Ability in Patients with Schizophrenia?" *Neuropsychologia* 50: 2736–44.

Martin, Brice et al. (2013). "Temporal Event Structure and Timing in Schizophrenia: Preserved Binding in a Longer 'now'" *Neuropsychologia* 51: 358–371.

Miall, R. et al. 1993. "Is the Cerebellum a Smith Predictor?" *Journal of motor behavior* 25: 203–16.

Minkowski, Eugène. 1933. *Le temps vécu* Presses Universitaires de France: Paris.

Mohr, Bettina et al. 2008. "Hemispheric Cooperation-a Crucial Factor in Schizophrenia? Neurophysiological Evidence" *NeuroImage* 41: 1102–10.

Nicol, Jeffrey R., and David I. Shore. 2007. "Perceptual Grouping Impairs Temporal Resolution" *Experimental Brain Research* 183: 141–8.

Phillips, Mary L., and Anthony S. Davis. 1994. "Understanding the Symptoms of Schizophrenia Using Visual Scan Paths" *British Journal of Psychiatry* 165: 673–5.

Pöppel, Ernst. 1997. "A Hierarchical Model of Temporal Perception" *Trends in Cognitive Science* 1: 56–61.

Pöppel, Ernst. 2009. "Pre-Semantically Defined Window for Cognitive Processing" *Philosophical Transactions of the Royal Society B Biological Sciences*, 364: 1887–96.

Saccuzzo, Dennis P., and David L. Braff. 1981. "Early Information Processing Deficit in Schizophrenia News Findings Using Schizophrenic Patients Centrally, Not Peripherally" *Archives of General Psychiatry*, 38: 175–9.

Schmidt, Heike et al. 2011. "Low-Level Temporal Coding Impairments in Psychosis: Preliminary Findings and Recommendations for Further Studies" *Journal of Abnormal Psychology* 120: 476–82.

Schwartz, Barry D., Daniel K. Winstead, and W.G. Walker. 1984. "A Corpus Callosal Deficit in Sequencial Analysis by Schizophrenics" *Biological Psychiatry* 19: 1667–76.

Schuck, John R., and Richard G. Lee. 1989. "Backward Masking, Information Processing, and Schizophrenia" *Schizophrenia Bulletin* 15: 491–500.

Silverstein, Steven M., and Brian P. Keane. 2011. "Perceptual Organization Impairment in Schizophrenia and Associated Brain Mechanisms: Review of Research from 2005 to 2010" *Schizophrenia Bulletin* 37: 690–9.

Spence, Charles, and Cesare Parise. 2010. "Prior-Entry: A Review" *Consciousness and Cognition* 19: 364–79.

Szelag, Elzbieta et al. 2004. "Individual Differences in Temporal Information Processing in Humans" *Acta Neurobiologica Experimentalis* 64: 349–66.

Trillenberg, Peter, Rebekka Lencer, and Wolfgang Heide. 2004. "Eye Movements and Psychiatric Disease" *Current Opinion in Neurology* 17: 43–7.

Turgeon, Martine et al. 2012. "Impaired Predictive Timing with Spared Time Interval Production in Individuals with Schizophrenia" *Psychiatry Research* 197: 13–8.

Uhlhaas, Peter, and Wolf Singer. 2010. "Abnormal Neural Oscillations and Synchrony in Schizophrenia" *Nature Reviews Neuroscience* 11: 100–13.

Van Assche, Mitsouko, and Anne Giersch. 2011. "Visual Organization Processes in Schizophrenia" *Schizophrenia Bulletin* 37: 394–404.

Van der Lubbe, Rob H.J., and Elger L. Abrahamse. 2011. "The Premotor Theory of Attention and the Simon Effect" *Acta Psychologica* 136: 259–64.

Van Wassenhove, Virginie. 2009. "Minding Time in an Amodal Representational Space" *Philosophical Transactions of the Royal Society B Biological Sciences* 364: 1815–30.

Varela, Francisco J. (1999). "The Specious Present: A Neurophenomenology of Time Consciousness" In *Naturalizing Phenomenology. Issues in Contemporary Phenomenology and Cognitive Science*, edited by Jean Petitot, 266–329. Stanford: Stanford University Press.

Vogeley, Kai, and Christian Kupke. 2007. "Disturbances of Time Consciousness from a Phenomenological and Neuroscientific Perspective" *Schizophrenia Bulletin* 33: 142–56.

Volz, Hans-Peter et al. 2001. "Time Estimation in Schizophrenia: An fMRI Study at Adjusted Levels of Difficulty" *Neuroreport* 12: 313–6.

Von Baer, Karl Ernst (1864). "Welche Auffassung der lebenden Natur ist die richtige? Und wie ist diese Auffassung auf die Entomologie anzuwenden?" In *Reden, gehalten in wissenschaftlichen Versammlungen une kleinere Aufsätze vermischten Inhalt*, edited by Karl Ernst von Baer, 237–84. St-Petersburg, Russia: H. Schmitzdorf.

Wittmann, Marc. 2011. "Moments in Time" *Frontiers in Integrative Neuroscience* 5(66): 1–9.

Wolpert, Daniel M., R. Chris Miall, and Mitsuo Kawato. 1998. "Internal Models in the Cerebellum" *Trends in Cognitive Science* 2: 338–47.

CHAPTER 6

Predictive Timing for Rhythmic Motor Actions in Schizophrenia

Yvonne Delevoye-Turrell, Hélène Wilquin†, and Mariama Dione**

1 Introduction

Time is a common word, used frequently in our everyday language. It appears in many expressions: *"I killed time playing cards," "Lets take time to talk," "I did two things at the same time."* However, even if used by all, the word Time refers back to a complex phenomenon that is difficult – even impossible – to define. To this a priori simple question: "What is time?" the theologist Saint Augustin answered: "If no one asks, I well know; but if someone poses me the question, and I try to reveal my thoughts, I realise that I in fact do not know" (Les Confessions de Saint Augustin, Livre XXI, chapitre XIV, traduit de Moreau, 1864, édition 1942). Time has the ineffable character that reveals its existence only through a construct of the mind (Kant, 1845). Hence, Kants' understanding of Time infers two distinct phenomena: a physical absolute time and the psychological subjective time, the later being the only one sensitive to a distorted mind.

The physical time, encapsulated in the real number "t," would correspond to that defined by Newton as *Absolute Time*: "Absolute time, without reference to anything external, flows uniformly" (1726 (1987), p. 408). Two concepts are here involved: absolute equality of time intervals ("uniform flow") and less obviously but equal essentially, absolute simultaneity, which is the more pervasive concept that underlies not only physics, but also the notion of past, present, and future. As such, the laws of physics impose to *Time* the idea of causality with the impossibility to act upon the past (i.e., to modulate events that have already taken place). The causality postulate may be one of the reasons why psychologists have revealed more interest in the study of *Subjective time*. The question being: "How does a subject experience the passage of Time? What is the experience of my own body as an actor in the past causing the occurrence of an event in the present time? Why can I, under certain situations, lose track of time with

* SCA Lab UMR CNRS 9193, University of Lille, rue du Barreau 59653 villeneuve d'Ascq.
† Aix-Marseille Université, LPCLS EA 3278, 13621, Aix en Provence, France.

confusions between actions that I am performing and those that are to come in the future?"

It is the case that Time is in the hearts of our daily motor activities. Time helps structure our sensations and thoughts; Time is required for the production of coordinated motor sequences and voluntary interactive behaviours. Because movements involve changes in muscle length over time, motor control and timing are inextricably related. In the present chapter, we will consider the specific case of Motor Timing in Schizophrenia, a pathology affecting the mind and also the subjective experience of the passage of time. We will present a brief overview of the literature on subjective timing in adults suffering from chronic schizophrenia, starting from the patients' verbatim to the initial empirical data that has been reported. Then, we will present our own work on rhythmic production in schizophrenia to show the importance of developing new paradigms, based on more objective and non verbal paradigms. We will further try to show that these paradigms have the advantage of providing a better insight on the psychological mechanisms that may be at the origin of the distorted experiences of subjective time and agency in schizophrenia. We will finally conclude on the possible neurocognitive mechanisms that may underlie timing processes required for normalised motor control but that may also be the basis for the emergence of an anchored experience of the present time, psychological experience that is required for the construct of self as a unity of being.

2 Subjective Time in Schizophrenia

The challenge in psychology has been to study the sensation of time, a sensation that has no specific sensory system or organ – contrary to sight, hearing, taste, smell, and touch. A multitude of experimental paradigms have thus been proposed to study temporal processing in mind for: (1) the experience of being through time, (2) perceptual timing, and (3) motor timing. Substantial reviews have been published throughout the years about these diverse approaches (e.g., Bindra and Waksberg, 1956; Grondin, 2010; Repp, 2005). In this section, we will discuss a series of studies that reported the presence of time abnormalities in schizophrenia, with both explicit and implicit measures of temporal processing in the experiential and the sensory domains. But before, we first briefly describe the clinical aspects of the pathology.

2.1 *Schizophrenia: The Pathology*

Schizophrenia is a chronic, severe, and disabling mental disorder that affects approximately 1% of the population at some point in their lifetime. The disorder often develops earlier in men, usually in the late teens or early twenties, while women typically develop the disorder in their twenties and early thirties. People with schizophrenia often experience positive symptoms, e.g., delusions (believing that other people are reading their minds, controlling their thoughts or plotting to harm them), hallucinations (usually hearing voices not heard by others), incoherence, and physical agitation. Patients are also often identified with significant negative symptoms, e.g., disturbed interpersonal relationships, fearfulness, and social withdrawn. In later years, schizophrenia has been considered to be associated systematically to a disorganisation aspect that would affect all aspects of cognitive interplay, with disorganised speech, thoughts, emotions, and motor behaviour. Treatment almost always includes medications, with newer drugs being more effective (especially for the positive symptoms) and associated to fewer side effects than older ones.

While there is evidence that genetic factors have a role in the development of the pathology, other unknown causes play a significant part as well, suggesting a multifactor disease (Bellak 1949; Harrow and Jobe, 2007; Tsuang et al., 2001). While some proposed causes have been proven false, such as bad parenting and poor will power, there are many theories that remain. One of the most famous and most debatable is the dopamine hypothesis. The proposed hypothesis states that the brain of schizophrenic patients produces more dopamine than normal brains. It is this increased dopamine that is believed to be responsible for the symptoms of the disease. However, there is much debate in the scientific community as to the exact mechanisms by which altered dopamine levels, especially in the prefrontal cortex, striatum, and limbic system, produce schizophrenia (for a review, see Howes and Kapur, 2009). Evidence has nevertheless been presented through the last decade indicating that brain structures directly linked to neural timing processes are also impaired in schizophrenia (Volz et al., 2001), with the cognitive dysmetria hypothesis being central to the pathology (Andreasen, 1999; Andreasen et al., 1998). To note the origin of the term "dysmetria," which comes from the Greek term "metron," which means, "to measure a (time) scale or distance."

2.2 *Experiential Distortions of Time: Studying Verbatim*

The idea that time distortions are associated to schizophrenia is far from being a new hypothesis. Indeed, in 1932, Aubrey Lewis did a very nice report on the verbatim of patients suffering from schizophrenia. In their reports, patients describe different time abnormalities, e.g., a feeling of intemporality, the sensation of having the power to influence forthcoming events, phenomenon of premonitions and of "déjà-vus." For example, a clear description of a patient revealed a sense of time interruption with the individual stating that "*Somebody was playing with the clocks*" (Lewis, 1932, p. 612). In this statement, the patient was in fact describing his feeling that time was passing extremely fast at certain moments, but was extremely slow at others with seconds seeming to last an eternity. Another patient reported: "he would shoot gun fires towards his watch in order to kill time, which he considered as his worst enemy" (Minkowski, 1933, p. 13). More recently, a French psychologist transcribed examples of time distortions in young patients with Ultra High Risk of developing schizophrenia and with first-episode psychosis (15–25 years of age) for her PhD work (Wilquin, 2010). In her work, she reported, for example, a young patient who said to have found himself in front of closed gates at his high school on three different consecutive days, in the middle of the night. In a surprising way, this patient evoked the fact that he was unable to read the time on his watch as if the hands had temporarily disappeared. Another patient described that time passed much faster now (i.e., since his first psychotic episode) than before; and when talking about motor actions, the patient would state: "I make a movement but I have the impression that my body stays put," "I walk but I feel as if my body remains immobile..." These two verbatim are striking revelations of the possibility that in schizophrenia, time distortions affect even the most basic brain mechanisms enabling the coordination of (i) those motor commands sent out to the limbs for movement execution with (ii) the sensory (reafferences) information coming back from the contracted muscles.

Overall, the verbatim of patients have been helpful in providing an insight in the nature of the time distortions experienced by patients suffering from schizophrenia. Starting in the 1950's, more empirical studies were conducted in patients in order to gain a better understanding of the neuropsychological mechanism that may cause the temporal processing dysfunctions.

2.3 *The Initial Empirical Methodologies*

Various experimental paradigms have been developed in the field of time psychology, using more or less strict psychophysical constraints and in different perceptual modalities. Four basic methods have been commonly adopted to investigate sensory timing capacities of interval-durations. In *verbal estimation* tasks, participants are required to verbally report the duration of a presented stimulus ("the sound lasted 2 seconds"). The *interval production* tasks require participants to produce a target duration (by pressing continuously a key press or by marking verbally the onset and the end of the duration; e.g., beep...beep) that has been presented by the experimenter using verbal temporal units (e.g., "Could you please produce an interval-duration of 2 seconds?"). Quite similarly, in *reproduction tasks*, participants are asked to reproduce a time interval that they were presented with, but the time interval is presented here by a continuous stimulus. Finally, many variants of a *comparative method* have been used. This approach, which resembles the closest to the classic psychophysical paradigms, consists in the sequential presentation of two intervals (a standard or *reference* and a *test* interval). Participants are asked to say whether the test interval is shorter or longer than the reference interval (forced choice response). All these paradigms have been adapted to the specificity of clinical populations, the auditory modality being the most commonly used with a limited number of trial repetitions in order to consider the fatigability parameter.

The Goldstone and Lhamon (1956) study is one of the first to have assessed through empirical measures the distorted perception of time intervals in schizophrenia. A sound was presented to the participants, who were required to indicate whether this auditory stimulus was longer or shorter than a second. Results from this method of time interval comparison, revealed that all participants had a net tendency to overestimate the duration of the interval. Nevertheless, the overestimation was significantly longer for the patients with schizophrenia compared to the controls. Patients were also characterised by more variability in their responses.

In the following section, we report complementary results confirming these findings in patients with schizophrenia using these various experimental paradigms to investigate more specifically the nature of time perception distortions.

2.4 *Time Perception in Schizophrenia*

Patients diagnosed with schizophrenia display timing deficits. The exact nature of these deficits, however, varies across studies (Brown et al., 2005; Clausen, 1950; Densen, 1977; Johnson and Petzel, 1971; Rammsayer, 1990; Tysk, 1983a,b, 1990; Volz et al., 2001; Wahl and Sieg, 1980).

A majority of reports have used verbal estimation tasks, most probably because of the simplicity of the approach (no equipment required). These studies have confirmed temporal disturbances in schizophrenia, with a systematic overestimation of duration (e.g., a 10-s stimulus elicits a verbal label larger than 10 s) in the seconds range (Johnson and Petzel, 1971; Tysk, 1983a; Densen, 1977; Wahl and Sieg, 1980). For example, Densen (1977) compared the results obtained in three different groups (10 patients with schizophrenia; 10 patients suffering from mental disorder other than schizophrenia; 10 controls) and reported a general tendency towards overestimation for intervals contained within the range of 5 to 120 s in the schizophrenic group of patients only and especially, for those short intervals of 5 s.

When using the producing and the reproducing paradigms, results described systematic underproductions of temporal durations ranging from several seconds to one minute in schizophrenia, i.e., participants responded before 10 s had elapsed when required to respond at 10 s (Clausen, 1950; Elvevåg et al., 2003; Tysk, 1990; Wahl and Sieg, 1980). The studies conducted by Tysk (1983b) are interesting as they tested large patient groups (N = 50) and showed that the timing deficits were found in the different subtypes of schizophrenia (following the DSM III classification) with an absence of symptom effects when contrasting positive and negative aspects (Tysk, 1990). Greater differences in time estimation of visual vs. auditory signals have also been reported in patients when compared to controls (Penney et al., 2000).

In the past twenty years, more contemporary paradigms, i.e., temporal discrimination tasks, have dominated research on time perception in schizophrenia (Carroll et al., 2008, 2009a; Davalos et al., 2003; Elvevåg et al., 2003; Rammsayer, 1990). For example, Elvevåg et al. (2003) employed brief stimuli so that the use of certain strategies, such as counting, would be reduced or eliminated (short time scales <2 s). In their study, participants were asked to perform two different tasks. In the temporal generalization task, participants were initially presented with a standard duration tone (500ms) and during the test phase they were asked to give interval duration judgments with standards of various durations (125, 250, 375, 625, 750, and 875ms). Results showed that patients made overall more mistakes but differences laid especially for those intervals greater than 500ms. These deficits could not be due to contrasting learning effects between patients and controls. A temporal bisection task was also used. Here, there were two standards, one long and one short, and participants were asked to judge whether test stimuli were more similar to the long or to the short standard. Once more, patients performed more mistakes but statistically reliable differences were found for those intervals greater than 600ms only. In both tasks however, patients were characterized by flatter gradients than controls, indicating less accurate and especially overall more variable timing capacities. Using a similar comparative method, Wilquin et al. (2010) and Ameller et al. (2011) also reported preserved perceptual detection of irregularities in otherwise regular sequences for interval shorter than 500ms, even if once more patients revealed overall more errors in their decision making.

Considering all in all, patients with schizophrenia seem to experience time as lengthened relative to objective time, especially for inter-tone intervals greater than 600ms, which is consistent with patients' subjective reports of an elongated experience of the passage of time in daily activities (Freedman, 1974). These findings have generally been interpreted as reflecting an increase in speed of a hypothetical "internal clock" (Johnson and Petzel, 1971; Tysk, 1990 – for theoretical basis to the question, see Gibbon and Church, 1984), an interpretation that has been assumed consistent with animal studies of the effects of dopamine agonists on timing (Rammsayer, 1990). We will come back to this concept of internal clock in Section 3 of the present chapter. However, while these results may reflect temporal deficits in schizophrenia, they may also be due to other (cognitive) dysfunctional processes. Specifically, global deficits in verbalisation, motivation, or even attention capacities may also affect performance levels in verbal estimation tasks (Brébion et al., 2000). Yet an additional possibility is that the overestimations and underproductions are a consequence of systematic memory distortions (Penney et al., 2000). Indeed, recent studies have observed significant correlations between the performance on time perception tasks and that found in neuropsychological outcomes typically affected in schizophrenia (e.g., Lee et al., 2009; Wilquin et al., 2010).

This limit of the verbal estimation paradigms, especially when working in clinical populations, has led to the use of non-verbal, non-decisional tasks in order to gain a more accurate (pure) assessment of timing deficits in schizophrenia.

In the following section, we report findings obtained for motor timing in adult patients suffering from schizophrenia in various motor sequential tasks in order to gain a better understanding of the nature of the timing processes that may be impaired. With this aim, we will now report those studies investigating timing deficits using synchronization-continuation finger-tapping tasks, a paradigm that does not require verbal or explicit report. Here, we have taken the position that motor timing and time perception rely on overlapping (and maybe identical) brain circuitries (Burr et al., 2007; Ivry and Hazeltine, 1995; see Buonomano and Laje, 2010 for a counter discussion).

3 Motor Timing to Assess Subjective Time in Schizophrenia

It is the case that a number of studies of human timing have indicated that accurate estimation of intervals in the order of several seconds requires sustained focusing of attention and/or memory (Fortin et al., 1993; Fortin, 1999; Zakay and Block, 1996). Furthermore, strategies such as counting can be recruited in making such judgments of interval durations. The involvement of attentional and strategic processes in these perceptual tasks is noteworthy given the well-documented reports in schizophrenia of systematic attentional impairments (e.g., Cornblatt and Keilp, 1994; Delevoye-Turrell et al., 2006; Nuechterlein and Dawson, 1984), and working memory deficits (e.g., Fleming et al., 1997; Stone et al., 1998). Hence, it is important to consider whether the systematic findings of perturbed timing processes in schizophrenia – reviewed above – reflect general impairments in cognitive processes rather than a specific difficulty with temporal processing (Rammsayer, 1999).

With this in mind, we have started to develop an original approach consisting in adapting the *synchronization-continuation finger-tapping paradigm* (Stevens, 1986) to pathological populations. Such a task requires participants to first tap in time with computer-generated tones separated by a fixed inter-tone interval (tone-paced tapping). After a series of tone-paced responses, the tones are discontinued and participants are required to continue. Using measuring devices like button presses, force transducers, and tactile screens, we have investigated the capacity of individuals to tap in rhythm to a metronome. Without explicit instructions to focus attention on the temporal information, participants perform the task as a simple rhythmic motor task. Importantly, the difficulty of the task can be manipulated experimentally by varying the complexity or length of the tapping sequence to produce; the rhythmic pattern can be maintained but task difficulty modulated by changing the spatial relationship between successive taps; and the pleasure experienced by the participants can be also modulated by changing the types of sounds the participants are required to synchronise to (tap to the sound of *marching feet* or to your *favourite pop tune* vs. tap to the *metronome* beat) – a nice approach to limit boredom.

We have used adapted synchronization-continuation tasks in order to work with different age groups of participants and with individuals suffering from different types of mental illnesses, with limited influences of associated cognitive abnormalities in memory, IQ or attention capacities. Thus, in the following sections, we report some of our recent findings obtained with these original motor timing tasks in patients suffering from schizophrenia (the "Touch 2 Know" tablet application).

Before, however, we wanted to emphasize in a first section the fact that accurate timing is crucial for motor control, especially for the fluency of motor sequencing. More precisely, we report data obtained in patients performing different types of sequential motor actions that require the precise timing of each sub-movement of the sequence in order to achieve motor goals.

3.1 *Sequential Motor Control in Schizophrenia*

Motor control, from catching a ball to playing a musical instrument or dancing a waltz, requires the production of complex spatiotemporal patterns of muscle activity. As such, most motor tasks, including speech production, require the carefully orchestrated movement times in the order of tens of milliseconds to a few seconds. In a task where a single effector is used to perform a sequence, each successive sub-movement is dependent on completion of the preceding one. It takes an appreciable amount of time for peripheral sensory information to reach the central nervous system, and in the presence of such delays, it is generally suggested that motor sequences may be timed in two fundamentally different ways depending on whether or not they require sensory information. First, sequencing may be based on concurrent processing of sensory information (Macefield and Johansson, 1994). For such triggered sequences, each element in the sequence will be initiated only after the arrival of the sensory feedback signalling the end of the preceding element. An alternative approach involves preplanning the entire sequence before its initiation as a whole (Billon et al., 1996). In this case, the sequence is represented as a single entity and performed without reference to feedback from external events in a pre-planned fashion. The correct preparation and execution of such planned sequences rely on well-defined predictive processes developed through previous experiences and depend on the brain's ability to organize through time and space information from multiple sources.

In the psychiatric literature of schizophrenia, many reaction time studies reported general slowness in schizophrenia (Cohen and Servan-Schreiber, 1992; Sereno and Holzman, 1996; for a review, see Nuechterlein, 1977). However, it is the case that the selected tasks often required sequencing over time, for example, detect a white dot and then detect a blue dot and then press the switch as soon as the blue dot disappears (Zahn et al., 1998). Hence, it is possible that slowness is not characteristic of patients with schizophrenia but arises in these patients because timed sequencing is necessary.

In a series of sequential motor tasks, we used a simple load cell to assess whether the motor slowness in schizophrenia was directly related to the need to sequence a series of actions through time (Delevoye-Turrell et al., 2003). Furthermore, through the implementation of different types of action sequences (triggered vs. planned), we were able to dissociate between lower and higher timing mechanisms for the deficit. Overall, results demonstrated that schizophrenia is associated to higher timing mechanisms only. Indeed, for triggered sequences, the participants' task was to let an object slip between fingertips and to augment grip force to arrest object fall. Results showed that the group mean time delays between two successive motor elements were similar in patients and controls, supporting the hypothesis of preserved lower order (subcortical) integrative mechanisms in schizophrenia (Delevoye-Turrell et al., 2007). For planned sequences, participants were required to press a series of flat button that were displayed on a table top, one after the other, as fast as possible at a stop signal (i.e., following a regular and fast rhythmic pattern). For the controls, the time delay between successive elements was short (40ms) affording a fluent and fast execution of the rhythmic pattern, suggesting that the timing of the motor elements was pre-programmed without the need of external sensory information (Billon et al., 1996). A different pattern of results was obtained in the patients with performances that were characterized by inter-movement delays that were: (a) longer for planned than for triggered sequences, (b) abnormally long compared with the controls, and (c) increased with sequence timing complexity. Overall, these data support the possibility that schizophrenia is characterized by a lack of fluency in the execution of planned sequences, which would be related to a higher order deficit in the temporal sequencing of multiple elements for motor fluency. This high order timing deficit may be a plausible explanation for the impaired ability to detect a mismatch between a self-generated movement and its consequences, but not impaired in their ability to automatically compensate for low level gain change in a continuous circle-drawing task (Knoblich et al., 2004), which may require two complementary timing mechanisms (Robertson et al., 1999; Zelaznik et al., 2000).

After presenting findings from sequential motor experiments in patients with schizophrenia, we continue in the next section, by reporting experimental studies tailor-made to assess motor timing distortions in schizophrenia during the planning and execution of rhythmic motor sequences.

3.2 *Motor Timing in Schizophrenia*

Elvevåg and collaborators (2004) were among the first that tested patients with schizophrenia in a synchronization-continuation finger-tapping task to an auditory beat, with a metronome at 30 or 60 beats per minute (bpm). After a 5-minute period, the metronome was then switched off and participants were to continue as before for 5 min, but now were required to generate the rhythm internally. Results revealed that all participants were rather accurate during the synchronisation task (unpublished data) but when asked to tap without the metronome, controls were better able to generate a constant tempo with minimal fluctuations than were patients. Interestingly, in the 30-bpm task patients disproportionately speeded up their responses over time. Note that the patients were reported being better when asked to perform faster rhythmic patterns (60-bpm) than slower ones (30-bpm) – we will come back to this point in the last section of this chapter.

Carroll et al. (2009b) used a similar synchronization-continuation finger-tapping task with an auditory tone paced at 500ms inter-tone intervals (ITI = 500ms). Thirty-two participants with schizophrenia displayed significantly faster tapping rates, with significantly shorter ITIs, for both tone- and self-paced portions of the task compared to the non-psychiatric group. They also displayed greater tapping variability during both tone- and self-paced portions of the task. In a second analysis, the Wing-Kristofferson model (Wing and Kristofferson, 1973) was applied to the self-paced data to determine whether increased tapping variance in schizophrenia should be attributed to clock or motor implementation factors. Consistent with the temporal representation demands of the self-paced task, clock variability was significantly higher in the schizophrenia group. Estimates of motor implementation variance did not differ between individuals with schizophrenia and controls. The partitioning of variance into clock and motor factors hence suggested that group differences in tapping variability were largely due to decreased temporal precision of the internal clock, in schizophrenia.

Still using a synchronization-continuation finger-tapping task, we wanted to go further by assessing the effects of task difficulty and attention demands for motor timing during motor sequencing. For this reason, we manipulated the complexity of the rhythm and/or the complexity of the motor task to perform motor sequences in synchrony with an external metronome (Wilquin et al., 2010). The participants' task was to squeeze a force transducer in its centre in beat with a metronome for a 12-s period, and then to continue producing the same rhythm in a continuation phase for another 12-s period (Figure 6.1-left). In order to manipulate task difficulty without modifying rhythm complexity, two pitch tones were used. Participants were required to produce strong squeeze forces for high pitch (16 N) and small squeeze forces for low pitch (8 N). These pitches were presented alternating providing the participants with the possibility to predict the rhythmic sequence to perform well before the start of each trial. In a second condition, we manipulated rhythm complexity without modifying task difficulty. Here, ITIs of 300 and 700ms were

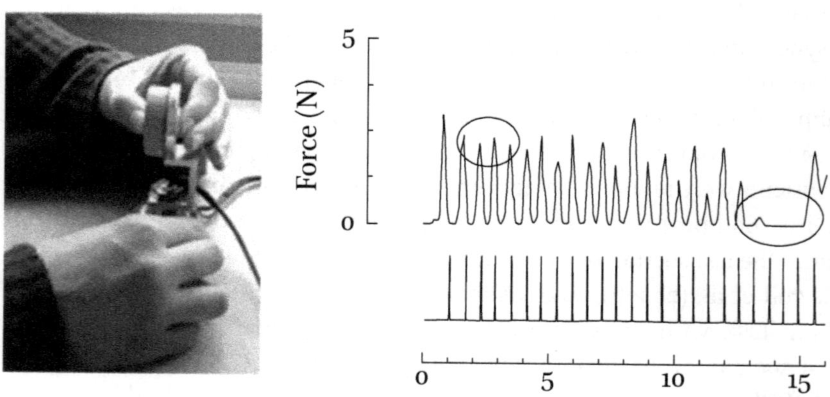

FIGURE 6.1 *Picture of the experimental apparatus used to measure squeeze forces produced on the load cell with the participants' dominant hand (example of a left-handed patient with schizophrenia) in function of a metronome* (left). *Example of an equivalent rhythm for which subjects were required to alternate force levels in function of high and low pitch tones as a function of time (in seconds). Note the difficulty in force alternation but also of the presence of an attentional lapse twelve seconds after the start of the trial* (right).

alternated with a single pitch tone for which participants were to use a 12 N squeeze force. Results revealed that in the simplest of conditions (equivalent force/equivalent time), patients and controls performed very few production errors and were highly accurate with mean errors smaller than 10ms, which is similar to that classically reported in the literature (for review, see Aschersleben, 2002). With motor task complexity (alternated force/equivalent time), patients were in difficulty and were characterised by more rejected trials. A more specific analysis of these errors confirmed that patients were characterised by more frequent arrests in rhythm execution (Figure 6.1-right), attentional lapses that are consistent with the clinical lapses frequently noted during semi-structured interviews of patients with schizophrenia (Ameller et al., 2011). Nevertheless, for those trials (alternated force/equivalent time) performed correctly and without lapses throughout the 24-s period, timing errors were similar in controls and patients, suggesting a problem of attention for action execution (Delevoye-Turrell et al., 2006) rather than a timing problem per se.

Interestingly, in both conditions for which the rhythm complexity was manipulated (equivalent force/alternated time; alternated force/alternated time), patients revealed on those trials correctly performed, significant timing errors both during the synchronisation and the continuation phases (Figure 6.2). These errors were not correlated with the treatment dosages or age; they were slightly correlated to the attention capacities as evaluated by the D2 test for attention (i.e., the participants are required to cross out as many *d"* symbols that are mixed up with *p*, *d*, and *d'* symbols within a timed period of 1 minute – see Brickenkamp, 1962). Following the theories developed for dynamical systems, sensorimotor synchronisation for equivalent time intervals would need the same timing mechanisms than that used for continuous movements (Zelaznik et al., 2005). It would require minimal attention resources because it would be highly automatic (Pressing, 1999) emerging from the inherent timing properties of the neural circuits processing motor planning and execution. For this low-level processing of motor timing, patients with schizophrenia would not be perturbed and thus, they show similar timing capacities than controls. For alternated interval timing, a more explicit control would be required in order to process each interval as a discrete time series (Pressing, 1998). Here, participants would be required to maintain cognitive representations of time interval durations, and order thanks to the existence of a kind of *internal clock*. Predictive timing would thus be involved, an anticipatory cognitive mechanism that would be impaired in schizophrenia. Because of the need of significant cognitive resources, alternated interval-timing tasks are more difficult to execute even for healthy controls than equivalent interval-timing tasks (Delignières et al., 2004; Zelaznik et al., 2002).

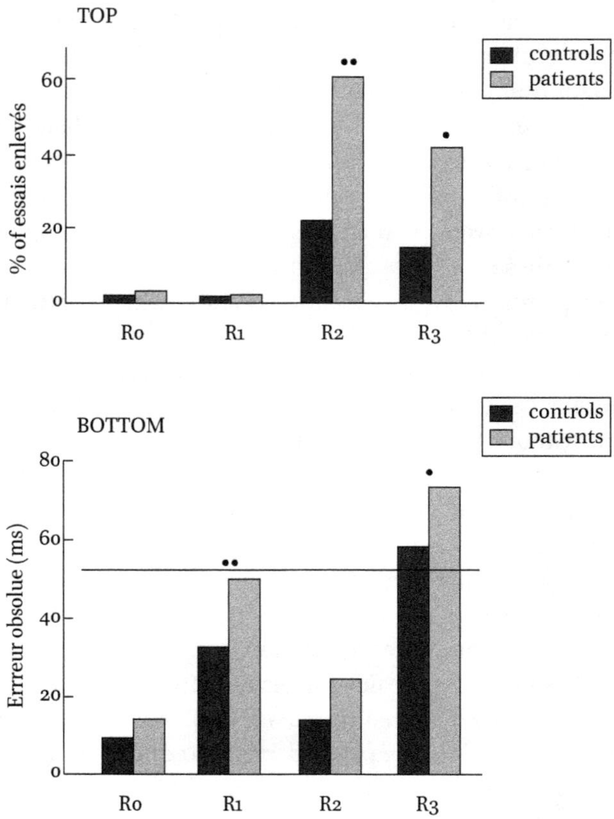

FIGURE 6.2 *Number of bad trials measured in healthy controls (black) and patients with schizophrenia (grey) in a synchronisation task requiring subjects to perform a series of squeezes on a load cell synchronisation with equivalent time intervals (R0; R2) and alternated time intervals (R1; R3) with equivalent force levels (R0; R1) or alternated force levels (R2; R3). Results showed that for both controls and patients, it was difficult to alternate force levels and maintain the timing motor task but patients failed more often than the controls (top). When assessing absolute timing errors on those trials performed correctly (bottom), results demonstrated that patients made significantly more timing errors than controls in those conditions for which alternated time intervals were to be performed. Overall, these results suggest a specific difficulty in predictive timing in schizophrenia.*

In the following section, we present two studies that were conducted specifically to test the hypothesis that the abnormal performance patterns reported both for time perception and motor timing in individuals with schizophrenia may be associated to a distorted internal clock for temporal processes. The inconclusive findings will lead us to propose in a fourth and final section a novel spatio-temporal paradigm that has given us the means to reveal in schizophrenia a specific problem of the anticipatory process that is necessary for predictive timing.

4 A Preserved Internal Clock in Schizophrenia

The overestimation and underproduction results described above for both time perception (Section 2.4) and motor timing (Section 3.2) suggest that the representation of time is severely distorted in schizophrenia. In the following section, we report two fundamentally different protocols that were developed in our laboratory to test the degree of normality of the pulse rate of a possible central oscillator (internal clock) for explicit and linear metric of time in schizophrenia.

4.1 *Time Decay of Body Schema*

Numerous contributions, mostly clinical reports and phenomenological considerations, have been published on body image pathology in schizophrenia (for reviews, see e.g., Fisher, 1986; Priebe and Röhricht, 2001). Schizophrenia patients are reported to suffer from symptoms of disembodiment such as not feeling at home in one's body anymore, feeling disintegrated, or feeling as if the body is torn apart, suggesting a problem in integrating different sources of information in a unified body representation. Explanations vary and range from neurological to psychological perspectives (Cumming, 1988). In the following study, we wanted to quantify the degree of abnormalities of body image in schizophrenia but especially assess its decay over time in order to get an insight on the phenomenology of the passage of time and its role in a possible distorted body image in schizophrenia.

For this purpose, we adapted the head "return" paradigm, with which one can dissociate the perceived orientation of a body part from its actual position without manipulating proprioceptive information. The so-called head "return" phenomenon was described by Gurfinkel et al. (1989) as the perception of a slow head movement toward the mid-sagittal plane, i.e., its neutral position, while the head is kept turned about 10 minutes with eyes closed. The head "return" phenomenon has since been studied by many researchers and it has been proposed that the illusory displacement of the body segment is due to the decay over time of the body image. In the present case, we studied an arm "return" phenomenon in patients with chronic schizophrenia to assess whether the decay over time would be faster than in healthy controls, under the effects of a faster ticking internal clock.

Twenty-nine patients with schizophrenia (10 females, mean age = 36, SD = 10.7) and thirty healthy controls (12 females, mean age = 34.5, SD = 14.4) were tested in a simple session (unpublished data from Gorog et al., 2005). All patients fulfilled the DSM-IV criteria for schizophrenia and three were drug-free. Participants were invited to seat in front of a table, facing a curtain that prevented them to see more than 30 cm in front of them. They were asked to passively let the experimenter put their (right) target hand behind the curtain, flat on the table, rotated with a 30° angle from the body's straight ahead axis. During the entire session, participants were required to maintain this body position, with shoulders and back straight, with minimal muscle contractions of upper limbs and no visual feedback of the hands.

At the experimenter's signal, the participants were required to move their (left) pointing hand in a symmetrical position from the target hand, relative to the body sagittal axis, and place their palm flat on the table. When set, the positions of both hands were measured on a bidirectional scale (X, Y) in centimetres (cm) with point zero being aligned with the participants' body sagittal axis. The experimenter was careful not to touch the participants' hand. With the measurements completed, the participants were required to bring their pointing hand back to the resting position, which was tactilely detectable with the help of a small rough surface that was placed on the table close to the participants' midline. They were asked to repeat this exact same procedure at five critical moments: 0, 1, 3, 5, and 8 minutes at the experimenter's signal.

Results revealed a significant Group effect for the error of initial hand position at To [$F(1,43) = 10.704$, $p < 0.01$]. Twenty-seven patients out of 29 were characterised with a greater error (ME = 12.1, SD = 10.6 cm) than that obtained in the control group (ME = 0.8, SD = 5.9 cm). Interestingly, errors at To for the patients were positively correlated with the negative score of the PANSS ($R = 0.43$, $p < 0.02$), suggesting greater initial distortions of body image for patients with more severe negative symptoms (even when controlling for treatment dosage). More importantly for the scope of the present chapter, the effects of Time on the decay of body image was highly significant [$F(1,43) = 15.05$, $p < 0.001$], with similar slopes for the linear regressions in both groups [$F(1,43) = 0.490$, $p = 0.481$]. This preserved "return" phenomenon in schizophrenia is illustrated in Figure 6.3.

Overall, these findings indicate that over the 8-minute period the effects of time gave the participants' the illusion of the hand progressively being closer to the body's sagittal axis. But importantly, the ticking clock decay was similar in the healthy controls and in the patients with schizophrenia.

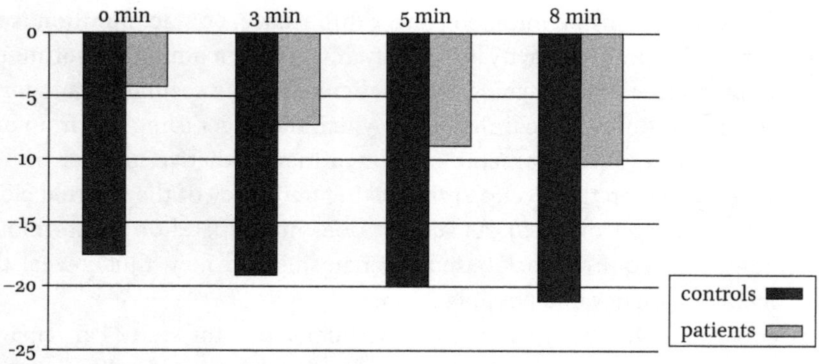

FIGURE 6.3 *Graphical representation of the errors in body schema observed in patients with schizophrenia (red) and controls (blue) during the arm "return" paradigm at four critical moments: 0, 3, 5, 8 minutes from the start of the experimental session. Note that reproduction errors of hand symmetrical position compared to the subjects' body sagittal axis increases at a similar rate for both subject groups suggesting a normalised ticking clock decay in schizophrenia.*

4.2 *Ticking Frequency of the Internal Clock*

Within the framework provided by pacemaker-accumulator models of interval timing (Gibbon et al., 1984), a faster internal clock than normal clock speed would result in a shift in an estimate of duration only when the experience of time on the test trial is measured with a faster clock than that used to measure the reference duration (Meck, 1996). Given the same contingencies in a verbal estimation task, the findings for patients with schizophrenia might be interpreted as due to differential clock speeds for reference and comparison durations, suggesting a less stable clock ticking frequency. To test this hypothesis directly, the spontaneous tapping task has been proposed for which participants are required to produce a regular sequence of finger taps (Vanneste et al., 2001). This measure has the advantage of requiring no subjective or explicit decision-making as well as no need for explicit memory capacities.

In a recent study, we measured spontaneous tapping tempo and used a touch screen in order to measure Inter Response Interval (IRI), i.e., the dependent variable traditionally measured in finger tapping studies. However, thanks to technological advances, we were able to decompose each IRI as the sum of two new variables that helped us dissociate timing distortions from motor control dysfunctions (Delevoye-Turrell et al., 2012). First, Contact Duration (CD) was defined as the time interval during which the finger is in contact with the screen. This indicator is associated to the time needed by the sensorimotor system to detect that the finger has made contact (finger down) and that the next element can thus be initiated (finger up). Hence, contact duration is an indicator of the brain capacity to quickly integrate incoming sensorimotor (haptic) afferences with outgoing motor efferences. The second parameter is the Flight Time (FT), i.e., the time during which the finger is moving from one target to the next. This parameter is planned in a predictive manner on the basis of an open loop that cycles at the ticking frequency of the internal clock (finger down…finger down…). As such, it does not depend on sensorimotor integration but requires clock-based mechanisms and may, thus, reveal the pure ticking frequency of a possible timer.

Twenty-four patients and 22 controls participated in the study. Participants were seated comfortably on a chair in front of a touch screen (Elo Touch, 23 cm 36 cm, 30 cm), which was placed on a narrow support at knee-height. Each trial started with the presentation of a visual image that was made of 6 black circles (Figure 6.4-left). The participants' task was to point each circle one after the other, clockwise, starting from the bottom-right circle. Participants were asked to use their preferred hand and to tap as regularly as possible, at their preferred rate. No sound was ever heard. A trial was ended after the production of 30 taps, and a total of 12 trials were recorded.

Results revealed that patients with schizophrenia were able to produce regular sequences of spontaneous taps but at a slower rate than that measured in age-matched controls (IRI of 610 vs. 487ms for patients and controls, respectively). These results are consistent with previous studies reporting (1) IRI of around 500ms in healthy adults (McAuley et al., 2006) and (2) general movement slowness in patients with schizophrenia (Morrens et al., 2007). Interestingly, tempo slowness was positively correlated to the slowdown of cognitive speed in patients, which suggested a central origin for the deficit (Figure 6.4-right).

However, through the use of a touch screen with an original time performance subdivision, our analyses revealed that an overall mean FT of 349ms was displayed across experimental groups (328 and 369ms for controls and chronic patients, respectively). Coherent with an absence of group effect, our results argue in favour of a normalised ticking frequency of the internal clock in chronic and stabilised patients with schizophrenia, with furthermore similar timing stability (FT variability) in both patients and healthy individuals. These results resemble those reported in the early 80s in studies investigating spontaneous eye-blink rates, with normalised spontaneous rates in stabilised patients under neuroleptic treatment (Karson, 1983). Our chronic patients were characterised nevertheless with significantly longer CDs than controls (see Table 6.1), which argues in favour of a specific problem in the fluent production of the finger down/up rhythmic movement in schizophrenia.

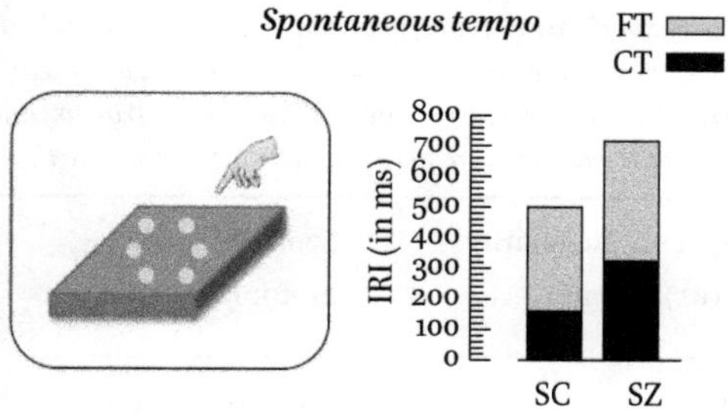

FIGURE 6.4 *Illustration of the experimental setup used to measure spontaneous tempo in controls and patients with schizophrenia* (left). *Results showed that tempo was slower in patients but this difference was solely due to a significant increase in contact duration* (CD) *of finger on the screen. Flight time* (FT) *was similar across groups arguing in favour of a normalised ticking frequency of the internal clock in schizophrenia* (right).

In conclusion, it is possible that the general psychomotor slowness and especially distorted temporal processes are not related to frequency abnormalities in a central oscillator that would play the role of a possible internal clock. We favour here the hypothesis of a specific deficit in the rhythmic timing of action sequences, i.e., a deficit in the sensorimotor integration of predicted incoming afferences with outgoing efferent commands.

5 Tempo Variations to Reveal Specific Timing Processes

In the field of visual perception, authors have suggested difficulties in discriminating simultaneity during the presentation of visual stimuli in schizophrenia, findings that may suggest a specific impairment in time event coding (Foucher et al., 2008; Giersch et al., 2009; Lalanne et al., 2010; see Chapter 5 of this book). With a difficulty to distinguish physical past and present events, the coding of the passage of time could be impaired with direct consequences on the capacity to correctly time and integrate sequences of multiple events. In the following section, we present the idea that the time-dependent properties of a task may induce two different timing processes for perceptual and motor timing. A new protocol in which the tempo is manipulated will be proposed to assess whether synchrony disturbances may be present only in certain time spectrum, suggesting a specific deficit in predictive timing in schizophrenia.

TABLE 6.1 *Mean values (standard deviation) obtained in healthy controls and patients with schizophrenia for rhythm production corresponding to spontaneous tempo, with the subdivision of inter-response-intervals (IRI) in Contact Durations (CD in ms) and Flight Times (FT in ms). Note that patients are as regular than controls in their rhythmic production.*

	Spontaneous Tempo (IRI)	Regularity ratio	Sensori-motor integration CD	Internal clock FT
Controls (N = 22)	487 (107)	1.004 (.007)	161 (74)	328 (78)
Patients (N = 30)	610 (155)	1.010 (.010)	246 (133)	369 (115)

5.1 *Two different Timing Mechanisms for Interval Timing*

Analyses of the neural basis of timing have generally focused on two general computational strategies: mechanisms based on neural *clocks* and mechanisms that emerge from the *dynamics* of neural networks. In general, these models must accomplish some variant of the same computational task by recoding the temporal information present in the (sensory) input into a spatial (motor) code. Hence, different cells must respond selectively to temporal features of the stimulus.

When considering the mechanisms of timing it is perhaps most intuitive to think in terms of clocks or interval timers. The basic computational unit of clock theories involves an oscillator and a counter (Creelman, 1962; Treisman, 1963). Conceptually, in this top-down approach, the oscillator beats at some constant frequency and each beat is then counted by some sort of neural integrator. These clock-like mechanisms could be involved in timing on the scale of seconds and minutes (Meck, 1996; Matell and Meck, 2000) but would clearly be, at the neurophysiological level, in difficulty for shorter intervals. We consider an alternative bottom-up approach to be interesting when considering short time intervals (<1 s). Here, starting from biologically realistic assumptions, authors have asked the extent to which temporal processing can be found as an emergent property. Then, it has been shown that cortical networks are inherently able to process temporal information because information about the recent input history is inherently captured by time-dependent changes in the state of the network (Buonomano and Merzenich, 1995; Buonomano, 2000; Maass et al., 2002). Given the high dimensionality and abundance of time-dependent properties of cortical networks, this type of model could provide a realistic means to decode complex temporal and spatial-temporal patterns of sensory information (Buonomano and Laje, 2010). As such, there would be no need of a centralized clock for temporal processing on the scale of tens to hundreds of milliseconds, but interval timing for fast tempi and complex temporal patterns would be accomplished through non-specialised network dynamics (Buonomano, 2000; Maass et al., 2002).

Ivry et al. (1988) provided one of the first neurophysiological evidence that central (clock) timing and emergent timing processes are independent. Patients with a lesion only in the medial region of a cerebellar hemisphere exhibited increases in emergent timing variance for the ipsilesional hand, compared to the contralesional hand, whereas those patients with localized lateral damage to a cerebellar hemisphere exhibited increases in central clock-like variability, only for the ipsilesional hand compared to the contralesional hand. Moreover, patients with cerebellar lesions also were impaired on a variety of sensory interval-timing processing tasks, suggesting that this structure operates as an internal timing system that is exploited for both motor and non motor tasks, which require the precise representation of temporal duration (reviewed in Ivry, 1997).

In an elegant and systematic analysis of timing aspects for motor control, Zelaznik and collaborators explored the idea that the timing of movement initiation processes involves an explicit representation of time, whereas movements that are continuous may not (Zelaznik et al., 2002; Robertson et al., 1999). They hypothesized furthermore that once initiated, continuous movements no longer involved explicit temporal control. Rather, the temporal regularities would be emergent, reflecting the implicit operation of a different control mode. For this, the authors used three different tasks in a classic synchronisation-continuation paradigm. Two tasks were used in the discrete condition: a finger-tapping movement and an intermittent circle drawing task, for which participants were required to insert a pause between each drawing cycle. In the continuous condition, participants were asked to perform the circle-drawing task without intermittence. Results showed that temporal consistency on the tapping and intermittent circle drawing tasks was significantly correlated in all of the experimental situations, and performances on these tasks were also correlated with acuity on a duration discrimination task. In contrast, the correlations were either weaker or absent between these tasks and the continuous circle drawing task. While all of the tasks could be considered as "timing" tasks, one fundamental difference here between the tasks in the discrete condition, on the one hand, and the task used in the continuous condition, on the other hand, was the need to explicitly time the end and beginning of each sub-movement, in the discrete condition.

In the above example, the two timing modes were revealed through the use of two different motor tasks. It is, thus, difficult to ensure that all other features of the motor planning and execution steps were identical and that the only differing parameter was the temporal aspect. In the following section, we present a unique synchronisation task that provided the means to investigate the accuracy and the stability of explicit and implicit timing mechanisms in both healthy controls and patients with schizophrenia.

5.2 *The Spatial-tapping Task and Schizophrenia*

When walking down the street, at a slow pace, it frequently happens that we control the end position of our foot. This may especially be the case in dirty streets to avoid dog-poo or mud holes. Now imagine walking to catch a bus that is on the verge of departing. Now, your steps will be executed at a higher tempo; you may not think of the cleanness of your shoes because your primary goal is to catch that bus. The Spatial Tapping task is based on this simple idea and requires participants to perform a synchronization task with a metronome that sets different tempo speeds. Contrary to all classic finger-tapping tasks, the participants will in addition need to *move through space* tapping in a counter clockwise order a series of 6 targets disposed around a circle (see Figure 6.4-left). As such, there is both a timing aspect and a spatial dimension to the task. We predicted that at slow tempi, participants would explicitly control the sequential timing of the initiation of each finger-tap; at fast tempi, the objective would be to move the arm around the global circle, keeping as best possible the overall timing of the cycle dynamics. In this later case, the timing of each individual tap would thus become emergent, slave to the overall timing of the arm movement. To test our hypothesis, we performed analyses on the endpoint distributions both in time and space in order to confirm the presence of two different temporal processes in function of imposed tempo.

In the motor control literature, the analysis of endpoint distributions of hand pointing actions has revealed the existence of different planning strategies depending upon task demands (Gordon et al., 1994). Inspired by the findings of Gordon and collaborators, Desmurget et al. (1998) demonstrated that the dispersion properties of endpoint distributions of single pointing movements – from a starting point to a visual target – could be influenced through instructions. When no other instructions were given than executing a pointing movement towards a visual target, Desmurget et al. observed that the distribution of the endpoint errors around the target was roughly circular. However, with the specific instruction to 'follow a straight line path', the endpoint distributions of the pointing actions were elliptical in shape, with the major axis being systematically oriented along a virtual line joining the starting point and the final visual target. Similar results have been published by others, suggesting that endpoint ellipses emerge as oriented when the motor plan is guided by an internal representation of grouped targets (Vindras and Viviani, 1998).

Recently, we applied the analysis of endpoint ellipses (orientation and area) to the results obtained in the finger-tapping synchronisation task to reveal distinct control (executive) strategies in relation to two different temporal control processes for various imposed tempi. Eighteen controls and 18 patients with schizophrenia participated in a unique experimental session. After sitting down in a silent room for 5-min (*brain-clearing phase for rhythmic cues*), participants were invited to produce a spontaneous series of 60 touchdowns on a tactile screen. The explicit instruction was as follows: *"Point each target, one after the other, at your preferred speed...one that seems the most comfortable to you. Try to be as regular as possible."* Customised software (written in MATLAB programming language) calculated the participants' spontaneous tempi and then, the software proposed a series of 11 trials at various tempi at and around, the middle-spontaneous tempo. On average, the spontaneous tempo was similar in patients and controls and was close to 600ms (patients: M = 658ms, SD = 131ms; controls: M = 579ms, SD = 127ms). Thus, for all individuals, the speed of the metronome was varied between Inter-Response-Intervals (IRI) of 300 and 1100ms and participants were invited to produce a series of 60 touchdowns at each tempo. The explicit instruction was as follows: *"Point each target, one after the other, following the speed set by the metronome. You must be as accurate as possible in space but when in difficulty, make sure you are always in time with the metronome."* Resting periods were given periodically to limit arm and cognitive fatigue.

Patients with schizophrenia and controls performed the task without apparent difficulty for 8 out of the 9 tempi. There were no significant differences for IRI errors between controls and patients at all tempi except for the fastest (300ms). For both groups, the fastest tempo was difficult to perform and was characterized by significant increases in IRI errors. These results are presented in Figure 6.5a. At the faster tempi of 400 and 300ms, many participants reported needing to be highly concentrated and to accept to "let go" in order to maintain rhythmic synchrony. Patients were characterized by larger spatial errors (area) throughout. However, longer contact times were revealed at slower tempi only (from ITI = 700 to 1100ms). Contrary to what would have been expected in patients characterized by general slowness, our patients with schizophrenia performed as the controls for the faster tempi (from ITI = 300 to 600ms – see Figure 6.5b).

It is possible that two different temporal processes are used for faster tempi (from ITI = 300 to 700ms), on the one hand, and slower tempi (from ITI = 700 to

1100ms), on the other hand. Following this idea, patients with schizophrenia would be impaired because of a specific difficulty in the timing of those slow tempi that required (1) the inhibition of spontaneous response and (2) a more cognitive control of timing for long intervals. A first indication of this would be the indication of contrasting cognitive strategies in these two different tempi intervals. The analysis of endpoint distributions (orientation bias) in the healthy controls confirmed that the spatial ellipses were significantly more oriented at faster tempi (ITI < 600ms), in reference to the straight line connecting successive targets, than at the slower tempi (i.e., smaller errors to the predicted model). As proposed by Desmurget (1998), we suggest that these results reveal the use of an internal representation of the grouped targets, which was used to plan and execute fast sequential actions without the needs to focus attention on each individual target; the temporal processes used here would be emergent and characteristic of the time needed for motor loops to initiate each step of the sequence. At slower tempi however, participants individuated each target more and as such, there was an absence of orientation of the distribution of the endpoint errors (see Figure 6.5c). The change from one type of temporal process to the other appears to be around the critical spontaneous tempo (i.e., 600ms). Interestingly, our patients with schizophrenia revealed similar results except around the critical spontaneous tempo and seemed to switch timing processes faster than the controls, with more oriented endpoint distributions than that observed in the controls, at the intervals of 600 and 700ms. It is possible that switching strategies occurred faster in the patients from predictive timing to emergent timing in order to engage faster in the emergent timing strategy that would be preserved. This may be also due to perturbed inhibitory functions that are frequently reported in schizophrenia.

Overall, our results suggest the existence of two fundamentally different timing processes with a change in the interplay around the critical moment of preferred tempo. In schizophrenia, the brain mechanism for emergent timing would be preserved. Indeed, patients revealed normalized performances both in time and space compared to adult controls, even if at these fast tempi all participants found the task difficult to achieve. The brain mechanism for predictive timing however seemed to be perturbed in our patients. With a difficulty to inhibit direct response, timing errors were significantly greater than that observed in controls; patients also were characterised with higher performance variability. This pattern of results strongly resembles those reported in the first two sections of this chapter; they also echo recent results published by our co-workers reporting a specific problem in predictive timing using a phase shift detection paradigm (Turgeon et al., 2012).

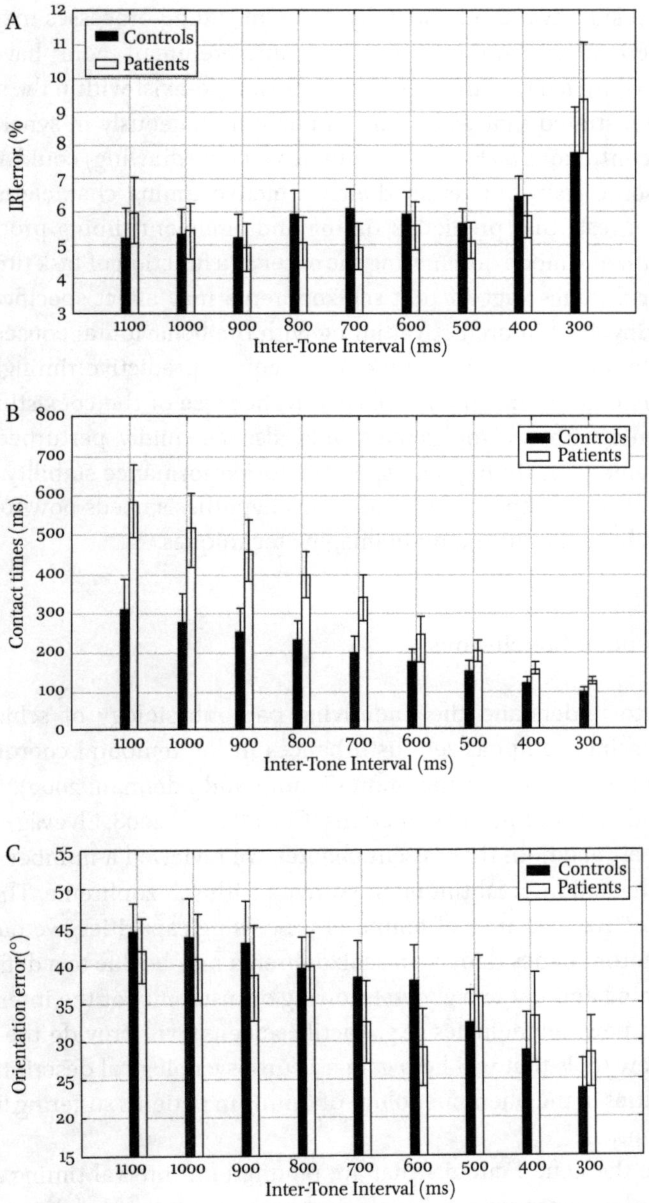

FIGURE 6.5 *Results obtained in the Spatio-tapping task in controls (grey) and patients with schizophrenia (black). The top figure (6.5a) illustrates the inter-tap interval error (ITI in ms) compared to the target inter response interval (IRI in ms) that was imposed by a regular metronome. The middle figure (6.5b) presents mean results obtained for the Contact times (CT in ms) measured for the 11 different tempi. Finally, the bottom figure (6.5c) presents the mean orientations of the endpoint error distributions calculated for oriented scatter plots towards the next target. Large values suggest an absence of orientation.*

In conclusion, two fundamentally different timing processes may exist for both perceptual and motor timing. Repp and Steinman (2010) have recently furthermore argued that these two mechanisms co-exist within the same task. They demonstrated that when carried out simultaneously in synchrony, discrete and continuous tasks (finger-tapping vs. circle drawing) could affect each other in some ways but retain their distinctive timing characteristics. This strongly suggests that predictive timing and emergent timing processes can coexist, with one mode dominating the other as a function of task timing properties. Our findings suggest that schizophrenia may affect specifically those processes involved in predictive timing with the behavioural consequence of significantly perturbing those tasks that require predictive timing, i.e., estimate interval durations in an explicit way; because of the coexisting properties, patients with schizophrenia would also be mildly perturbed in tasks requiring emergent-timing but especially for performance stability, indicator of the event-based aspect of the task. This hypothesis needs now to be tested using both behavioural and brain imaging techniques.

6 General Conclusions

Attempts to understand the underlying pathophysiology of schizophrenia have increasingly emphasized disturbances in the temporal coordination of information processing in the brain (Tononi and Edelman, 2000), leading to sensory and motor timing dysfunctions (Carroll et al., 2008; Elvevåg et al., 2003; Penney et al., 2000). In the present chapter, we reviewed a number of studies that investigated interval timing in patients with schizophrenia. Through the use of objective measures of temporal processing capacities, we have shown that the motor fluency deficits in schizophrenia may be due to a deficit in predictive timing and not to a general slowing or instability of the internal clock. The use of new technologies (e.g., tactile screens) will provide the means to develop new tools that will help gain a neuropsychological description of the abnormalities in the phenomenology of timing in patients suffering from brain dysfunctions.

Because the neural circuits that are required for interval timing and those thought to be compromised in schizophrenia overlap, it has been proposed that a focus on temporal information processing offers a window into understanding the cognitive deficits of schizophrenia and how deficits may contribute to a variety of clinical symptoms (Ward et al., 2012). In our laboratory, we are specifically interested in understating the functional overlap that may exist between predictive timing and anticipatory motor control, which seem to be two processes that are necessary for a normalised sense of agency (Jardri et al., 2009) and a stable construct of self as a unity of being (Wilquin and Delevoye-Turrell, 2012).

References

Ameller, A. et al. 2011. "Production de rythmes dans la schizophrénie : Rôle central de l'attention" *L'Encéphale*.

Andreasen, N.C., S. Paradiso, and D.S. O'Leary 1998. "'Cognitive Dysmetria' as an Integrative Theory of Schizophrenia: A Dysfunction in Cortical-Subcortical-Cerebellar Circuitry?" *Schizophrenia Bulletin* 24(2): 203–18.

Andreasen, Nancy C. et al. 1999. "Defining the Phenotype of Schizophrenia: Cognitive Dysmetria and Its Neural Mechanisms" *Biological Psychiatry* 46(7): 908–20.

Aschersleben, Gisa. 2002. "Temporal Control of Movements in Sensorimotor Synchronization" *Brain and Cognition* 48(1): 66–79.

Augustin. 1942. *Les confessions de saint Augustin*.

Bellak, L. 1949. "A Multiple-Factor Psychosomatic Theory of Schizophrenia; An Attempt at a Consistent Conceptualization" *The Psychiatric Quarterly* 23(4): 738–55.

Billon, M.A., Semjen, J. Cole, and G. Gauthier. 1996. "The Role of Sensory Information in the Production of Periodic Finger-Tapping Sequences" *Experimental Brain Research*. 110(1): 117–30.

Bindra, D., and H. Waksberg. 1956. "Methods and Terminology in Studies of Time Estimation" *Psychological Bulletin* 53(2): 155–9.

Brébion, G. et al. 2000. "Depression, Psychomotor Retardation, Negative Symptoms, and Memory in Schizophrenia" *Neuropsychiatry, Neuropsychology, and Behavioral Neurology* 13(3): 177–83.

Brickenkamp, R. 1962. Testd2. Aufmerksamkeits-Belastungs-Test. Hogrefe Verlag, Göttingen.

Brown, S.M. et al. 2005. "Eyeblink Conditioning Deficits Indicate Timing and Cerebellar Abnormalities in Schizophrenia" *Brain and Cognition* 58(1): 94–108.

Buonomano, D.V. 2000. "Decoding Temporal Information: A Model Based on Short-Term Synaptic Plasticity" *The Journal of Neuroscience: The Official Journal of the Society for Neuroscience* 20(3): 1129–41.

Buonomano, D.V., and M.M. Merzenich. 1995. "Temporal Information Transformed into a Spatial Code by a Neural Network with Realistic Properties" *Science* 267(5200): 1028–30.

Buonomano, Dean V., and Rodrigo Laje. 2010. "Population Clocks: Motor Timing with Neural Dynamics" *Trends in Cognitive Sciences* 14(12): 520–7.

Burr, David, Arianna Tozzi, and M. Concetta Morrone. 2007. "Neural Mechanisms for Timing Visual Events are Spatially Selective in Real-World Coordinates" *Nature Neuroscience* 10(4): 423–5.

Carroll, Christine A. et al. 2008. "Temporal Processing Dysfunction in Schizophrenia" *Brain and Cognition* 67(2): 150–61.

Carroll, Christine A. et al. 2009a. "Timing Dysfunctions in Schizophrenia Span from Millisecond to Several-Second Durations" *Brain and Cognition* 70(2): 181–90.

Carroll, Christine A. et al. 2009b. "Timing Dysfunctions in Schizophrenia as Measured by a Repetitive Finger Tapping Task". *Brain and Cognition* 71(3): 345–53.

Clausen, J. 1950. "An Evaluation of Experimental Methods of Time Judgment" *Journal of Experimental Psychology* 40(6): 756–61.

Cohen, J.D., and D. Servan-Schreiber. 1992. "Context, Cortex, and Dopamine: A Connectionist Approach to Behavior and Biology in Schizophrenia" *Psychological Review* 99(1): 45–77.

Cornblatt, B.A., and J.G. Keilp. 1994. "Impaired Attention, Genetics, and the Pathophysiology of Schizophrenia" *Schizophrenia Bulletin* 20(1): 31–46.

Creelman, C. Douglas. 1962. "Human Discrimination of Auditory Duration" *The Journal of the Acoustical Society of America* 34(5): 582–93.

Cumming, W.J. 1988. "The Neurobiology of the Body Schema" *The British Journal of Psychiatry* 2: 7–11.

Davalos, Deana B., Michael A. Kisley, and Randal G. Ross. 2003. "Effects of Interval Duration on Temporal Processing in Schizophrenia" *Brain and Cognition* 52(3): 295–301.

Delevoye-Turrell, Yvonne, Anne Giersch, and Jean-Marie Danion. 2003. "Abnormal Sequencing of Motor Actions in Patients with Schizophrenia: Evidence from Grip Force Adjustments During Object Manipulation" *The American Journal of Psychiatry* 160(1): 134–41.

Delevoye-Turrell, Yvonne Nathalie, Pierre Thomas, and Anne Giersch. 2006. "Attention for Movement Production: Abnormal Profiles in Schizophrenia" *Schizophrenia Research* 84(2–3): 430–2.

Delevoye-Turrell, Yvonne et al. 2007. "Motor Fluency Deficits in the Sequencing of Actions in Schizophrenia" *Journal of Abnormal Psychology* 116(1): 56–64.

Delevoye-Turrell, Yvonne, Hélène Wilquin, and Anne Giersch. 2012. "A Ticking Clock for the Production of Sequential Actions: Where Does the Problem Lie in Schizophrenia?" *Schizophrenia Research* 135(1–3): 51–4.

Delignières, Didier, Loïc Lemoine, and Kjerstin Torre. 2004. "Time Intervals Production in Tapping and Oscillatory Motion" *Human Movement Science* 23(2): 87–103.

Densen, M.E. 1977. "Time Perception and Schizophrenia" *Perceptual and Motor Skills* 44(2): 436–8.

Desmurget, M. et al. 1998. "From Eye to Hand: Planning Goal-Directed Movements" *Neuroscience and Biobehavioral Reviews* 22(6): 761–88.
Elvevåg, B. et al. 2003. "Duration Judgements in Patients with Schizophrenia" *Psychological Medicine* 33(7): 1249–61.
Elvevåg, B. et al. 2004. "Identification of Tone Duration, Line Length, and Letter Position: An Experimental Approach to Timing and Working Memory Deficits in Schizophrenia" *Journal of Abnormal Psychology* 113(4): 509–21.
Fisher, Seymour, Stephen G. Bryant, and Ronica M. Kluge. 1986. "New Approaches to Postmarketing Surveillance" *Psychopharmacology* 90(3).
Fleming, K. et al. 1997. "Visuospatial Working Memory in Patients with Schizophrenia" *Biological Psychiatry* 41(1): 43–9.
Fortin, C., and N. Massé. 1999. "Order Information in Short-Term Memory and Time Estimation" *Memory and Cognition* 27(1): 54–62.
Fortin, C. et al. 1993. "Time Estimation and Concurrent Nontemporal Processing: Specific Interference from Short-Term-Memory Demands" *Perception and Psychophysics* 53(5): 536–48.
Freedman, B.J. 1974. "The Subjective Experience of Perceptual and Cognitive Disturbances in Schizophrenia. A Review of Autobiographical Accounts" *Archives of General Psychiatry* 30(3): 333–340.
Gibbon, J., and Church, R.M. 1984. "Sources of Variance in an Information Processing Theory of Timing. " In H.L. Roitblat, T.G. Bever, and H.S. Terrace (Eds), *Animal Cognition* 465–88. Hillsdale, NJ: Erlbaum.
Giersch, Anne et al. 2009. "Extended Visual Simultaneity Thresholds in Patients with Schizophrenia" *Schizophrenia Bulletin* 35(4): 816–25.
Goldstone, S., and W.T. Lhamon. 1956. "The Time Sense: Estimation of One Second Durations by Schizophrenic Patients" *A.M A. Archives of Neurology and Psychiatry* 76(6): 625–9.
Gordon, J., M.F. Ghilardi, and C. Ghez. 1994. "Accuracy of Planar Reaching Movements. I. Independence of Direction and Extent Variability" *Experimental Brain Research* 99(1): 97–111.
Grondin, S. 2010. "Timing and Time Perception: A Review of Recent Behavioral and Neuroscience Findings and Theoretical Directions" *Attention, Perception, and Psychophysics* 72(3): 561–82.
Gurfinkel, V.S. et al. 1989. "[Changes in the Direction of Vestibulomotor Responses in the Process of Adaptation to Prolonged Static Head Turning in Man.]" *Neĭrofiziologiia Neurophysiology* 21(2): 210–7.
Harrow, Martin, and Thomas H. Jobe. 2007. "Factors Involved in Outcome and Recovery in Schizophrenia Patients Not on Antipsychotic Medications: A 15-Year Multifollow-Up Study" *The Journal of Nervous and Mental Disease* 195(5): 406–14.

Howes, Oliver D., and Shitij Kapur. 2009. "The Dopamine Hypothesis of Schizophrenia: Version III – The Final Common Pathway" *Schizophrenia Bulletin* 35(3): 549–62.

Ivry, R. 1997. "Cerebellar Timing Systems" *International Review of Neurobiology* 41: 555–73.

Ivry, R.B., and R.E. Hazeltine. 1995. "Perception and Production of Temporal Intervals Across a Range of Durations: Evidence for a Common Timing Mechanism" *Journal of Experimental Psychology. Human Perception and Performance* 21(1): 3–18.

Ivry, R.B., S.W. Keele, and H.C. Diener. 1988. "Dissociation of the Lateral and Medial Cerebellum in Movement Timing and Movement Execution" *Experimental Brain Research* 73(1): 167–80.

Jardri, Renaud et al. 2009. "Clinical Practice of rTMS Reveals a Functional Dissociation between Agency and Hallucinations in Schizophrenia" *Neuropsychologia* 47(1): 132–8.

Johnson, J.E., and T.P. Petzel. 1971. "Temporal Orientation and Time Estimation in Chronic Schizophrenics" *Journal of Clinical Psychology* 27(2): 194–6.

Kant, E. 1845. *Critique de la raison pure*. Ladrange.

Karson, C.N. 1983. "Spontaneous Eye-Blink Rates and Dopaminergic Systems" *Brain: A Journal of Neurology* 106(3): 643–53.

Knoblich, Günther, Frank Stottmeister, and Tilo Kircher. 2004. "Self-Monitoring in Patients with Schizophrenia" *Psychological Medicine* 34(8): 1561–9.

Lalanne, Laurence, Mitsouko van Assche, and Anne Giersch. 2010. "When Predictive Mechanisms Go Wrong: Disordered Visual Synchrony Thresholds in Schizophrenia" *Schizophrenia Bulletin*.

Lee, Kwang-Hyuk et al. 2009. "Time Perception and Its Neuropsychological Correlates in Patients with Schizophrenia and in Healthy Volunteers" *Psychiatry Research* 166(2–3): 174–83.

Lewis, Aubrey. 1932. "The Experience of Time in Mental Disorder" *Proceedings of the Royal Society of Medicine* 25(5): 611–20.

Maass, Wolfgang, Thomas Natschläger, and Henry Markram. 2002. "Real-Time Computing without Stable States: A New Framework for Neural Computation Based on Perturbations" *Neural Computation* 14(11): 2531–60.

Macefield, V.G., and R.S. Johansson. 1994. "Electrical Signs of Cortical Involvement in the Automatic Control of Grip Force" *Neuroreport* 5(17): 2229–32.

McAuley, J. Devin et al. 2006. "The Time of Our Lives: Life Span Development of Timing and Event Tracking" *Journal of Experimental PsychologyGeneral.* 135(3): 348–67.

Meck, W H. 1996. "Neuropharmacology of Timing and Time Perception" *Cognitive Brain Research* 3(3–4): 227–42.

Minkowski, Eugène. 1933. *Le temps vécu: études phénoménologiques and psychopathologiques*. Collection de l'évolution psychiatrique.

Morrens, Manuel, Wouter Hulstijn, and Bernard Sabbe. 2007. "Psychomotor Slowing in Schizophrenia" *Schizophrenia Bulletin* 33(4): 1038–53.

Newton, Isaac. 1987. *Philosophiae naturalis principia mathematica.*

Nuechterlein, K.H. 1977. "Reaction Time and Attention in Schizophrenia: A Critical Evaluation of the Data and Theories" *Schizophrenia Bulletin* 3(3): 373–428.

Nuechterlein, K.H, and M.E. Dawson. 1984. "Information Processing and Attentional Functioning in the Developmental Course of Schizophrenic Disorders" *Schizophrenia Bulletin* 10(2): 160–203.

Penney, T.B., J. Gibbon, and W.H. Meck. 2000. "Differential Effects of Auditory and Visual Signals on Clock Speed and Temporal Memory" *Journal of Experimental Psychology. Human Perception and Performance* 26(6): 1770–87.

Pressing, J. 1998. "Referential Behaviour Theory: A Framework for Multiple Perspectives on Motor Control. Motor Behavior and Human Skill: A Multidisciplinary Approach" 357–84.

Pressing, Jeff. 1999. "The Referential Dynamics of Cognition and Action" *Psychological Review* 106(4): 714–47.

Priebe, Stefan, and Frank Röhricht. 2001. "Specific Body Image Pathology in Acute Schizophrenia" *Psychiatry Research* 101(3): 289–301.

Rammsayer, T. 1990. "Temporal Discrimination in Schizophrenic and Affective Disorders: Evidence for a Dopamine-Dependent Internal Clock" *The International Journal of Neuroscience* 53(2–4): 111–20.

Rammsayer, T.H. 1999. "Neuropharmacological Evidence for Different Timing Mechanisms in Humans" *The Quarterly Journal of Experimental Psychology. B, Comparative and Physiological Psychology* 52(3): 273–86.

Repp, Bruno H. 2005. "Sensorimotor Synchronization: A Review of the Tapping Literature" *Psychonomic Bulletin and Review* 12(6): 969–92.

Repp, Bruno H., and Susan R. Steinman. 2010. "Simultaneous Event-Based and Emergent Timing: Synchronization, Continuation, and Phase Correction" *Journal of Motor Behavior* 42(2): 111–26.

Robertson, Shannon D. et al. 1999. "Correlations for Timing Consistency among Tapping and Drawing Tasks: Evidence against a Single Timing Process for Motor Control ". *Journal of Experimental Psychology: Human Perception and Performance* 25(5): 1316–30.

Stevens, L.T. 1986. "On the Time-Sense" *Mind* 43: 393.

Stone, M. et al. 1998. "Working and Strategic Memory Deficits in Schizophrenia" *Neuropsychology* 12(2): 278–88.

Tononi, G., and G.M. Edelman. 2000. "Schizophrenia and the Mechanisms of Conscious Integration" *Brain Research. Brain Research Reviews* 31(2–3): 391–400.

Treisman, M. 1963. "Temporal Discrimination and the Indifference Interval. Implications for a Model of the 'Internal Clock'" *Psychological Monographs* 77(13): 1–31.

Tsuang, M.T., W.S. Stone, and S.V. Faraone. 2001. "Genes, Environment and Schizophrenia" *The British journal of psychiatry. Supplement* 40: S18-24.

Turgeon, Martine et al. 2012. "Impaired Predictive Timing with Spared Time Interval Production in Individual with Schizophrenia" *Psychiatry Research* 197(1–2): 13–8.

Tysk, L. 1983a. "Estimation of Time and the Sub Classification of Schizophrenic Disorders" *Perceptual and Motor Skills* 57(3): 911–8.

Tysk, L. 1983b. "Time Estimation by Healthy Participants and Schizophrenic Patients: A Methodological Study" *Perceptual and Motor Skills* 56(3): 983–8.

Tysk, L. 1990. "Estimation of Time by Patients with Positive and Negative Schizophrenia" *Perceptual and Motor Skills* 71(3): 826.

Vanneste, S., V. Pouthas, and J.H. Wearden. 2001. "Temporal Control of Rhythmic Performance: A Comparison between Young and Old Adults" *Experimental Aging Research* 27(1): 83–102.

Vindras, P, and P. Viviani. 1998. "Frames of Reference and Control Parameters in Visuomanual Pointing" *Journal of Experimental Psychology. Human Perception and Performance* 24(2): 569–91.

Volz, H.P. et al. 2001. "Time Estimation in Schizophrenia: An fMRI Study at Adjusted Levels of Difficulty" *Neuroreport* 12(2): 313–6.

Wahl, O.F., and D. Sieg. 1980. "Time Estimation among Schizophrenics" *Perceptual and Motor Skills* 50(2): 535–41.

Ward, Ryan D. et al. 2012. "Timing as a Window on Cognition in Schizophrenia" *Neuropharmacology* 62(3): 1175–81.

Wilquin, Hélène, and Yvonne Delevoye-Turrell. 2012. "Motor Agency: A New and Highly Sensitive Measure to Reveal Agency Disturbances in Early Psychosis" *PloS One* 7(2): e30449.

Wilquin, H. et al. 2010. "Production de rythmes dans la schizophrénie: Un déficit de l'alternance de temps" *L'Année psychologique* 110(4): 517–39.

Wing, A., and Kristofferson, A. 1973. "The Timing of Interresponse Intervals" *Attention, Perception, and Psychophysics* 13(3): 455–60.

Zahn, T.P. et al. 1998. "Manual and Saccadic Reaction time with Constant and Variable Preparatory Intervals in Schizophrenia" *Journal of Abnormal Psychology* 107(2): 328–37.

Zakay, Dan, and Richard A. Block. 1996. "The Role of Attention in Time Estimation Processes" In *Time, Internal Clocks and Movement*, 115:143–64. North-Holland.

Zelaznik, Howard N., Rebecca M.C. Spencer, and Richard B Ivry. 2002. "Dissociation of Explicit and Implicit Timing in Repetitive Tapping and Drawing Movements"

Journal of Experimental Psychology. Human Perception and Performance 28(3): 575–88.

Zelaznik, H., Spencer, R., and Doffin, J. 2000. "Temporal Precision in Tapping and Circle Drawing Movements at Preferred Rates is Not Correlated: Further Evidence against Timing as a General-Purpose Ability" *Journal of motor behavior* 32(2): 193–9.

Zelaznik, Howard N. et al. 2005. "Timing Variability in Circle Drawing and Tapping: Probing the Relationship between Event and Emergent Timing" *Journal of Motor Behavior* 37(5): 395–403.

CHAPTER 7

Interactions of Timing and Motivational Impairments in Schizophrenia

Ryan D. Ward, Billur Avlar**, and Peter D. Balsam***†

1 Introduction

Cognitive impairments are a hallmark of schizophrenia and include disturbances in working memory, attention, and executive function (Kerns et al., 2008). Some investigators have proposed that dysfunctional cognition is the core deficit in schizophrenia (Elvevåg and Goldberg, 2000; Heinrichs, 2005). For example, Andreason and colleagues have suggested that the primary phenotype in schizophrenia is "cognitive dysmetria" which they define as "a disruption in the fluid coordination of mental activity that is the hallmark of normal cognition" (Andreason et al., 1999). Such a dysfunction in some fundamental aspect of cognition could result in a detrimental cascade of effects, ultimately giving rise to more serious cognitive impairments, as well as contributing to the positive and negative symptoms of schizophrenia.

What crucial aspect of cognition could be responsible for such a pervasive impairment? Some have suggested that the fundamental process that may be impaired is temporal information processing (Andreason et al., 1999; Elvevag et al., 2004). Distorted timing is well documented in patients with schizophrenia. Patients have been described as being disoriented in time: not knowing the time of day, day of the week, month or even year (Lewis, 1932). Patients also report alterations in the perception of time. Sometimes time slows (even to the point of standing still) and sometimes it rushes by at high speed (Freedman, 1974; Lewis, 1932). In addition to these more anecdotal reports of temporal distortions, a growing literature has demonstrated empirically the existence of distorted temporal information processing in a variety of experimental paradigms in individuals diagnosed with schizophrenia (e.g., see Penney et al., 2005, for review). For example, patients have been reported to overestimate

* Department of Psychology, University of Otago, New Zealand.
** Department of Psychology, Columbia University, USA.
† Department of Psychology, Barnard College, USA.

 | DOI 10.1163/9789004230699_008

interval duration when they are asked to verbally report the duration of a presented stimulus (e.g., Clausen, 1950; Densen, 1977; Johnson and Petzel, 1971; Lhamon and Goldstone, 1956; Tysk, 1983, 1984a, 1990). When asked to reproduce a standard interval of some specified duration, patients have been reported to underestimate time (e.g., Clausen, 1950; Johnson and Petzel, 1971; Tysk, 1983, 1990; Wahl and Sieg, 1980). While the specific nature of the timing deficit is different depending on the experimental paradigm employed, when all data are considered together, the most reliable finding across paradigms and studies is that patients are more variable in their timing of temporal intervals than controls (Carroll et al., 2008, 2009; Davalos et al., 2003; Elvevag et al., 2003, 2004; Todd, 2006; Yang et al., 2004). Thus, distortions in temporal information processing are reliably associated with the disease. Furthermore, interval timing deficits were evident in patients who were at risk for developing schizophrenia but not major affective disorder (Penney et al., 2005) providing additional evidence that schizophrenia and timing are deeply linked.

1.1 *Dopaminergic Dysfunction in Schizophrenia*

Although the specific neural underpinnings of schizophrenia are not fully described, one of the most influential hypotheses is that the disease reflects pathological changes in activity of the dopaminergic systems (e.g., Howes and Kapur, 2009). Hyperactivity of the striatal dopamine (DA) system in patients has been consistently observed through a variety of methods (see Kellendonk, 2009, for a review). For example, a number of postmortem studies have reported an increase in subcortical DA in patients (e.g., Davis et al., 1991). In addition, studies using positron emission tomography (PET) to assess dopaminergic neurotransmission have reported increased uptake of the radioligand fluorodopa in the striatum of patients with schizophrenia (e.g., Frankle, 2007) as well as in individuals in the prodromal phase of the disease (Howes et al., 2009), indicating increased DA synthesis. Other PET studies have shown increased DA neurotransmission in schizophrenia, as evidenced by increased amphetamine-induced displacement of D2 binding in patients (see Frankle, 2007, for review).

In addition to increased DA synthesis and neurotransmission, other studies have shown that striatal D2 receptors are also altered in schizophrenia. One of the most general and replicable findings is hyperactivity of DA D2 receptors in the striatum of patients and all effective antipsychotic drugs antagonize D2 receptors (e.g., Seeman et al., 1976). Post mortem studies report an upregulation in striatal D2 receptors in drug-free patients (Davis et al., 1991). These data have been confirmed in PET studies, which demonstrate about a 12% average increase in DA D2 receptor density in the striatum of drug-free and drug-naïve patients (see Laruelle, 1998, for meta-analysis and review). Patients also display increased occupancy of striatal D2 receptors by DA (e.g., Abi-Dargham et al., 2000). Thus, the majority of evidence confirms that there is a hyperfunction of the striatal DA D2 system in schizophrenia, suggesting it is an important part of the pathophysiology of the disease. Furthermore, PFC hypoactivity has also been reported (Barch, 2005; Glahn et al., 2005) and the PFC has long been a target of schizophrenia research, given the well documented reliance of working memory, which is compromised in schizophrenia, on normal PFC function (e.g., Goldman-Rakic and Selemon, 1997; Goldman-Rakic et al., 2004; Lewis et al., 1999). Abnormal PFC and striatal functioning is thought to compromise the integrity of cortico-striatal circuits, leading to functional impairments (see Simpson, Kellendonk, and Kandel, 2010, for review and discussion on how excessive striatal D2 activity could lead to PFC dysfunction).

1.2 *Modeling Increased Striatal D2 Receptor Activity in Schizophrenia*

One of the ways to examine the causal relationship between D2 receptors and the deficit in interval timing observed in patients is to utilize transgenic animal models that may mimic the endophenotypes in schizophrenia. In an effort to model the increased occupancy and density of striatal DA D2 receptors in patients, Kellendonk et al. (2006) generated transgenic mice which selectively overexpress DA D2 receptors in the striatum. To accomplish this, they employed the tetracycline controlled gene expression system (see Aiba and Nakao, 2007, for discussion). This system utilizes a recombinant protein, the tetracycline transactivator (tTA), which is a transcription factor that binds to a specific DNA sequence, the tetracycline response element (tetO). The transgene is linked downstream of tetO in this system, so that transcription of the transgene only occurs upon binding of tTA to tetO. The antibiotic tetracycline as well as the tetracycline derivative doxycycline bind tTA and render it incapable of binding to tetO, thus preventing expression of target genes. This allows for temporal control of transgene expression; by supplementing the animals diet with doxycycline the transgene is silenced.

Kellendonk et al. (2006) used a line of transgenic mice in which the expression of tTA is directed by a region and cell type specific promoter to a particular brain area and cell type of interest (Mayford et al., 1996). They generated a second line of transgenic mice, in which the DNA coding sequence for the human dopamine D2 receptor was downstream of the tetO promoter. When mice from each of these lines are bred together, mice carrying both transgenes are produced, in which the transgenic D2 receptor gene is expressed when tTA binds to its tetO promoter, thus resulting in region and cell specific expression of transgenic D2 receptors (Figure 7.1).

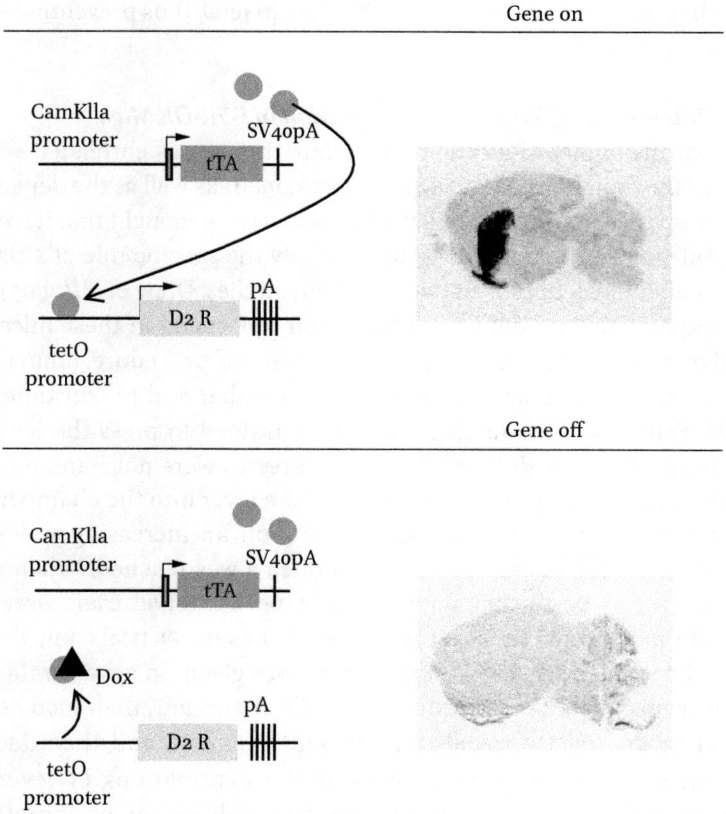

FIGURE 7.1 *The upper panels show the genetic system used to generate the D2R-OE mice. In one mouse line, the human D2 receptor gene is expressed under control of the tetO promoter. In a second mouse line, the tTA element is expressed under control of the CamKIIa promoter. In a mouse in which both transgenes are present, the tTA element binds to the tetO promoter, driving expression of D2 receptors. In the D2R-OE mice, expression of excess D2 receptors is confined to the striatum and olfactory tubercule (shown via in situ hybridization in the upper right panel). Feeding the mouse doxycycline prevents tTA from binding to the tetO promoter (bottom left panel), thereby eliminating expression of the transgenic D2 receptors (bottom right panel). Figure adapted from Kellendonk (2009).*

While the endogenous expression pattern of the CamKIIα gene is in neurons of the entire forebrain, this promoter, in combination with the tetO-D2R transgene, resulted in transgenic D2 receptor expression restricted to the striatum and olfactory tubercule. This model system has a 15% increase in overall striatal D2 receptors, fortuitously mimicking the level of increase found in patients. The overexpression can be turned off by feeding the mice doxycycline, which prevents tTA from binding to tetO, thus preventing expression of the D2 receptor gene.

1.3 *Interaction of Timing and Motivation in D2R-OE Mice*

Given that the ability of a drug to modulate timing was correlated with the affinity of the drug for the D2 receptor (Meck, 1986) as well as the dependence of timing on the striatum (Matell and Meck, 2004) we thought that it would be worthwhile to use the D2R-OE mouse to study the specific role of striatal D2 signaling on interval timing. In the first of our studies, Drew et al. (2007) found severe impairments in temporal information processing in these mice when trained on a task called the peak procedure. In this procedure, animals were first trained to press a lever in an operant chamber and to consume liquid rewards. Following this training, mice were trained to press the lever on a fixed interval (FI) schedule in which lever presses were not reinforced until after a FI had elapsed from the insertion of the lever into the chamber at the start of a trial. Mice went through training on an increasing series of FI schedules until the final target duration of 24 s was reached. When performance on the FI 24 s schedule was stable, peak interval trials were introduced. On peak trials, the lever was extended as on FI trials, but the lever remained extended for 96 s and no reward was given. In a well-trained animal, the average rate of responding generally begins low, then increases and peaks at or around the usual time of reinforcement, and then decreases (Roberts, 1981). Quantitative analysis of the distributions of lever press responses on the peak trials can be used to provide estimates of motivation to respond in the task, accuracy, and precision of interval timing. The peak height of the response rate distribution indexes motivation, and is affected by manipulations that target motivational states, such as feeding the animal prior to the session, changing the magnitude of the reward, or devaluing the reward via pairings with lithium chloride (Galtress and Kirkpatrick, 2009; Ludvig et al., 2007; Roberts, 1981). The location of the maximal rate of responding gives an index of how accurately the mice were able to time the temporal interval, and the spread of the response rate distribution is a measure of the precision (variability) of interval timing. Figure 7.2 shows that when performance was characterized in this way, D2R-OE mice responded at overall lower rates, indicating lack of motivation.

In addition, the response rate distributions of D2R-OE mice were flatter, and peaked later than the target interval of 24 s, indicating decreased precision and accuracy of timing. Turning off the transgene with doxycycline improved motivation (increased response rate), and partially rescued the deficit in the accuracy and precision of timing. These data suggest the possibility of a motivational

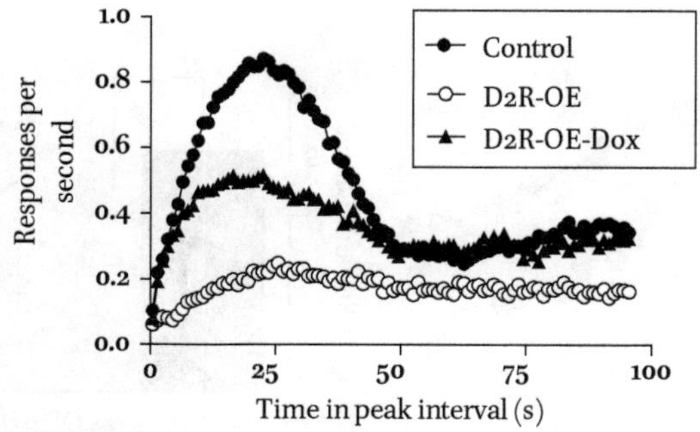

FIGURE 7.2 *Temporal information processing during the peak procedure. Responses per second as a function of time during peak interval trials (see text for details). D2R-OE mice have lower response rates, and are less accurate and precise in their timing of the 24 s target interval. Turning off the transgene (D2R-OE-Dox) increases response rates and rescues accuracy of interval timing. Data taken from Drew et al. (2007).*

component to the timing deficits, which is rescued by turning off the transgene.

Response rates in the peak procedure provided an indirect measure of motivation. To more directly assess motivation, the progressive ratio task, which measures the amount of effort a subject will expend to earn a reward, was used (Drew et al., 2007). In this task, the mouse earns reward after completing a criterion number of lever presses. The criterion was set at two lever presses for the first trial and then doubled with each successive trial. In this procedure, as shown in Figure 7.3, D2R-OE mice ceased responding significantly earlier, earned fewer rewards, and made fewer responses. These deficits were ameliorated when the transgene was turned off, indicating that the progressive ratio impairment in D2R-OE mice results from acute overexpression of the D2 receptor.

We investigated the interaction between timing and motivation in the peak procedure by reducing the probability of reward on FI trials (Ward et al., 2009). Our reasoning was that if the timing deficits in D2R-OE mice resulted from motivational impairments, we should be able to mimic the impairment in controls by reducing motivation during the peak procedure. The peak procedure was employed exactly as described above, except that the percentage of rewarded FI trials was decreased across conditions from 100 to 5%. Each reward percentage condition was in place for at least five sessions until the response rate data appeared stable. As shown in Figure 7.4A, manipulation of reward probability impacted performance of both control and D2R-OE mice, with poorer performance as reward probability was reduced.

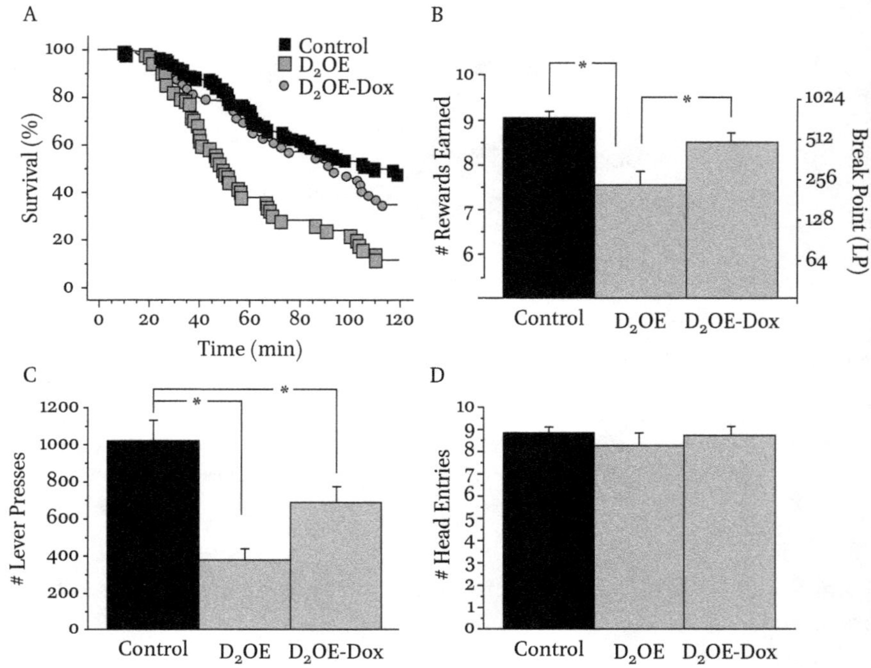

FIGURE 7.3 *Assessment of motivation in the progressive ratio schedule in D2R-OE mice. A. Survival functions showing the percentage of mice in all groups that were still responding on the schedule as a function of time in the progressive ratio session. D2R-OE mice quit responding significantly sooner than controls, and turning off the transgene (D2OE-Dox) rescues the progressive ratio performance. B. Number of rewards earned and break point (last ratio completed) for all groups of mice. D2R-OE mice earned significantly fewer rewards and had significantly lower break points than control and D2R-OE-Dox mice. C. Total number of lever presses emitted by all groups of mice on the progressive ratio schedule. D2R-OE mice made significantly fewer lever presses than control mice. Turning off the transgene improved performance. D. Number of head entries made by all groups of mice following a reward delivery. There was no difference in this measure across groups, indicating that all mice consumed the reward once it was earned. Reprinted with permission from Drew et al. (2007).*

Most interestingly however, when the level of motivation as indexed by peak height was similar between control and D2R-OE mice, timing accuracy and precision were also similar, strongly suggesting a motivational component to the timing deficits in D2R-OE mice.

These results are consistent with a large body of evidence implicating the DA D2 receptor in both motivation and temporal information processing. As far as motivation is concerned, DAergic signaling has been shown to be critical to the performance of motivated behavior in a number of paradigms (see

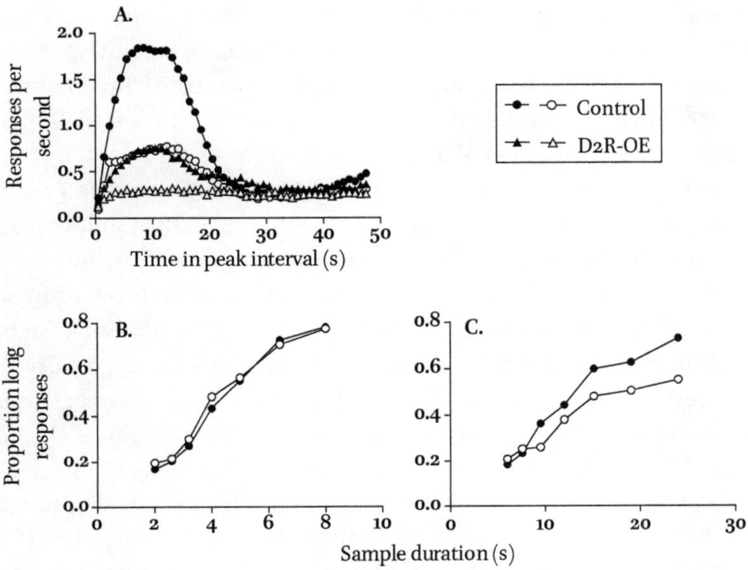

FIGURE 7.4 *A. Peak interval performance of both control and D2R-OE mice is sensitive to manipulation of motivation. Motivation was manipulated by changing the percentage of rewarded fixed interval trials (see text for details). Circles show performance under conditions of 100% rewarded FI trials, while triangles show performance under 10% rewarded FI trials. Closed symbols show control performance, while open symbols show performance of D2R-OE mice. Decreasing motivation in control mice results in performance that is indistinguishable from that of D2OE mice. B. Performance of control and D2OE mice on the bisection procedure (see text for details) with anchor durations of 2 and 8 s. The figure shows the proportion of trials on which the mice chose the response lever corresponding to a "long" sample duration as a function of sample duration. There are no differences in performance between genotypes. C. Performance of control and D2R-OE mice on the bisection procedure with anchor durations of 6 and 24 s. D2R-OE mice are selectively impaired on longer duration sample trials, suggesting a deficit in working memory or sustained attention. Data from Ward et al. (2009).*

Salamone et al., 2007, 2009, 2012, for reviews). For example, both DA antagonists and depletion of nucleus accumbens DA decrease operant responding for food, particularly when the lever press requirement becomes more effortful (e.g., Aberman and Salamone 1999; Ishiwari et al., 2004). The specific role of DA in reward motivated behavior has been clarified using a procedure in which animals are given the choice to work (lever press) for a more preferred reward, or consume a freely available less preferred reward. In this procedure, animals with intact DA signaling press the lever for the preferred reward, and consume little of the free chow. However, antagonism of DA receptors leads to a shift in behavior away from the more effortful lever press towards the freely available chow (Cousins and Salamone, 1994; Farrar et al., 2010; Salamone et al., 1991, 1996; Salamone and Correa, 2002). Data like these are clarifying the role of DA in motivated behavior and emphasize the importance of DA signaling in contributing to the computation of work-related response costs.

Many studies indicate that temporal information processing can be distorted by manipulations that target the DA system (Buhusi and Meck, 2005; Maricq and Church, 1983). Although theoretical accounts of the underlying mechanisms of these effects differ, administration of both DA agonists and antagonists has been shown repeatedly to produce disruption of temporal information processing in a variety of experimental protocols (see Coull et al., 2011 for review). In particular, empirical evidence indicates that temporal information processing is particularly sensitive to manipulation of D2 receptor function. For example, Meck (1986) demonstrated that the dose of a neuroleptic that was required to distort rats' perception of a time interval by 10-15% was negatively correlated with the drugs' affinity for DA D2 receptors. Thus, D2 receptor activity is critical in accurate temporal information processing.

1.4 *Rescuing Motivation Rescues Timing in D2R-OE Mice*

The data from the peak procedure in which the probability of reward was manipulated indicate that manipulation of motivation impacts timing, and strongly suggest that the timing impairment in D2R-OE mice has a motivational component. However, in the peak procedure it is not possible to separate deficits in timing from those in motivation, because the index of timing is response rate, and response rate is greatly impacted by motivation. To further examine the nature of the timing impairment, we tested the D2R-OE mice on a timing task, which is not dependent on response rate as the measure of timing (Ward et al., 2009). In the temporal bisection task (Church and Deluty, 1977) at trial onset a sample (e.g., tone) is presented for either a short (e.g., 2 s) or long (e.g., 8 s) duration. Following sample presentation, response levers are presented and responses on one lever are only rewarded following presentation of short samples while responses on the other lever are only rewarded following presentation of long samples. Once performance with these anchor durations is learned, mice can be tested on trials in which intermediate duration samples are presented. Accuracy and precision of interval timing performance can be assessed by examining the proportion of choices to the lever corresponding to a 'long' sample duration as a function of sample duration. Once performance is stable, the proportion of responses to the long choice option is generally an increasing sigmoidal function of sample duration (Church and Deluty, 1977). The slope of this function indicates precision of interval timing, while the point on the function that corresponds to 50% long choices, or the point of subjective equality, gives an estimate of the accuracy of interval timing. Because only a single response is required in each trial, performance in the bisection task is not as affected by motivation to respond as the peak interval procedure.

The bottom panels of Figure 7.4 show the results of these experiments. The results indicated that D2R-OE mice have difficulty processing temporal information only at relatively long durations. With the relatively short anchor durations of 2 and 8 s (Figure 7.4B), for both control and D2R-OE mice the proportion of responses corresponding to the "long" response option increases with increasing sample duration, but there is no difference between control and D2R-OE mice, indicating that timing of these relatively short durations is intact. However, when the anchor durations were 6 and 24 s (Figure 7.4C), a difference emerged. At the longer durations, the proportion of "long" responses was significantly lower for D2R-OE mice than controls, indicating a selective impairment when required to process temporal information over longer intervals. This selective impairment in timing of relatively long durations could be due to a deficit in working memory or sustained attention, psychological processes that are critical to accurately process temporal information over longer intervals. However, when we tested the D2R-OE mice in operant paradigms, which assessed retention of information in working memory and sustained attention, we found no deficit. This suggested to us the possibility that there may be a motivational component to the timing impairment.

We had previously demonstrated that the motivational impairment could be rescued by turning off the transgene by feeding the mice doxycycline (Drew et al., 2007; Simpson et al., 2011). We next tested whether the timing impairment in the bisection procedure could also be rescued by turning off the transgene. Indeed, turning off the transgene rescued the timing impairment in the bisection procedure (Avlar et al., in preparation) suggesting a motivational basis for this deficit. The specific contribution of motivation to the timing deficits in the bisection procedure must be complex. The data from the peak procedure in which response rates of D2R-OE mice were significantly lower than controls indicates the presence of a basal motivational impairment. However, when the influence of response rates on the index of timing was removed by testing the mice in the bisection procedure, there was no general timing impairment with brief stimuli (Figure 7.4C). Only when the target duration became relatively long (>15 s) did the impairment manifest itself. On the face of it, this selective impairment in processing of longer temporal durations suggests a difficulty in working memory or attention. The fact that sustained attention and working memory maintenance are not impaired in the D2R-OE mice when assessed separately, however, suggests the possibility that the deficit results from an interaction of these cognitive processes when they are both required to work in concert, as in processing of longer temporal delays.

If there is a motivational component to the temporal information processing impairments in D2R-OE mice, we should be able to improve performance using an experimental manipulation that improves motivation. Therefore, manipulating the reward magnitude in a temporal discrimination task is particularly beneficial, because it gives us an opportunity to assess the interval timing independent of lever pressing vigor. Preliminary results from our lab (Avlar et al., in preparation) showed that altering the levels of motivation by increasing the reward magnitude for the longer cue durations, also improved the timing of both control and D2R-OE mice, although D2R-OE mice required a longer period of exposure to the increased reward magnitude before the improvements were manifest.

2 How Does Motivation Impact Timing?

The results from our assessment of timing and motivation in D2R-OE mice indicated that motivation is critical in accurate temporal information processing. To conceptually dissect this interaction, Figure 7.5 shows the psychological processes thought to be involved in interval timing.

Although different accounts differ in the specifics, the general component processes are common across theories. In order to accurately process temporal information, an organism must perceive and attend to the passage of time, store a representation of time in memory, and then make decisions about appropriate behavioral responses based on a comparison of the currently elapsing duration with previously experienced durations. First, some environmental event must trigger a timing mechanism. Information about the duration of the currently elapsing interval is stored and continuously updated in working memory. When the stimulus event ends or another biologically significant event occurs, the duration of the interval is transferred to long term memory. On subsequent occasions when the same or similar stimulus events are encountered, the elapsing duration is compared to a stored duration in long term memory. If the comparison crosses response thresholds, an appropriate behavioral action is selected. Thus, accurate interval timing requires proper functioning of a cascade of related psychological processes, including perception, attention, working memory, long term memory, and decision processes. In addition, these processes are impacted by motivation. A large literature has demonstrated that manipulations of aspects of reward, which putatively

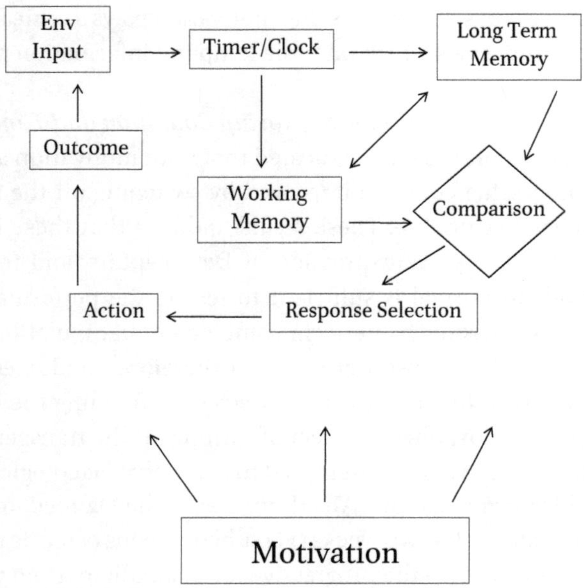

FIGURE 7.5 *Conceptual framework showing the psychological processes which are thought to underlie temporal information processing, and their interaction with motivation* (*see text for details*).

impact motivation during cognitive tasks, impact performance on these tasks. For example, in the peak procedure, when animals are pre-fed a large amount before the daily session, the response rate function flattened and shifted to the right, and this effect was interpreted as evidence of a change in clock speed (Roberts, 1981; see also Galtress and Kirkpatrick, 2009; Ward and Odum, 2006, 2007). Others have demonstrated similar effects with manipulations of reward magnitude. Increases and decreases of reward magnitude produce leftward, and rightward shifts in the response rate function, respectively, and these shifts have also been interpreted as changes in the speed of the clock (Grace and Nevin, 2000; Kacelnik and Brunner, 2002; Ludvig, Conover, and Shizgal, 2007; see also Bizo and White, 1995; Killeen et al., 1999; Ward and Odum, 2007). In addition to effects on clock speed, motivation has also been theorized to impact the latency to initiate the timing mechanism (Gibbon and Church, 1984; Lejuene, Macar, and Zakay, 1999), the vigilance with which the elapsing interval is attended to (Fortin and Masse, 2000; Ward and Odum, 2007; Zakay, 1989), and decision related computations (e.g., Bendiksby and Platt, 2006; Gold and Shadlen, 2001; Leon and Shadlen, 1999; Platt and Glimcher, 1999; Sugrue et al., 2004), among others. Determining the impact of motivational manipulations on specific psychological processes or their interaction is challenging, but what is clear from these types of studies is that motivation plays a critical modulatory role in cognitive processes which underlie temporal information processing.

2.1 *Motivation as a Target for Improving Cognition and Timing*

Our work with the D2R-OE mice indicated that both motivation and temporal information processing could be improved by switching off the transgene by feeding the mice doxycycline. These results indicate that these impairments are caused by the acute overexpression of D2 receptors, and returning D2R expression levels to normal is sufficient to rescue the performance deficits. Thus, it should be possible to perform some acute manipulation and rescue the performance. We (Simpson et al., 2011) therefore conducted a series of studies in an effort to find successful strategies for rescuing the motivational deficit. Because the most obvious effect of turning off the transgene is normalization of D2R activity, we first attempted to use a pharmacological strategy to recapitulate this normalization. We administered the D2 receptor antagonist haloperidol chronically for two weeks (via subcutaneous osmotic mini-pumps) and assessed motivation in the progressive ratio paradigm. Even with the lowest dose used, we were unable to improve performance via haloperidol. In fact, at the higher doses, haloperidol produced performance decrements. Interestingly, these results are in line with clinical reports, which indicate that haloperidol, although effective in treating positive symptoms in schizophrenia, has no efficacy for treatment of motivational deficits. In fact, as we observed and as corroborated by numerous published studies, haloperidol has been shown to decrease motivation as assessed in a number of paradigms. The lack of efficacy of haloperidol could possibly be due to the difficulty in targeting and antagonizing only the excess D2 receptors, or it could be due to haloperidol's actions at other receptors. We, therefore, searched for an alternative strategy for modulating circuit function. We performed an unbiased gene chip assay, which determined the expression levels of thousands of genes. Because the motivational deficit was reversed when the transgene was turned off, we focused on changes in gene expression, which were also reversed when the transgene was turned off. Given the demonstrated role of the serotonin system in psychiatric diseases in which motivation is compromised, we were particularly interested in changes in this system. We surveyed the changes in the serotonin system and found increased expression of the serotonin 2C (5-HT2C) receptor in the striatum of D2R-OE mice, which was normalized when the transgene was turned off. To see if this overexpression of the 5-HT2C receptor contributed to the motivational impairment of the D2R-OE mice, we administered the selective 5-HT2C antagonist SB 242084. Systemic administration of this drug rescued performance in a progressive ratio schedule.

Not only did the 5-HT2C antagonist rescue motivation in D2R-OE mice, it also improved motivation in control mice. This suggests that the impact of the drug was not through a mechanism specific to the deficit in D2R-OE mice. Rather, the effect was likely due to a general role of 5-HT2C receptors in motivation. These results indicate that motivation in the D2R-OE mice can be rescued by 5-HT2C receptor antagonism and suggest a novel therapeutic strategy for patients.

Consistent with the idea that motivation modulates the accuracy and precision of timing, preliminary data from our lab is in line with the hypothesis that timing deficits can be ameliorated by increasing motivation via 5-HT2C antagonism. In a temporal bisection task, administration of SB242084 improved the precision and accuracy of the timing in both controls and D2R-OE mice (Avlar et al., in preparation) in the same way such improvements might be anticipated when reward factors are directly manipulated to increase motivation (Galtress and Kirkpatrick, 2010; Ward et al., 2009). Of course, improvement in timing could be due to the 5-HT2C receptor antagonist having a direct pro-cognitive effect (e.g., altered memory or attention) but the evidence so far is consistent with the hypothesis that the improvements arise from an indirect effect on cognition via modulation of motivation. We suggest that therapeutic strategies that target motivation may be a viable strategy for improving cognition in schizophrenia.

These results may also be relevant to depression. Lack of motivation is also characteristic of depression, and there are data to suggest that the 5-HT2C receptor may be altered in depression. For example, post mortem studies of suicide victims with a history of depression showed that 5-HT2C pre-mRNA editing was altered in prefrontal cortex (Gurevich et al., 2002). In addition, using a mouse model, Mombereau et al. (2010) reported that complete editing of 5-HT2C receptor mRNA resulted in an antidepressant-like phenotype as measured in a forced swimming test. There is also some evidence that temporal information processing is disrupted in depression (Bschor et al., 2004; Gil and Droit-Volet, 2009; Grinker et al., 1973; Sevigny et al., 2003; Tysk, 1984). Given the demonstrated role of the 5-HT2C receptor in motivation (Pentkowski et al., 2010; Simpson et al., 2011), perhaps targeting this receptor would be therapeutically beneficial in depression as well.

3 Conclusions

The dysmetria hypothesis of the symptoms of schizophrenia is an appealing one. It is easy to imagine how improper timing in the flow of information leads to many of the symptoms of the disorder (see Ward et al., 2012, for discussion) including disorganized behavior, delusions, and misjudgments of causality (Andreason et al., 1999; Carroll et al., 2008; Metcalfe et al., 2012). For example, consider the case of delusions. Disruption in temporal perception could lead to failure to correctly perceive the temporal order of temporally contiguous events. This in turn could lead to a lack of attribution of causality to one's own actions, and instead causally attributing events in one's life to other sources (aliens, spies, etc.). Of course some of the symptoms of schizophrenia may also be directly affected by difficulties in other aspects of information processing such as memory and decision-making, or as the results of our experiments with the D2R-OE mice suggest, some of the cognitive deficits may be a consequence of a dysfunction in motivation. The developmental overexpression of striatal D2 receptors is sufficient to produce both motivational and timing impairments. Our data indicate that motivational and cognitive processes interact in producing the timing deficits. The level of motivation modulates timing accuracy and precision in both D2R-OE and control mice to such an extent that as motivation is decreased in control mice, a timing phenotype emerges that is identical to that of D2R-OE mice. Thus, it seems plausible that at least in the subset of patients who exhibit negative symptoms, the deficit in motivation contributes to their cognitive deficits including those in timing.

In recent years, there has been a renewal of research effort into the negative symptoms of schizophrenia. While positive symptoms (disordered thoughts, delusions, hallucinations) are perhaps the most prototypical symptoms of schizophrenia, severity of positive symptoms is not correlated with functional impairment in the disease. Cognitive (deficits in perception, attention, working memory, executive functioning) and negative (blunted affect, social withdrawal, lack of motivation) symptoms, on the other hand, are highly predictive of functional outcomes (Beng-Choon et al., 1998). There has been significant research effort focused on developing pharmacological interventions for the cognitive deficits in schizophrenia, but so far an effective therapeutic strategy remains elusive (Keefe et al., 2007). Aside from the general complexity of the neurobiological mechanisms of the disease, one reason that current pharmacological strategies are ineffective in producing improvements in functional outcomes may be that a critical motivational component is not being addressed. While there have been some modest successes in treating cognitive impairment, there are no efficacious therapies for treating the motivational impairments. In fact, current therapeutic strategies for treating positive symptoms based on some level of D2 receptor antagonism likely worsen motivational impairments (Salamone et al., 2009; Simpson et al., 2011). By focusing on the interaction of motivation and cognition, the work with the D2R-OE mouse model directly points to some novel treatment strategies. The motivational impairments are rescued and there is a concomitant improvement in timing when D2 receptor expression is returned to normal levels. This suggests that targeting motivational circuits might be an effective means of improving both motivation and cognition. Consistent with this hypothesis we found that a 5-HT2C receptor antagonist improves motivation (Simpson et al., 2011) and here describe preliminary data showing that timing is also improved by this intervention.

More generally, the difficulty in finding effective treatments for psychiatric diseases with heterogeneous symptom manifestation illustrates the need for caution when trying to understand psychiatric disease as comprising distinct clusters of clinical symptoms. Of course, parsing psychiatric syndromes into symptom clusters, as in the case with the three symptom clusters in schizophrenia (positive, negative, cognitive) aids in being able to speak in a coherent way about patients and in forming a coherent clinical picture of the disease. This view, however, becomes problematic when it is treated as indicative of an actual partitioning of various psychological and biological processes that may be impaired in disease states. This mindset has perhaps been perpetuated by the advances in functional magnetic resonance imaging (fMRI) technology, which make it possible to visualize brain activity of participants while they are engaged in various experimental protocols. Many of these studies then correlate brain activity in a certain region of interest with severity of patient symptoms as indexed on a clinical scale. This practice reinforces the tendency to interpret fMRI data as a visualization of specific and distinct symptom types (or cognitive processes which underlie these symptoms) physically instantiated in the brain. These types of interpretations lead to the compartmentalization of the brain into areas based on some purported functional specificity. Although some functional selectivity has been supported empirically, in general, the notion that specific brain regions or networks subserve specific subsets of cognitive processes is the topic of considerable debate among cognitive neuroscientists (Barrett, 2012; Friston and Price, 2011; Mahon and Cantlon, 2011; Uttal, 2001, 2011).

The situation becomes even more challenging when one considers the fact that even specific symptoms are not monolithic constructs, but nuanced and complex phenomena. For example, it is easy to say that motivation is impaired in schizophrenia, but what is motivation? The outward behavioral manifestation of a lack of motivated behavior could result from impairments and interactions in numerous aspects of reward and cognitive processing (Barch and Dowd, 2010; Salamone et al., 2007; Ward et al, 2012) which would have to be dissected in order to understand the specific nature of the deficit. Thus, attempting to treat certain symptoms while failing to consider their interactions with other symptoms is not likely to lead to the most efficacious therapeutic strategies.

In closing, we suggest that is important to consider the *inseparability* of motivation, cognition, emotion, perception, decision-making, and action selection. All these features of behavioral regulation evolved to produce a reasonably successful behavioral output in the right place and at the right time. We identify psychiatric disorders because the output of these systems becomes dangerous or disruptive. Consequently, it seems foundational to understand that behavioral outputs are simultaneously regulated by this entire set of interacting processes. In this context, we should not expect to find a single cause for complex psychiatric disorders, nor perhaps a single cure. Although this approach increases the complexity of the search to understand causes of psychiatric disease, it also holds a promise for contributing to significant advances in helping patients deal with their difficulties.

References

Aberman, J. and Salamone, J. 1999. Nucleus Accumbens Dopamine Depletions Make Rats More Sensitive to High Ratio Requirements but do Not Impair Primary Food Reinforcement. *Neuroscience*, 92(2), 545–552.

Abi-Dargham, Anissa et al. 2000. "Increased Baseline Occupancy of D2 Receptors by Dopamine in Schizophrenia" *Proceedings of the National Academy of Sciences of United States of America* 97: 8104–9.

Aiba, Atsu, and Harumi Nakao. 2007. "Conditional Mutant Mice Using Tetracycline-Controlled Gene Expression System in the Brain" *Neuroscience Research* 58: 113–17.

Andreason, Nancy C. et al. 1999. "Defining the Phenotype of Schizophrenia: Cognitive Dysmetria and Its Neural Mechanisms" *Biological Psychiatry* 46: 908–20.

Barch, Deanna M. 2005. "The Cognitive Neuroscience of Schizophrenia" *Annual Review of Clinical Psychology* 1: 321–53.

Barch, Deanna M., and Erin C. Dowd. 2010. "Goal Representations and Motivational Drive Inschizophrenia: The Role of Prefrontal-Striatal Interactions" *Schizophrenia Bulletin* 36: 919–34.

Bschor, T. et al. 2004. "Time Experience and Time Judgment in Major Depression, Mania and Healthy Subjects. A Controlled Study of 93 Subjects" *Acta Psychiatrica Scandinavia* 109: 222–9.

Bendiksby, Michael. S., and Michael L. Platt. 2006. "Neural Correlates of Reward and Attention in Macaque Area LIP" *Neuropsychologia* 44: 2411–20.

Ho Beng-Choon et al. 1998. "Two-year Outcome in First-Episode Schizophrenia: Predictive Value of Symptoms for Quality of Life" *American Journal of Psychiatry* 155: 1196–201.

Barrett, H. Clark. 2012. "A Hierarchical Model of the Evolution of Human Brain Specializations" *Proceedings of the National Academy of Sciences of United States of America* 109: 10733–40.

Bizo, Lewis A., and K. Geoffrey White. 1995. "Reinforcement Context and Pacemaker Rate in the Behavioral Theory of Timing" *Animal Learning and Behavior* 23: 376–82.

Buhusi, Catalin V., and Warren H. Meck. 2005. "What Makes Us Tick? Functional and Neural Mechanisms of Interval Timing" *Nature Reviews Neuroscience* 6: 755–65.

Carroll, Christine A. et al. 2008. "Temporal Processing Dysfunction in Schizophrenia" *Brain and Cognition* 67: 150–61.

Carroll, Christine A. et al. 2009. "Timing Dysfunctions in Schizophrenia as Measured by a Repetitive Finger Tapping Task" *Brain and Cognition* 71: 345–53.

Church, Russell M., and Marvin Z. Deluty. 1977. "Bisection of Temporal Intervals" *Journal of Experimental Psychology: Animal Behavior Processes* 3: 216–28.

Clausen, Johs. 1950. "An Evaluation of Experimental Methods of Time Judgment" *Journal of Experimental Psychology* 40: 760–1.

Cousins, Michael S., and John D. Salamone. 1994. "Nucleus Accumbens Dopamine Depletions in Rats Affect Relative Response Allocation in a Novel Cost/Benefit Procedure" *Pharmacology Biochemistry and Behavior* 49: 85–91.

Coull, Jennifer T., Ruey-Kuang Cheng, and Warren H. Meck. 2011. "Neuroanatomical and Neurochemical Substrates of timing" *Neuropsychopharmacology* 36: 3–25.

Davalos, Deana B., Michael A. Kinsley, and Randal G. Ross. 2003. "Effects of Interval Duration on Temporal Processing in Schizophrenia" *Brain and Cognition* 52: 295–301.

Densen, Michelle E. 1977. "Time Perception and Schizophrenia" *Perceptual and Motor Skills* 44: 436–8.

Davis, Kenneth L. et al. 1991. "Dopamine in Schizophrenia: A Review and Reconceptualization" *American Journal of Psychiatry* 148: 1474–86.

Drew, Michael R. et al. 2007. "Transient Overexpression of Striatal D2 Receptors Impairs Operant Motivation and Interval Timing" *Journal of Neuroscience* 27: 7731–39.

Elvevåg, Brita, and Terry E. Goldberg. 2000. "Cognitive Impairment in Schizophrenia is the Core of the Disorder" *Critical Reviews in Neurobiology* 14: 1–21.

Elvevåg, Brita et al. 2003. "Duration Judgments in Patients with Schizophrenia" *Psychological Medicine* 33: 1249–61.

Elvevåg, Brita et al. 2004. "Identification of Tone Duration, Line Length, and Letter Position: An Experimental Approach to Timing and Working Memory Deficits in Schizophrenia" *Journal of Abnormal Psychology* 113: 509–21.

Farrar, A.M. et al. 2010. "Nucleus Accumbens and Effort-Related Functions: Behavioral and Neural Markers of the Interaction between Adenosine A2A and Dopamine D2 receptors" *Neuroscience* 166: 1056–67.

Fortin, Claudette, and Nathalie Massé. 2000. "Expecting a Break in Time Estimation: Attentional Time-Sharing Without Concurrent Processing" *Journal of Experimental Psychology: Human Perception and Performance* 26: 1788–96.

Frankle, Gordon W. 2007. "Neuroreceptor Imaging Studies in Schizophrenia" *Harvard Review of Psychiatry* 15: 212–32.

Freedman, Barbara J. 1974. "The Subjective Experience of Perceptual and Cognitive Disturbances in Schizophrenia" *Archives of General Psychiatry* 30: 333–40.

Friston, Karl J., and Cathy J. Price. 2011. "Modules and Brain Mapping" *Cognitive Neuropsychology* 28: 241–250.

Galtress, T., and Kirkpatrick, K. 2010. "Reward Magnitude Effects on Temporal Discrimination" *Learning and Motivation, 41*(2), 108–124.

Galtress, Tiffany, and Kimberly Kirkpatrick. 2009. "Reward Value Effects on Timing in the Peak Procedure" *Learning and Motivation* 40: 109–31.

Gibbon, John, and Russell M. Church. 1984. "Sources of Variance in an Information Processing Theory of Timing" In *Animal cognition*, edited by Herbert L. Roitblat, Thomas G. Bever, Herbert S. Terrace, 465–88. Hillsdale, NJ: Erlbaum.

Gil, Sandrine, and Sylvie Droit-Volet. 2009. "Time Perception, Depression and Sadness" *Behavioural Processes* 80: 169–76.

Glahn, David C. et al. 2005. "Beyond Hypofrontality: A Quantitative Meta-Analysis of Functional Neuroimaging Studies of Working Memory in Schizophrenia" *Human Brain Mapping* 25: 60–9.

Gold, Joshua I., and Michael N. Shadlen. 2001. "Neural Computations that Underlie Decisions about Sensory Stimuli" *Trends in Cognitive Sciences* 5: 10–6.

Goldman-Rakic, Patricia S., and Lynn D. Selemon. 1997. "Functional and Anatomical Aspects of Prefrontal Pathology in Schizophrenia" *Schizophrenia Bulletin* 23: 437–58.

Goldman-Rakic, Patricia S. et al. 2004. "Targeting the Dopamine D1 Receptor in Schizophrenia: Evidence from Cognitive Dysfunction" *Psychopharmocology* 174: 3–16.

Grace, Randolph C., and John A. Nevin. 2000. "Response Strength and Temporal Control in Fixed-Interval Schedules" *Animal Learning & Behavior* 28: 313–31.

Grinker, Joel et al. 1973. "Time Perception as a Function of Weight Reduction: A Differentiation Based on Age at Onset of Obesity" *Psychosomatic Medicine* 35: 104–11.

Gurevich, Ilona et al. 2002. "Altered Editing of Serotonin 2C Receptor Pre-mRNA in the Prefrontal Cortex of Depressed Suicide Victims" *Neuron* 34: 349–56.

Heinrichs, R. Walter. 2005. "The Primacy of Cognition in Schizophrenia" *The American Psychologist, 60*(3), 229–42.

Howes, Oliver D., and Shitij Kapur. 2009. "The Dopamine Hypothesis of Schizophrenia: Version III-the Final Common Pathway" *Schizophrenia Bulletin* 35: 549–62.

Howes, Oliver D. et al. 2009. "Elevated Striatal Dopamine Function Linked to Prodromal Signs of Schizophrenia" *Archives of General Psychiatry* 66: 13–20.

Ishiwari, K. et al 2004. "Accumbens Dopamine and the Regulation of Effort in Food-Seeking Behavior: Modulation of Work Output by Different Ratio or Force Requirements" *Behavioural Brain Research, 151*(1–2), 83–91.

Johnson, James E., and Thomas P. Petzel. 1971. "Temporal Orientation and Time Estimation in Chronic Schizophrenics" *Journal of Clinical Psychology* 27: 194–96.

Kacelnik, Alex, and Dani Brunner. 2002. "Timing and Foraging: Gibbon's Scalar Expectancy Theory and Optimal Patch Exploitation" *Learning and Motivation* 33: 177–95.

Keefe, Richard S. et al. 2007. "Neurocognitive Effects of Antipsychotic Medications in Patients with Chronic Schizophrenia in the CATIE Trial" *Archives of General Psychiatry* 64: 633–47.

Kellendonk, Christoph et al. 2006. "Transient and Selective Overexpression of Dopamine D2 Receptors in the Striatum Causes Persistent Abnormalities in Prefrontal Cortex Functioning" *Neuron* 49: 603–15.

Kellendonk, Christoph. 2009. "Modeling Excess Striatal D2 Receptors in Mice" In *Genetic Models of Schizophrenia,* edited by Akira Sawa, 59–65. Amsterdam: Elsevier.

Kerns, John G. et al. 2008. "Executive Functioning Component Mechanisms and Schizophrenia." *Biological Psychiatry* 64: 26–33.

Killeen, Peter R., Scott Hall, and Lewis A. Bizo. 1999. "A Clock Not Wound Runs Down." *Behavioural Processes* 45: 129–39.

Laruelle, Marc. 1998. "Imaging Dopamine Transmission in Schizophrenia. A Review and Meta-Analysis." *Quarterly Journal of Nuclear Medicine* 42: 211–21.

Lejuene, Helga, Francoise Macar, and Dan Zakay. 1999. "Attention and Timing: Dual-Task Performance in Pigeons." *Behavioural Processes* 45: 141–57.

Leon, Matthew I., and Michael N. Shadlen. 1999. "Effect of Expected Reward Magnitude on the Response of Neurons in the Dorsolateral Prefrontal Cortex of the Macaque." *Neuron* 24: 415–25.

Lewis, David A. et al. 1999. "Altered GABA Neurotransmission and Prefrontal Cortical Dysfunction in Schizophrenia." *Biological Psychiatry* 46: 616–26.

Lewis, Aubrey. 1932. "The Experience of Time in Mental Disorder." *Proceedings of the Royal Society of Medicine* 25: 611–20.

Lhamon, William, and Sanford Goldstone. 1956. "The Time Sense: Estimation of One Second Durations by Schizophrenic Patients." *Archives of Neurological Psychiatry* 76: 625–29.

Ludvig, Elliot. A., Kent Conover, and Peter Shizgal. 2007. "The Effects of Reinforcer Magnitude on Timing in Rats." *Journal of Experimental Analysis of Behavior* 87: 201–18.

Mahon, Bradford Z., and Jessica F. Cantlon. 2011. "The Specialization of Function: Cognitive and Neural Perspectives." *Cognitive Neuropsychology* 28: 147–55.

Maricq, Andres V., and Russell M. Church. 1983. "The Differential Effects of Methamphetamine and Haloperidol on Time Estimation in the Rat." *Psychopharmacology* 79: 10–15.
Matell, Matthew S., and Meck, Warren H. 2004. "Cortico-striatal Circuits and Interval Timing: Coincidence Detection of Oscillatory Processes" *Brain Research. Cognitive Brain Research, 21*(2).
Mayford, Mark et al. 1996. "Control of Memory Formation through Regulated Expression of a CamkII Transgene." *Science* 274: 1678–83.
Meck, Warren H. 1986. "Affinity for the Dopamine D2 Receptor Predicts Neuroleptic Potency in Decreasing the Speed of an Internal Clock." *Pharmacology Biochemistry and Behavior* 25: 1185–9.
Metcalfe, Janet et al. 2012. "Action Monitoring and Metacognition of Agency in Participants with Schizophrenia and Healthy Controls." *Proceedings of the Royal Academy of Sciences* 367: 1391–400.
Mombereau, Cedric et al. 2010. "Functional Relevance of Serotonin 2C Receptor mRNA Editing in Antidepressant- and Anxiety-like Behaviors." *Neuropharmacology* 59: 468–73.
Penney, Trevor B. et al. 2005. "Interval Timing Deficits in Individuals at High Risk for Schizophrenia." *Brain Cognition* 58: 109–18.
Pentkowski, Nathan S. et al. 2010. "Stimulation of Medial Prefrontal Cortex Serotonin 2C (5-HT2C) Receptors Attenuates Cocaine-Seeking Behavior." *Neuropsychopharmacology* 35: 2037–48.
Platt, Michael L., and Paul W. Glimcher. 1999. "Neural Correlates of Decision Variables in Parietal Cortex." *Nature* 400: 233–38.
Roberts, Seth. 1981. "Isolation of an Internal Clock." *Journal of Experimental Psychology Animal Behavior Processes* 7: 242–68.
Salamone, John D., and Correa, Merce (2012). "The Mysterious Motivational Functions of Mesolimbic Dopamine" *Neuron, 76*(3).
Salamone, John D., and Merce Correa. 2002. "Motivational Views of Reinforcement: Implications for Understanding the Behavioral Functions of Nucleus Accumbens Dopamine." *Behavioral Brain Research* 137: 3–25.
Salamone, J.D. et al. 1991. "Haloperidol and Nucleus Accumbens Dopamine Depletion Suppress Lever Pressing for Food but Increase Free Food Consumption in a Novel Food Choice Procedure." *Psychopharmacology* 104: 515–21.
Salamone, J.D. et al. 1996. "Different Behavioral Effects of Haloperidol, Clozapine and Thioridazine in a Concurrent Lever Pressing and Feeding Procedure." *Psychopharmacology* 125: 105–12.
Salamone, John D. et al. 2007. "Effort-Related Functions of Nucleus Accumbens Dopamine and Associated Forebrain Circuits." *Psychopharmacology* 191: 461–82.
Salamone, John. D. et al. 2009. "Dopamine, Behavioral Economics, and Effort." *Frontiers in Behavioral Neuroscience* 3: 13.
Seeman, P. et al. 1976. "Antipsychotic Drug Doses and Neuroleptic/dopamine Receptors." *Nature* 261: 717–19.

Sévigny, Marie-Claude, James Everett, and Simon Grondin. 2003. "Depression, Attention, and Time Estimation." *Brain and Cognition* 53: 351–3.

Simpson, Eleanor H., Christoph Kellendonk, and Eric R. Kandel. 2010. "A Possible Role for the Striatum in the Pathogenesis of the Cognitive Symptoms of Schizophrenia." *Neuron* 65: 585–96.

Simpson, Eleanor H. et al. 2011. "Pharmacologic Rescue of Motivational Deficit in an Animal Model of the Negative Symptoms of Schizophrenia." *Biological Psychiatry* 69: 928–35.

Sugrue, Leo P., Greg S. Corrado, and William T. Newsome. 2004. "Matching Behavior and the Representation of Value in the Parietal Cortex." *Science* 304: 1782–7.

Todd, Juanita. 2006. "Impaired Detection of Silent Interval Change in Schizophrenia." *Neuroreport* 17: 785–9.

Tysk, Lennart. 1983. "Estimation of Time and the Subclassification of Schizophrenic Disorders." *Perceptual and Motor Skills* 57: 911–18.

Tysk, Lennart. 1984a. "A Longitudinal Study of Time Estimation in Psychotic Disorders." *Perceptual and Motor Skills* 59: 779–89.

———. 1984b. "Time Perception and Affective Disorders." *Perceptual and Motor Skills* 58: 455–64.

Tysk, Lennart. 1990. "Estimation of Time by Patients with Positive and Negative Schizophrenia." *Perceptual and Motor Skills* 71: 826.

Uttal, William R. 2001. *The New Phrenology: The Limits of Localizing Cognitive Processes in the Brain.* Cambridge: MIT Press.

Uttal, William R. 2011. *Mind and Brain: A Critical Appraisal of Cognitive Neuroscience.* Cambridge: MIT Press.

Wahl, Otto F., and David Sieg. 1980. "Time estimation among schizophrenics." *Perceptual and Motor Skills* 50: 535–41.

Ward, Ryan D., and Amy L. Odum. 2006. "Effects of Prefeeding, Intercomponent Interval Food, and Extinction on Temporal Discrimination and Pacemaker Rate." *Behavioural Processes* 71: 297–306.

Ward, Ryan D., and Amy L. Odum. 2007. "Disruption of Temporal Discrimination and the Choose-Short Effect." *Learning and Behavior* 35: 60–70.

Ward, Ryan D. et al. 2009. "Impaired Timing Precision Produced by Striatal D2 Receptor Overexpression is Mediated by Cognitive and Motivational Deficits." *Behavioral Neuroscience* 123: 720–30.

Ward, Ryan D. et al. 2012. "Dissociation of Hedonic Reaction to Reward and Incentive Motivation in an Animal Model of the Negative Symptoms of Schizophrenia." *Neuropsychopharmacology* 371: 699–707.

Yang, Yen K. et al. 2004. "Association between Cognitive Performance and Striatal Dopamine Binding is Higher Intiming and Motor Tasks in Patients with Schizophrenia." *Psychiatry Research* 131: 209–16.

Zakay, Dan. 1989. "Subjective and Attentional Resource Allocation." In *Time and Human Cognition*, edited by Dan Zakay, 365–97. North-Holland: Elsevier Science.

CHAPTER 8

Timing in Neurogenerative Disorders of the Basal Ganglia

*Deborah L. Harrington*** and Stephen M. Rao*†

1 Introduction

Time is an integral facet of behavior that structures our perceptions, experiences, and memories. Although there is no clinical disorder that is specifically characterized by timing disturbances, some are known to alter certain interrelated cognitive functions, such as planning. A host of neurological and psychiatric diseases that affect basal ganglia, cerebellar, and cerebral cortex functioning affect timing within the scale of tenths of milliseconds on up to seconds or even minutes. As such, there has been a lively debate over the neural sources of temporal processing deficits in various disease processes. Presently, there is substantial support for the centrality of the striatum and dopamine neurotransmission in explicit timing (Balci et al., 2012; Hohn et al., 2011; Lake and Meck, 2012; Meck, 2006a,b; Rammsayer, 1993). This chapter discusses timing disturbances in two disorders of the basal ganglia in which dopaminergic functioning is decreased, namely Parkinson's disease (PD) and prodromal Huntington disease (HD). There is a growing interest in these disorders since temporal processing deficits contribute to the breakdown in the spatiotemporal organization of movement in PD (Vercruysse et al., 2012) and timing ability is a marker of proximity to a diagnosis of manifest HD (Harrington et al., 2012; Rowe et al., 2010). This chapter places an emphasis on functional imaging investigations of timing disturbances in PD and prodromal HD, as this has received little attention in the literature to date. To set the stage for our discussion, we first review two influential timing models that underscore the crucial roles of key component processes and neurophysiological mechanisms. The framework provided by these models, together with advancements in neuroanatomical connectivity, has guided research into the

* VA San Diego Healthcare System, San Diego, CA, USA.
** Department of Radiology, University of California, San Diego, La Jolla, CA, USA.
† Schey Center for Cognitive Neuroimaging, Neurological Institute, Cleveland Clinic, Cleveland, OH, USA.

neurocognitive mechanisms of timing and its breakdown. We then give an overview of timing disturbances in PD and the effects of dopamine therapy, followed by research into prodromal phases of HD. This section is not intended to provide a thorough treatment of this topic as there are several excellent, recent reviews of temporal processing disturbances in PD (Allman and Meck, 2012; Koch, Oliveri, and Caltagirone, 2009). We then discuss emerging functional imaging research, which is beginning to reveal brain circuits that govern timing deficits in these disorders. Lastly, we consider the clinical relevance of temporal processing dysfunction in these diseases and future avenues for research.

2 Models of Timing

We lead with a brief description of two influential models that have driven research in neurologically intact adults and in clinical disorders of temporal processing. The information processing model has provided a strong framework for studying component processes of interval timing, which can have distinct effects on timing accuracy and variability. In contrast, a leading neurophysiological model delineates the physiological and neuroanatomical properties of timing networks, thereby fostering an understanding of the functional significance of abnormalities in corticostriatal networks that underlie timing disturbances.

2.1 *Information Processing Model*

Over the last two decades there has been an explosion of research into the neuroanatomical underpinnings of interval timing in disorders of the basal ganglia. Initially, a driving force behind much of this work was scalar expectancy theory (SET; Gibbon, 1977; Gibbon, Church, and Meck, 1984), which defined sources of timing variability that were derived from clock, memory, and decision processes (Figure 8.1A). Empirical findings in pharmacological and lesion studies in animals further suggested that the clock process depended on dopamine neurotransmission and the striatum (Buhusi and Meck, 2002; Maricq and Church, 1983; Meck, 1996). In SET, timing is implemented via a pacemaker mechanism, which represents time through the accumulation of pulses. Pulses are turned on and off by a mode switch and then passed onto an accumulator to be counted. Perceived duration is intimately related to the interplay between the pacemaker and the level of attention given to the passage of time. Attention affects the mode switch, which controls the flexible starting and stopping of pulses from the pacemaker, thereby enabling anticipation of predictable events. Attention also affects the accumulation of pulses, such that when attention is diverted there is a shortening of perceived duration, suggesting that pulses are lost. The current representation of time is stored in working memory, where it is maintained in accord with current goals and over time, more enduring interval representations are stored in long-term memory. A decision process compares pulse counts from the accumulator with ones in memory to determine when or how to respond. At this stage, strategic or attentional factors can influence decision processes by biasing response thresholds that determine when the current time is sufficiently close to the remembered time. Altogether, the perception of time emerges from the interplay among these component processes. For this reason, SET continues to be an influential framework for unraveling the potential influence of various cognitive processes on disturbances in temporal processing. Potential sources of timing deficits are inferred through different patterns of accuracy and variability, which has been discussed in detail by others (Allman and Meck, 2012; Wearden, 1999).

2.2 *Neurophysiological Model*

As knowledge grew about the neurobiological bases of timing, the challenge of unraveling the functional significance of temporal processing deficits in neurological disorders became more apparent. This challenge can be better appreciated upon consideration of the complexity of corticostriatal connectivity (Figure 8. 1B). Cortical networks subserving different behaviors have separate, partially segregated pathways to the striatum that return to the same areas via direct and indirect pathways. There is also a hyper-direct pathway from the cortex to the subthalamic nucleus (STN), which circumvents the striatum. Dopaminergic input to the striatum and cerebral cortex is via the substantia nigra pars compacta (SNc; nigrostriatal pathway) and the ventral tegmental area (VTA; mesolimbic and mesocortical pathways), respectively. The cerebellum also has separate parallel connections with the cerebral cortex (Akkal, Dum, and Strick, 2007; Clower, Dum, and Strick, 2005). Moreover, the cerebellum and the basal ganglia communicate with each other by way of output nuclei from each region (dentate nuclei) and the STN (Figure 8.1C; Bostan, Dum, and Strick, 2010; Hoshi et al., 2005). Thus, timing might emerge from complex interactions within corticostriatal, cortico-cortical, and/or striatal-cerebellar networks.

The Striatal Beat Frequency Model (SBF; Matell and Meck, 2004), which evolved from an elegant body of work in rodents, captured the inherent interactive nature of timing networks (Matell, Meck, and Nicolelis, 2003; Matell et al., 2011; Meck et al., 2012). By this model (Figure 8.1B), the striatum receives

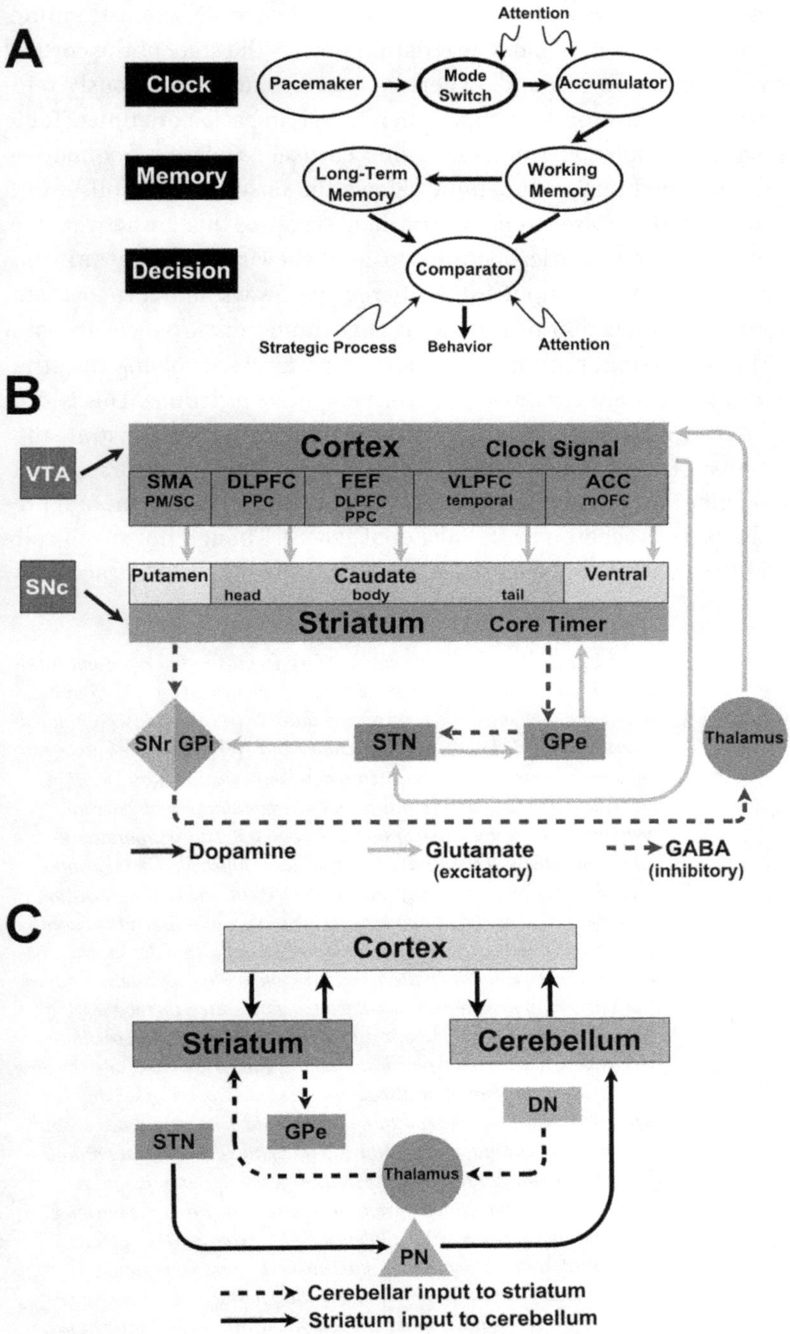

FIGURE 8.1 *Interval timing models and brain networks. A: Illustration of the three component processes of an information processing model of timing, scalar expectancy theory*

cortical oscillatory activity, which evolves as a function of event duration, thereby serving as a clock signal. The striatum detects the state of the cortical oscillatory patterns and responds when a pattern matches previously reinforced or intrinsically important states, thereby serving as a core timer. Tonic dopamine and glutamate are thought to alter cortical oscillation frequencies, whereas nigrostriatal phasic dopamine signals the onset and the offset of a timed event. Dopamine also strengthens striatal synapses that are activated by the frequency pattern of cortical neurons to a timed event. So by the SBF, timing depends on dopamine input from different pathways and corticostriatal interactions. An implication of the SBF is that timing disturbances in basal ganglia disorders should arise from changes in pathways involving the striatum, but also other brain circuits that govern cognitive functions. This is supported by psychopharmacological challenges in rodents that demonstrate that cognitive enhancers (atomoxetine and physostigmine) and disruptors (scopolamine and chlordiazepoxide) respectively improve and worsen temporal precision in the peak procedure task (Balci et al., 2008). Though not specifically addressed by the SBF, diminished striatal functioning might also engage brain

FIGURE 8.1 (CONT.) *(SET). The clock component consists of a pacemaker that represents time through the accumulation of pulses. Pulses are turned on and off by a switch and then passed into an accumulator to be counted. Accumulated pulses are encoded into working memory and over time, more enduring interval representations are stored in long-term memory. Decision processes compare pulse counts from the accumulator with ones in memory to determine when or how to respond. B: The neuroanatomically connectivity of the striatum and cortex is illustrated in the context of the Striatal Beat Frequency Model (SBF). By this model, the striatum receives cortical oscillatory activity, which evolves as a function of event duration, thereby serving as a clock signal. The striatum detects the state of the cortical oscillatory patterns and responds when a pattern matches previously reinforced or intrinsically important states, thereby serving as a core timer. Tonic dopamine and glutamate alter cortical oscillation frequencies; nigrostriatal phasic dopamine signals the onset and the offset of a timed event. Dopamine also strengthens striatal synapses that are activated by the frequency pattern of cortical neurons to a timed event. C: Illustration of anatomical pathways that enable two-way communication between the striatum and cerebellum. ACC = anterior cingulate cortex; DLPFC = dorsolateral prefrontal cortex; DN = dentate nucleus; FEF = frontal eye fields; GPe = globus pallidus externa; GPi = globus pallidus interna; PPC = posterior parietal cortex; PM = premotor; PN = pontine nucleus; SMA = supplementary motor area; SC = sensory cortex; SNc = substantia nigra pars compacta; SNr = substantia nigra pars reticulata; STN = subthalamic nucleus; VLPFC = ventrolateral prefrontal cortex; VTA = ventral tegmental area.*

circuits that compensate via degeneracy mechanisms, wherein different anatomical networks are engaged (Merchant, Harrington, and Meck, 2013). Degeneracy in timing systems is plausible since time-related cell activity is not only found in the basal ganglia, but also the cerebellum, thalamus, posterior parietal cortex, prefrontal cortex, and the supplementary motor area (SMA and preSMA; Merchant et al., 2013). Timing in different behavioral contexts is also associated with different neural architectures. For example, implicit and explicit timing are thought to respectively depend on the cerebellum and striatum (Coull, Cheng, and Meck, 2011). Similarly, cortico-cortical systems are more engaged when timed movements that are externally paced, whereas the striatum is engaged when movements are self-paced or internally timed (Rao et al., 1997; Taniwaki et al., 2003). Thus, different neural systems may be capable of timing. At the same time, it is far from clear whether degeneracy mechanisms actually compensate (i.e., improve performance) for diminished timing. We will return to this issue when discussing neuroimaging studies of timing in PD and prodromal HD.

3 Temporal Processing Deficits in Basal Ganglia Disorders

Temporal processing in PD and prodromal HD is of keen interest due to the centrality of the striatum and dopamine neurotransmission in timing. Though traditionally considered disorders of movement, timing disturbances in both diseases are not simply due to motor symptoms, since they are found in contexts that minimize motor output. Temporal processing has been far more studied in PD than prodromal HD. In this section, behavioral studies will be reviewed in PD, followed by those in prodromal HD. For each disorder, we first briefly review the underlying neuropathology. Then clinical features are discussed that are relevant to elucidating the neurobehavioral mechanisms of temporal processing deficits and interpreting discrepancies that have been reported across studies.

3.1 *Parkinson's Disease*

3.1.1 Neuropathology and Symptoms

The principal neuropathological finding in PD is dopamine cell loss in nigrostriatal pathway and to lesser extent, the mesocortical/limbic pathway. PD not only presents with classic motor symptoms (tremor, rigidity, slowness, postural instability, freezing), but it has a progressive impact on cognition. In PD without dementia, cognitive decline is often subtle and attributed to frontostriatal dysfunction, as it most commonly affects executive functions, attention, and working memory. By SET, some of these functions may interact with the clock, thereby complicating the interpretation of the source for timing deficits. It is also important to recognize that there is considerable clinical heterogeneity of cognitive symptoms in PD (Carbon, Edwards, and Eidelberg, 2003; Lewis et al., 2003; Williams-Gray et al., 2009). This heterogeneity likely relates to individual differences in pathological changes and their severity, although this is not well understood nor is the effect of dopamine therapy on various cognitive functions. Another important consideration is that there are day-to-day fluctuations in symptoms and individual differences in responsiveness to medication therapy. Thus, a gamut of clinical phenotypes may underlie some discrepancies reported across studies with regard to whether timing is impaired in PD or affected by dopamine therapy.

3.1.2 Temporal Processing in Parkinson's Disease and the Effect of Dopamine Therapy

Most studies have reported temporal processing deficits in PD across a variety of tasks including time production or reproduction (Elsinger et al., 2003; Harrington, Haaland, and Hermanowicz, 1998; Jones et al., 2011; Jones et al., 2008; Koch et al., 2008; O'Boyle, Freeman, and Cody, 1996; Pastor et al., 1992a; Pastor et al., 1992b; Perbal et al., 2005; Wing, Keele, and Margolin, 1984), time estimation (Koch et al., 2004a, 2004b; Malapani, Deweer, and Gibbon, 2002; Pastor et al., 1992b; Smith et al., 2007; Wild-Wall et al., 2008), and time discrimination (Harrington et al., 2011b; Harrington , Haaland, and Hermanowicz, 1998; Rammsayer and Classen, 1997; Riesen and Schnider, 2001). Basal ganglia lesions also disrupt the detection and production of intervals (Schwartze et al., 2011). However, the mechanisms underlying these deficits are unclear. By SET, a slowdown in the clock process in PD should alter timing accuracy, thereby producing an underestimation of time. Although this has been reported for time estimation (Malapani et al., 2002; Pastor et al., 1992; Smith et al., 2007; Wild-Wall et al., 2008), measures of accuracy that are straightforward to interpret within SET are often not available for other tasks. For example, repetitively timed movements in PD can be over produced, especially for subsecond intervals (Elsinger et al., 2003; Harrington, Haaland, and Hermanowicz, 1998; Jones et al., 2011; Pastor et al., 1992a), but this may not be due to a timing disturbance per se. Rather, the finding may relate to a clinical feature of PD, namely motor festination, wherein individuals attempt to correct for the decreased amplitude of repetitive movements by speeding them up (e.g., gait, handwriting). Most studies of timing reproduction or production also report greater timing variability in PD, which by SET may reflect changes both in the clock and other component processes. Though time discrimination paradigms circumvent the potentially confounding effects of motor symptoms on timing, increased temporal difference thresholds in PD (i.e., the least amount of time that can be perceived between two intervals) may also relate to changes in the clock and/or other component processes. As an aside, several reviews suggest that the basal ganglia are more important for timing within the range of seconds, whereas the cerebellum subserves timing functions with the subsecond range (Koch et al., 2009; Lewis and Miall, 2003). Although this has received support in some studies of PD (Koch et al., 2008; Smith et al., 2007), others have found timing deficits in the range of subseconds (Elsinger et al., 2003; Harrington, Haaland, and Hermanowicz, 1998; O'Boyle et al., 1996; Rammsayer and Classen, 1997; Riesen and Schnider, 2001). The reasons for the discrepancies are not entirely clear and will require carefully conducted psychophysical studies across a range of timing tasks to sort this out. It is also important to recognize that while dichotomies can provide useful frameworks to advance research, they often oversimplify the organization of neurocognition.

As for the effect of dopamine replacement therapy, it can improve (O'Boyle et al., 1996; Pastor et al., 1992a, 1992b), have no effect (Elsinger et al., 2003; Harrington et al., 2011; Jahanshahi et al., 2010; Koch et al., 2008), or hinder timing performance (Jones et al., 2008). At first glance, the absence of an effect of dopamine on timing in PD seems at odds with the SBF model and the effects of dopamine agonists and antagonists on timing (Buhusi and Meck, 2002; Lake and Meck, 2012; Meck, 1996; Meck et al., 2012). However, dopamine replacement therapy is not complete in PD, which concurs with timing dysfunction even when patients are tested on their medication therapy (Harrington et al., 2011; Harrington, Haaland, and Hermanowicz, 1998; Rammsayer and Classen, 1997; Riesen and Schnider, 2001). There are also lingering effects of medications when they are temporarily stopped for up to 24 hours, because they are long acting and there are individual differences in treatment responsiveness. The effect of dopamine therapy on cognition is not well understood, but it can hinder performance if there is too much dopamine in functionally-intact corticostriatal circuits that support a cognitive function (Cools, 2006). Altogether, these factors can make it difficult to evaluate the influence of dopamine on temporal processing in the PD model.

An important consideration is that timing deficits have not always been found in PD. For example, normal performance was reported on a test of motor timing (Spencer and Ivry, 2005) and on several different tests of time perception (Wearden et al., 2008). The reasons for this are unknown, but may be due to the insensitivity of tasks when time discriminations are rather easy (i.e., 25% or more than the duration of anchor stimuli) and/or when feedback is regularly given (Wearden et al., 2008). In addition, timing deficits in PD correlate with disease severity (Artieda et al., 1992). Several studies have tested early-stage patients (Spencer and Ivry, 2005), who despite considerable dopamine cell loss may have the capacity to compensate for timing difficulties. The cerebellum may be one compensatory route (Kotz and Schwartze, 2011), possibly because it predicts and finely tunes behavioral states based on efferent copy of sensory and motor information. Cortical systems might support compensatory processing in PD as well. Apart from these issues, the finding that there are subgroups of patients who do and do not exhibit timing disturbances (Merchant et al., 2008) resonates with the substantial heterogeneity of clinical phenotypes in the disease, day-to-day fluctuations in symptoms, and individual differences in response to dopamine therapy. Neurodegenerative disorders of the basal ganglia eventually alter cortical functioning, which may be another source of interval timing disturbances. This prospect was suggested in an early report that damage to the right hemisphere of the prefrontal (dorsolateral prefrontal and premotor cortices) and the inferior parietal cortex disrupted time perception (Harrington, Haaland, and Knight, 1998). Additionally, damage to the right, but not the left hemisphere, produced elevated temporal discrimination thresholds that correlated with a decreased ability to flexibly reorient attention. This finding comports with a report that time estimation is improved in PD after repetitive transcranial magnetic stimulation (rTMS) to the right dorsolateral prefrontal cortex, but not the SMA (Koch et al., 2004b). Moreover, rTMS to the supramarginal gyrus of the right hemisphere in healthy adults causes a distortion in perceived duration (Wiener et al., 2012; Wiener et al., 2010). Thus, frontoparietal systems, which govern attention functions, interact with timekeeping processes (Lustig, Matell, and Meck, 2005). Unfortunately, systematic investigations into cortical regions that are essential for timing in humans have been hampered by difficulties in obtaining sufficient samples of patients with focal cortical damage. For this reason, functional imaging has become increasingly important for understanding brain circuits that govern disturbances in temporal processing.

3.2 *Huntington Disease*

3.2.1 Neuropathology and Symptoms

Huntington disease (HD) is an autosomal dominant neurodegenerative disorder caused by a polyglutamine (CAG) expansion in the IT15 gene, which leads to the production of a mutant form of the protein huntingtin. HD usually manifests in midlife, but the age of onset varies considerably and partly depends on the CAG repeat length. The mutant huntingtin protein is expressed throughout the brain, yet the medium spiny neurons of the caudate nucleus and to a lesser extent, the putamen, are the most vulnerable early in the disease process. Pathological changes in cortical gray- and white-matter also gradually develop as individuals approach onset of the illness (Aylward et al., 2012; Nopoulos et al., 2010). The HD diagnosis is based on the appearance of unequivocal extrapyramidal motor symptoms (e.g., chorea, oculomotor problems, dysarthria, dystonia, gait disturbance, postural instability, rigidity, bradykinesia).

There is an inverse correlation between age of onset and CAG repeat length, such that greater repeat lengths signify a stronger genetic burden and earlier onset of illness. Because this relationship accounts for only 47 to 73% of the variability in age of illness onset (Brinkman et al., 1997; Ranen et al., 1995), there has been a concerted effort to identify early markers of the disease during the prodromal phase, when potential treatments are more likely to be effective. In prodromal HD, the striatum begins to atrophy decades before diagnosis, rendering it a good model of basal ganglia dysfunction. Pathological changes are accompanied by subtle motor and psychiatric symptoms, but also cognitive decline (e.g., processing speed, executive functions, and sensory-perceptual processes; Harrington et al., 2012; Stout et al., 2011) that typically is not clinically significant until an individual approaches diagnosis. Like PD, however, subtle cognitive changes may interact with the clock process, thereby complicating the interpretation of the source for temporal processing deficits.

3.2.2 Temporal Processing in Huntington Disease

Interval timing in HD has been far less studied than in PD. Deficits in timed repetitive movements were first reported in manifest HD using the classic paced tapping task wherein finger movements are initially entrained to a series of isochronous tones (synchronization phase), followed by a period of tapping at the same pace without the tones (continuation phase; Freeman et al., 1996). Abnormal timing accuracy and variability was found, irrespective of whether cues (tones) were present or absent. Recent studies have replicated this finding in manifest HD using similar motor timing tasks (Bechtel et al., 2010; Thompson et al., 2010). Deficits in time estimation and discrimination have also been found in manifest HD (Beste et al., 2007), which suggests that motor disturbances cannot entirely explain impaired motor timing. Still, sources of temporal processing dysfunction remain difficult to ascertain due to the significant cognitive disturbances in manifest HD.

A better model of basal ganglia dysfunction is prodromal HD, since motor and cognitive changes are typically subtle. In prodromal HD, paced-tapping variability, but not accuracy, is abnormal for both subsecond and suprasecond intervals during synchronization and continuation phases of the task (Hinton et al., 2007; Zimbelman et al., 2007). Timing variability also increased nonlinearly with estimated years to onset, demonstrating that timing dysfunction accelerates as individuals approach a clinical diagnosis (Hinton et al., 2007). This finding was replicated for subsecond intervals by large multisite studies, which also controlled for subtle motor symptoms (Bechtel et al., 2010; Rowe et al., 2010; Stout et al., 2011). In addition, a longitudinal study of paced timing performance demonstrated that timing variability gradually increased over a two to four year period, but at a faster rate in individuals who were closer to diagnosis (Rowe et al., 2010). Thus, there is consensus that rhythmically timed movements progressively deteriorate with proximity to diagnosis.

To date, only two studies have investigated temporal processing on tasks that minimize motor output requirements. One study found time discrimination deficits (reduced accuracy) for suprasecond intervals (1200ms) in individuals close (<12 years), but not far (>12 years) from diagnosis (Paulsen et al., 2004). Another study found that time discrimination was normal in prodromal HD relative to a control group (Beste et al., 2007). However, higher error rates on the task correlated with estimated proximity to diagnosis, suggesting that the absence of group differences may have been due to the small (N = 12) heterogeneous sample of prodromal HD participants. Clearly, more studies of time perception tasks are needed to determine if temporal processing deficits in prodromal HD are independent of changes in the motor system.

Another issue is that pathological changes in the cortex (i.e., cortical thinning, white-matter atrophy, altered white-matter diffusivity) are seen more than a decade before a diagnosis of manifest PD (Dumas et al., 2012; Nopoulos et al., 2010; Paulsen et al., 2010). These changes likely cause subtle cognitive decline, which progresses with proximity to diagnosis. As in PD, cognitive dysfunction may interact with timekeeping processes, thereby rendering it difficult to identify the neurocognitive mechanisms of temporal processing dysfunction. The advent of functional imaging has therefore been a welcome development.

4 Neuroanatomical Underpinnings of Temporal Processing Dysfunction in Basal Ganglia Disorders

The neural underpinnings of timing dysfunction in PD and prodromal HD are not well understood due to the dearth of functioning imaging investigations. To date, there have been only four functional imaging studies of motor timing dysfunction in PD, all of which have used the paced-tapping task (Cerasa et al.,

2006; Elsinger et al., 2003; Jahanshahi et al., 2010; Yu et al., 2007). Despite differences among studies in the analysis methods and the control conditions, some results are remarkably consistent, particularly as they concern the functional role of the cerebellum in timing. However, there are also discrepancies among studies in regional patterns of timing-related dysfunction during the synchronization and the continuation phases of the task or whether PD is characterized by hypo- or hyperactivation. This may be partly a consequence of the small samples PD participants in all studies (10 or fewer), as there is considerable heterogeneity of clinical phenotypes in PD. Although the brain circuits that govern time perception deficits in PD have been far less studied (Dusek et al., 2012; Harrington et al., 2011), it is clear that patterns of regional brain dysfunction in PD depend on the extent to which timing and other cognitive processes are emphasized within a behavioral context. Another important development concerns investigations into disturbances in timing-related connectivity of the striatum (Harrington et al., 2011b; Jahanshahi et al., 2010), which appears more sensitive to the influence of dopamine therapy than conventional regional analyses of brain activation. As for prodromal HD, two studies have been conducted to date, one of timed movements (paced-tapping task; Zimbelman et al., 2007) and the other of time discrimination (Paulsen et al., 2004). Although more research is needed, timing deficits in prodromal HD appear to be governed by some similar brain circuits as in PD.

4.1 *Functional Imaging Studies of Parkinson's Disease*

4.1.1 Timed Movements

The neuroanatomical basis of motor timing deficits in PD was first explored by Elsinger and colleagues in an functional magnetic resonance imaging (fMRI) study of participants who were tested on (ON) and off (OFF) their medication therapy (Elsinger et al., 2003). Paced tapping (600ms isochronous interval) was studied for both the synchronization and continuation phases. Timing was impaired in the PD group during both phases, irrespective of medication. Though the fMRI analyses did not directly compare the control and PD groups or the medication conditions, group differences were implied by two main patterns of regional activation. Regardless of medication or phase of the task, cerebellar activation was found in the control, but not the PD group, suggesting an absence of cerebellar compensation. Second, during the continuation phase, medication reinstated motor circuit activation, similar to the control group. This result suggested that internally timed movements in PD depended on the motor circuit.

Cerasa and colleagues subsequently used fMRI to study paced tapping (750ms isochronous interval) during the synchronization and continuation phases in PD OFF participants relative to controls (Cerasa et al., 2006). Timing was impaired in PD during both phases of the task, despite different patterns of abnormal activation in the two phases. During the synchronization phase, the PD group showed hyperactivity in the motor circuit (putamen, thalamus, SMA), the inferior frontal gyrus and insula, and the cerebellum. During the continuation phase, the PD group showed hyperactivity in the thalamus and cerebellum. Thus, in contrast to Elsinger and colleagues, externally guided rather than internally controlled timing was related to motor-circuit dysfunction. Moreover, potential compensatory responses were observed in the inferior frontal gyrus and insula (synchronization phase) and in the cerebellum (both phases).

Yu and colleagues partially replicated and extended these findings in an fMRI study of synchronized timing (900 and 2400ms isochronous intervals), which compared activation in PD OFF participants with controls (Yu et al., 2007). Despite no group differences in timing accuracy or variability, the PD group exhibited hypoactivation of the motor circuit, rather than hyperactivation (Cerasa et al., 2006). However, consistent with Cerasa and colleagues, hyperactivation was found in the bilateral cerebellum, but also the contralateral motor area. Importantly, ipsilateral cerebellar activation negatively correlated with contralateral putamen activation, whereas motor cortex activation correlated positive with the severity of rigidity. The authors speculated that over activation of the cerebellum might normalize timing performance, since this structure supports timed movements (Harrington et al., 2004a; Ivry and Keele, 1989). As illustrated in Figure 8.1C, this could be achieved via direct cerebellar input into the striatum (Bostan et al., 2010; Hoshi et al., 2005). The prospect that the cerebellum may support timing in PD due to degeneracy in timing systems is intriguing (Kotz and Schwartze, 2011). However, hyperactivation may also reflect a loss in the topographic specificity of basal ganglia output or generalized spreading of activity due to a loss in regional specialization (Bergman et al., 1998; Bronfeld and Bar-Gad, 2011; Pessiglione et al., 2005), which may be the result of increased neural nose (Matell and Meck, 2004).

In a positron emission tomography (PET) study, Jahanshahi and colleagues (Jahanshahi et al., 2010) also found hyperactivation of the cerebellum OFF medication during synchronization and continuation (1000ms isochronous interval), despite normal timing performance, irrespective of medication. In addition, hypoactivity of frontal, temporal, and parietal regions was generally alleviated by medication, whereas hypoactivity of the caudate was not. To explore whether medication altered interactions between the caudate and the whole brain, the psychophysiological interaction (PPI) method was used to test the effective connectivity of the caudate with the whole brain. In the PPI analysis, the time course of a seed region (caudate) is correlated with the time courses of other brain voxels to determine if it differs as a function of medication. The main results showed that effective connectivity between the caudate and the cerebellum was stronger OFF than ON medication, which is compatible with the prospect of timing-related compensatory activation by the cerebellum (Cerasa et al., 2006; Yu et al., 2007). Interestingly, effective connectivity of the caudate with elements of the motor circuit (putamen and SMA) was also stronger OFF medication, which may relate to excessive synchronicity in corticostriatal circuits after dopamine depletion in animal models (Costa et al., 2006). In contrast, stronger connectivity of the caudate with prefrontal areas was found ON than OFF medication, possibly reflecting greater frontostriatal modulation of performance by cognitive-control networks. Altogether, the results suggest that timing in PD is associated with altered patterns of striatal connectivity with the prefrontal cortex and classic motor systems, the strength of which depends on dopamine therapy.

Although the prospect of degeneracy in timing mechanisms is intriguing, the above interpretations should be tempered by the fact that the mechanisms of effective connectivity are not well understood, nor is their relationship to regional patterns of hyper- and hypoactivity. More research is needed to determine if strengthening or weakening of connectivity in PD has a direct influence on performance or signifies dedifferentiation of activation due to reduced regional specialization. It is also unknown whether cerebellar hyperactivity or changes in striatal-cerebellar connectivity are due to disturbances in timing per se or motor-control processes. For example, the cerebellum is also hyperactive in PD during the performance of self-initiated and automatic movements (Wu and Hallett, 2005; Wu et al., 2010), presumably due to deficits in motor control. Moreover, motor symptoms in PD correlate with altered effective connectivity of the cerebellum, basal ganglia, and SMA during movement (Wu et al., 2011; Wu, Chan, and Hallett, 2010). Thus, patterns of connectivity in classic motor areas may partly relate to disturbances in motor-control processes due to the significant motor-output component of the paced-tapping task.

4.1.2 Perceptual Timing

With this in mind, we used fMRI to investigate brain systems that govern time perception deficits in PD and to examine the effect of dopamine therapy on neurocognition (Harrington et al., 2011b). Participants included 21 volunteers with PD and 19 older adult controls. Figure 8.2A illustrates the time discrimination task, in which a standard interval (SI) was presented, followed by a delay period and then a comparison interval (CI). The participant judged if the CI was longer or shorter in duration than the SI. Because temporal processing

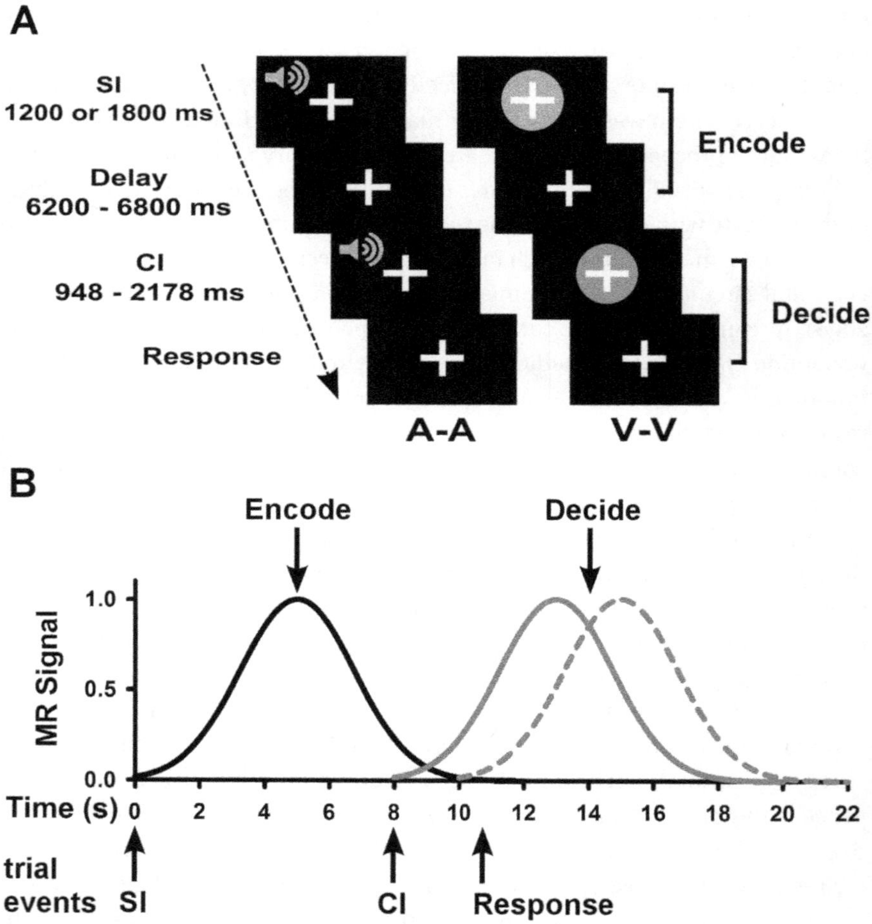

FIGURE 8.2 *Time perception paradigm. A: Illustration of the trial events in the time perception task. A standard and a comparison interval were successively presented and separated by a delay. The standard (SI) was 1200 or 1800ms and respectively pegged to a delay of 6800 or 6200ms. Three shorter and three longer comparison intervals (CI) were ±7% increments of each SI. Intervals were designated by filled tones or a blue sphere. B: The three hypothetical time-course functions illustrate the expected MR signal associated with encoding the standard interval (black curve), encoding the comparison interval (solid gray curve), and making a response (dotted gray curve). Arrows leading from each trial event designate their onset. The hemodynamic response peaks 4 to 6 s after the onset of the events. An image of the entire brain is acquired every 2 s. The fixation cross is displayed throughout the task. Figure adapted from Harrington and colleagues (2011b).*

unavoidably engages a host of cognitive processes, we sought to deconstruct component processes by separating activation associated with encoding the SI from activation associated with encoding the CI and making a decision. This can be done by inserting a long delay period (6200 and 6800ms) between the end of the SI and the onset of the CI (Harrington et al., 2004b). Figure 8.2B illustrates a hypothetical hemodynamic response function (HRF), which peaks 4 to 6 s after the onset of an event (e.g., SI, CI, response). By acquiring an image of the whole brain every 2 s, the HRF for the first 12 s of the trial constituted brain activity associated with encoding the SI and holding it in memory (encoding phase) and the last 12 s of the trial constituted brain activity associated with encoding the CI and judging its duration relative to the SI (decision phase). Both phases were assumed to engage timing. We reasoned that the encoding phase would also engage working memory, whereas the decision phase would engage executive or decisional processes involved in comparing the two intervals. Activation was compared to rest (i.e., fixation plus ambient noise) to better evaluate potential dysfunction in cognitive and sensory systems that normally support timing (Bueti and Macaluso, 2011).

Time perception performance during fMRI was impaired in PD, irrespective of the signal modality of intervals and dopamine therapy. In addition, volumetric analyses of anatomical MRIS revealed no significant cortical or basal ganglia atrophy in the PD group. Figure 8.3 displays areas of activation that were associated with temporal processing in both groups during the two phases of the trial. Areas color-coded blue showed normal activation in the PD OFF condition, whereas regions color-coded red showed significant group differences in activation. Despite similar patterns of regional activation during both phases of the trial, the figure shows that abnormal regional activation in PD largely depended on context-specific processes. An exception was striatal activation (caudate and putamen), which typically was hypoactive in the PD OFF condition during both phases. In the encoding period, hypoactivation was also found in the motor circuit (putamen, preSMA/SMA and cingulate, premotor cortex), an attention/working memory network (caudate, middle-frontal gyrus, inferior parietal cortex, lateral cerebellum), a memory center (parahippocampus), an attention and integration center (insula), and a sensory processing hub (vermis). Most of these regions are commonly associated with timing in young adults (Bueti and Macaluso, 2011; Coull, Nazarian, and Vidal, 2008; Coull et al., 2004; Harrington et al., 2004b; Harrington et al., 2010; Koch et al., 2004b; Melgire et al., 2005; Pouthas et al., 2005; Rao, Mayer, and Harrington, 2001; Wiener et al., 2012; Wittmann et al., 2010). In contrast, in the decision phase, we found hypoactivation in the PD OFF condition only in elements of a memory retrieval hub (posterior cingulate, parahippocampus) and in the

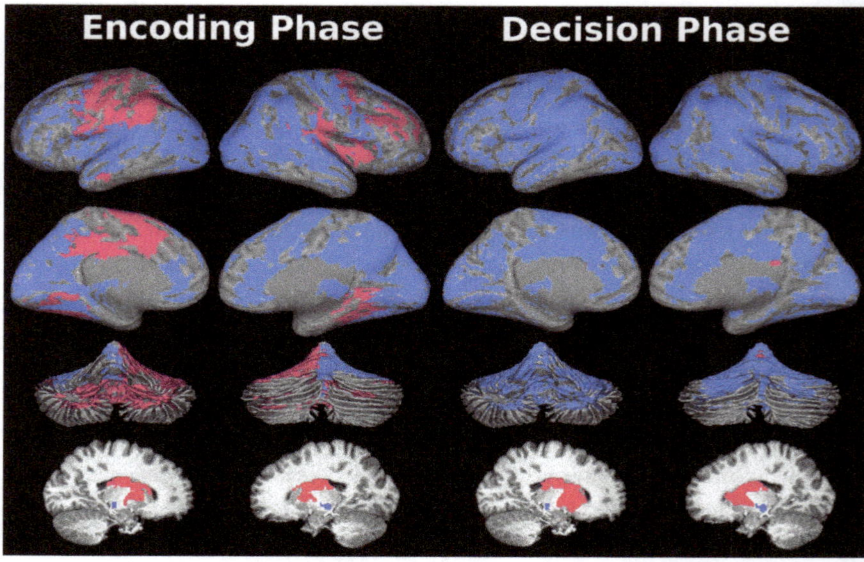

FIGURE 8.3 *Regional analyses of brain activation in PD for the encoding and decision phases. Functional regions of interest (ROI) were derived by conjoining timing-related activation from the control and PD groups. Blue areas blue signify regions in which activation did not differ between the groups. Red areas showed hypoactivity in the PD OFF group relative to the control group. Brain activation is projected onto the lateral (row 1) and medial (row 2) surfaces of the left and right hemispheres, the anterior and posterior surfaces of the cerebellum (row 3), and the left and right basal ganglia (row 4). Brain sections are displayed in neurological view. Figure adapted from Harrington and colleagues (2011b).*

vermis. These results resonate with the distinct patterns of timing dysfunction in PD that have been attributed to the storage and retrieval to temporal memories (Koch et al., 2004a; Malapani et al., 2002). Our results suggest the possibility that memory storage and retrieval are related to dysfunction in different brain circuits in PD. Interestingly, the cerebellum was hypoactive in PD OFF during both phases of the time perception task. This may suggest that cerebellar hyperactivity during motor timing (Cerasa et al., 2006; Jahanshahi et al., 2010; Yu et al., 2007) relates to motor control functions, rather than timing per se. Additional research is needed to test this proposal.

As for dopamine therapy, it had rather circumscribed effects on regional activation. In the encoding phase, it normalized activation in elements of the limbic system (insula, parahippocampus) and the vermis, but otherwise had no effect on cortical or striatal activation. In the decision phase, dopamine therapy slightly improved putamen activation and normalized vermis activity. Normalization of vermis activity during both phases may come about via the striatal-cerebellar pathways (Figure 8.1C), although this was not sufficient to improve timing performance.

The rather limited effect of dopamine therapy on brain activation may be related to a host of factors discussed above. Another important consideration is that conventional regional analyses of brain activation are insensitive to changes in brain networks (Rowe, 2010). Since temporal processing depends on communication of the striatum with other brain regions, it is crucial to study striatal connectivity. We therefore used the PPI method to explore whether dopamine therapy altered the effective connectivity of the striatum (caudate and putamen) with the cortex and the cerebellum. These analyses were conducted separately for activation during the encoding and decision phases of a trial. The results showed that striatal connectivity was modulated by dopamine only in the decision phase, which concurred with its effect on striatal activation during this phase only. This finding does not mean that corticostriatal connectivity is absent during the encoding phase, only that it is not altered by medication. Figure 8.4 shows that striatal connectivity was stronger OFF than ON medication with areas of the motor circuit (green), the frontoparietal attention network (purple), and the insula (blue). This effect is consistent with reports that dopamine depletion produces excessive spontaneous synchronicity in corticostriatal circuits (Costa et al., 2006; Gatev, Darbin, and Wichmann, 2006; Hammond, Bergman, and Brown, 2007), which may hinder flexible updating and integration by the striatum in contexts that call for cognitive flexibility (Lustig et al., 2005). The finding also comports with findings of stronger resting-state connectivity in PD OFF than ON medication (Baudrexel et al., 2011; Kwak et al., 2010; Stoffers et al., 2008). In contrast, left putamen connectivity was stronger ON than OFF medication with the left superior frontal gyrus (red), possibly signifying compensatory processing in some frontostriatal networks. Interestingly, the putamen and caudate did not show medication-modulated effective connectivity with the cerebellum, as the caudate does for motor timing (Jahanshahi et al., 2010).

Although it is not clear why dopamine altered striatal connectivity in the decision, but not the encoding phase, a similar result was recently reported in an fMRI study that examined regional activation during the encoding and the reproduction of intervals ranging between 5 and 11.89 s (Dusek et al., 2012). One clue may come from the work of Cools and colleagues (Cools, 2006) who proposed that mesocortical dopamine stabilizes cortical oscillatory activity, which in our time perception task is important during the encoding phase where interval durations must be maintained in working memory. An implication of this model is that cortico-cortico connectivity might benefit from dopamine therapy during interval encoding, which we did not explore. In contrast, nigrostriatal dopamine is thought to improve updating and integrative functions of basal ganglia networks, which is emphasized during the decision phase of our task. Though speculative, this model might advance an understanding of seemingly discrepant effects of dopamine therapy across different tasks, because it considers the demands placed on cognitive stability and flexibility control-processes, which respectively engage the cortex and striatum.

4.2 *Functional Imaging Studies of Huntington Disease*

4.2.1 Timed Movements

Since striatal atrophy is seen decades before a manifest diagnosis, it is of keen interest to delineate the brain networks that are associated with motor timing deficits in prodromal HD and to ascertain the relationship to disease burden. Structural changes in the striatum, but also the cortex, are known to correlate with motor timing proficiency. In a combined sample of prodromal HD and early manifest HD individuals, striatal atrophy and cortical thinning on the lateral and medial surfaces of the rostral-frontal and occipital-parietal cortices correlated with changes in paced-tapping accuracy to a metronome (550ms isochronous interval; Bechtel et al., 2010). Though the strong association with thinning in the occipital cortex was unexpected, the finding may reflect disease progression rather than a functional association, because occipital cortex thinning begins decades before a diagnosis and progresses over time (Nopoulos et al., 2010). At the same time, structural changes are not necessarily functionally significant, since some degree of atrophy or thinning may not alter functioning.

For these reasons, functional imaging investigations are vital. To date, there has been only one functional imaging investigation into the brain circuits that govern motor timing deficits in prodromal HD (Zimbelman et al., 2007). Using the paced-tapping task (600ms isochronous interval), we sought to identify the brain systems associated with motor timing deficits and to determine if neurocognitive dysfunction was related to disease burden. Prodromal HD participants were divided into two groups of 13 participants each based on their genetic testing (CAG repeat length) and age. One group was estimated to be further (FAR) from diagnosis (more than 12 years to onset) and the other group was closer (CLOSE) to a diagnosis (less than 12 years to estimated onset). Both groups were compared to a healthy control group. Timing variability was greater in the CLOSE, but not the FAR group during the synchronization and the continuation phases of the task. Volumetric analyses showed significant caudate atrophy in the FAR group and significant caudate and putamen atrophy in the CLOSE group. No group differences were found in cortical atrophy.

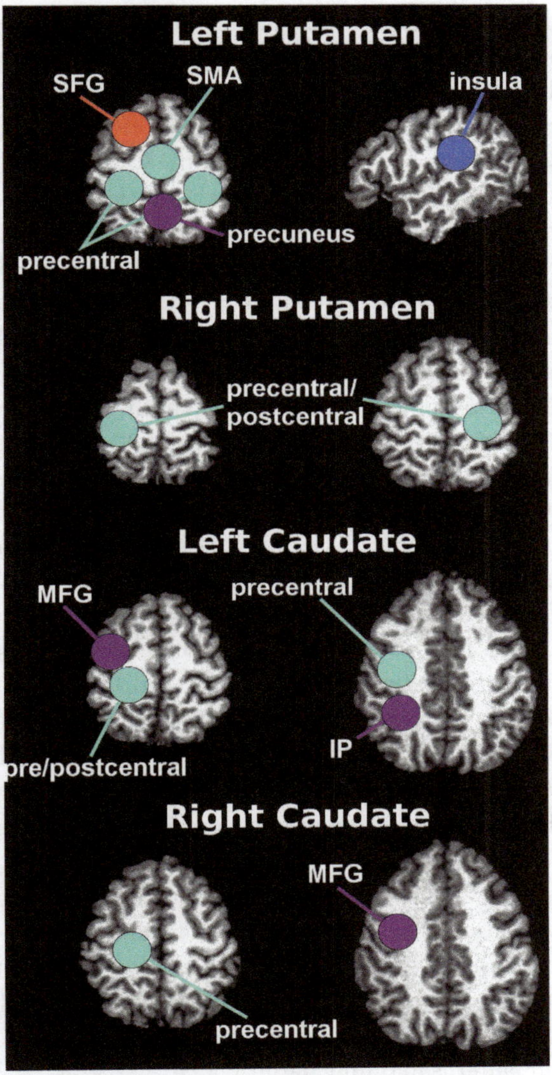

FIGURE 8.4 *Cortical regions showing dopamine-modulated connectivity with the striatum during the decision phase. For each striatal seed region, effective connectivity of theputamen and caudate is illustrated on axial or sagittal brain sections, which are displayed in neurological view. Striatal connectivity was stronger OFF than ON medication with elements of the motor circuit (green areas), the frontoparietal working-memory network (purple areas), and the limbic system (insula). Striatal connectivity was stronger ON than OFF medication with the superior frontal gyrus.* IP *= inferior parietal cortex;* MFG *= middle frontal gyrus;* SFG *= superior frontal gyrus;* SMA *= supplementary motor area. Figure adapted from Harrington and colleagues (2011b).*

For the fMRI analyses, the synchronization and continuation phases were combined and compared to a rest condition. Figure 8.5 (top) displays the regions that showed significant activation in the timing task for one or more groups. The graphs show the MR signal change in selected regions in each group. The main results revealed hypoactivation in the CLOSE group relative to the control and the FAR groups in the motor circuit (left putamen, SMA, preSMA, cingulate motor area; Figure 8.5, FAR = Controls > CLOSE), the left anterior insula, and the right inferior frontal gyrus. All of these brain centers are known to exhibit timing-related activation in young adults (Bueti, van Dongen, and Walsh, 2008; Coull et al., 2008; Harrington et al., 2010). In addition, the FAR group showed two patterns of functional changes. First, there was a stepwise reduction in activation with disease burden in the right anterior cingulate and right anterior insula (Figure 8.5, Controls > FAR > CLOSE), suggesting early functional decline in components of the limbic system that support attention and executive functions. The FAR group also exhibited hyperactivation relative to the control and the CLOSE groups in sensory and in motor areas including the sensorimotor cortex, the precentral gyrus, medial frontal gyrus, superior temporal gyrus and the right cerebellum (Figure 8.5, FAR > Controls = CLOSE). This result may signify compensatory responses in early stages of neurodegeneration, similar to PD during motor timing (Jahanshahi et al., 2010; Yu et al., 2007). Hyperactivation could be an intermediate phenotype of cell dysfunction, which begins long before cell death (Tobin and Signer, 2000). Because the close group did not exhibit hyperactivation in any brain regions, this purported compensatory mechanism may weaken as the neurodegenerative process advances. Indeed, a nonlinear trajectory of activation across the continuum of cognitive impairment (i.e., increases and decreases in brain activation) has been observed for mild cognitive impairment and Alzheimer's disease (Celone et al., 2006). Nevertheless, it is also important to consider that the findings may be an early sign of dedifferentiation. To tease apart these explanations, longitudinal studies are needed that relate brain-activation patterns to timing performance.

In contrast to PD (Harrington et al., 2011b), significant caudate and/or putamen atrophy was found in the prodromal HD groups. Striatal atrophy is one of the better predictors of disease prognosis in prodromal HD (Aylward et al., 2012). Thus, one question is whether measures of cortical functioning better distinguish controls from the different prodromal HD groups, beyond other known predictors of disease prognosis including striatal atrophy, executive functioning on neuropsychological tests, and timing (accuracy and variability). We therefore performed a hierarchical discriminant analysis to examine group classification accuracy using striatal atrophy, task performance, and

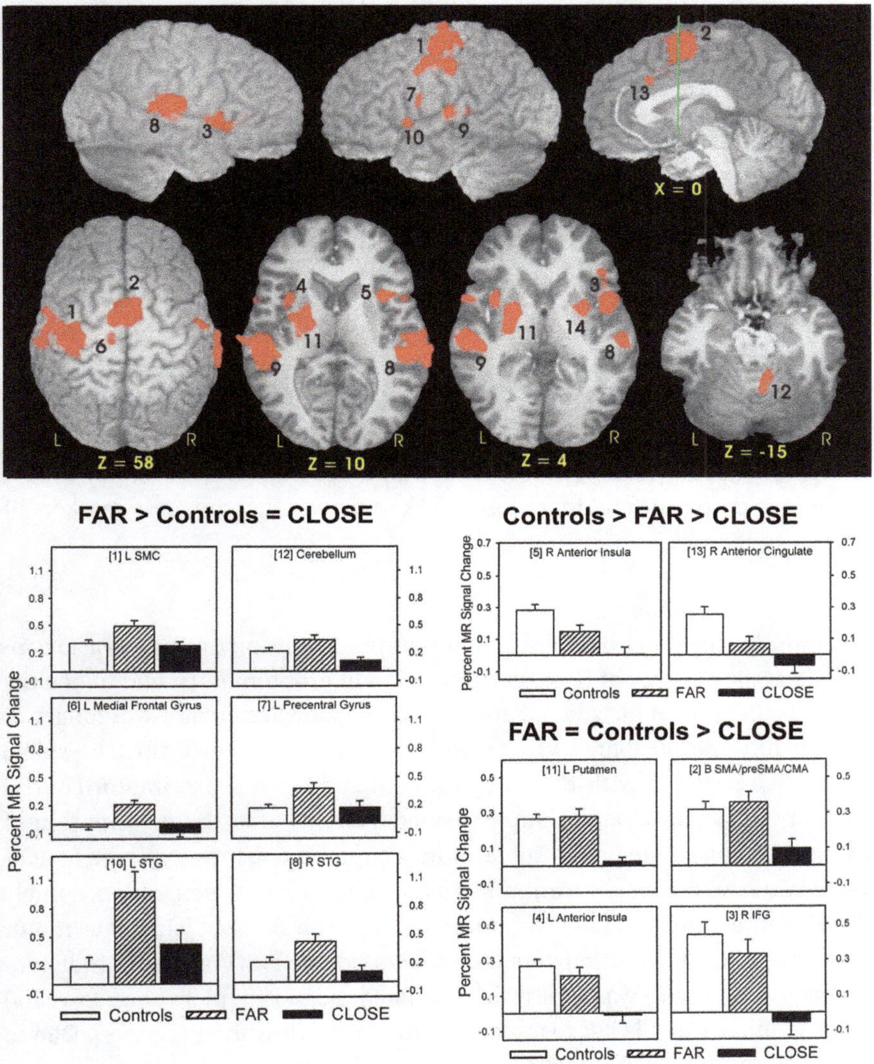

FIGURE 8.5 *Regions showing abnormal activation during motor timing in prodromal HD. Top: Map of the functional regions of interest, which were derived from a conjunction map that was used to test for group differences in signal intensity. Green line in the sagittal image is perpendicular to the anterior-posterior commissure. Z coordinate = mm superior to the anterior commissure -posterior commissure line. Bottom: The graphs display the percent MRI signal change for the control, FAR, and CLOSE groups. Three different patterns of activation were uncovered: 1) FAR > Controls = CLOSE; 2) Control > FAR > CLOSE; and 3) FAR = Controls > CLOSE. B = bilateral hemispheres; L = left hemisphere; R = right hemisphere. Figure adapted from Zimbelman and colleagues (2007).*

fMRI variables as predictors. Classification accuracy was 60.7% for caudate and putamen volumes; the cognitive and timing measures did not add to the classification accuracy. However, MRI signal intensity in the left medial frontal gyrus, right cerebellum, and right anterior insula increased classification accuracy to 86.8% (83.3, 84.6, and 92.3% for the control, FAR, and CLOSE groups, respectively). Recall that in the FAR group two of these regions showed hyperactivity (left medial frontal gyrus and right cerebellum), whereas the right anterior insula was hypoactive. Thus, the addition of these variables improved classification the most for the FAR group (46.2% without fMRI; 84.6% with fMRI). Though hyperactivity in motor regions may relate to early compensatory mechanisms, the insula is associated with timing-related activation in young adults (Harrington et al., 2010; Kosillo and Smith, 2010; Wittmann et al., 2010a,b) and mediates the synthesis of temporal information across the senses (Harrington et al., 2011a). Altogether, our findings indicate that motor timing probes for corticostriatal functioning in key centers that are early markers of disease prognosis.

4.2.2 Perceptual Timing

To minimize the potential role of motor output factors on timing, we also conducted an fMRI study of time discrimination in prodromal HD and in healthy controls (Paulsen et al., 2004). Participants judged whether a CI was longer or shorter in duration than a SI (1200ms) as they underwent fMRI. The single intensity associated with this task was compared to a sensorimotor control task, in which participants simply pressed a button after the presentation of two isochronous tone pairs. The fMRI measure of interest was volume of activated tissue within regions commonly associated with time perception, namely the thalamus, the caudate, the putamen, and the preSMA/cingulate motor area. Prodromal HD participants were separated into two groups of 7 individuals each. One group was FAR from estimated diagnosis (more than 12 years) and the other was CLOSE to a manifest diagnosis (less than 12 years). Due to the small sample sizes, direct comparisons between the groups were not conducted for the fMRI analyses. The behavioral results revealed that time discrimination accuracy was reduced in the CLOSE group, but normal in the FAR group relative to the controls. Significant atrophy was also found in the CLOSE, but not the FAR group.

Figure 8.6 displays the spatial extent of activation in the regions of interest for each group, wherein the activated volume was greater for the timing than the control task. Despite normal timing performance and an absence of striatal atrophy, individuals in the FAR group demonstrated a larger spatial extent of activation in the preSMA and the cingulate motor area relative to the control group (Figure 8.6, bottom graph). We speculated that this finding may reflect compensation for early striatal dysfunction. Although this result cannot be directly compared to findings from our study of motor timing (Zimbelman et al., 2007), presumed compensatory activation in the FAR group may differ for time perception than for motor timing (sensorimotor areas) due to differences between the timing tasks in the emphasis placed on context-specific processes. In addition, the spatial extent of thalamus and striatal activation was reduced in the FAR group, whereas the spatial extent of activation in all regions of interest was reduced in the CLOSE group. These preliminary results would be strengthened by larger samples of prodromal HD participants and by conducting direct group comparisons on MR signal intensity across all regions that demonstrate timing-related activation. However, they suggest that time perception deficits in prodromal HD are related to dysfunction in elements of the motor circuit, just as they are in PD for motor timing and time perception (Cerasa et al., 2006; Elsinger et al., 2003; Harrington et al., 2011b; Yu et al., 2007).

5 Clinical Implications and Future Directions

Knowledge of the neural mechanisms associated with temporal processing disturbances in PD and HD is beginning to inform an understanding of normal timing, wherein converging findings support the view that timing emerges from interactions of the striatum with extensive brain networks that govern cognitive-control and sensory processing (Bueti, 2011; Merchant et al., 2013). There is also mounting evidence that disturbances in timing have clinical significance. In PD, freezing of gait and finger movements benefit from predictable pacing cues (McIntosh et al., 1997; Spildooren et al., 2012; Thaut et al., 1996; Vercruysse et al., 2012). When cues are withdrawn, motor control deteriorates resulting in smaller movement amplitudes, a hastening or more variable frequency of movements, and decreased coordination stability. Isochronous auditory or visual cues also improve gait in PD without freezing (Georgiou et al., 1993; Lee et al., 2012; Majsak et al., 1998; Nieuwboer et al., 2009). The motor circuit normally governs the internal generation of beat processing (Grahn, 2009; Grahn and Rowe, 2009), which is impaired in PD (Grahn and Brett, 2009). One proposal is that the effective utilization of external, predictable temporal cues in PD may be due to bypassing the frontostriatal system in favor of cerebellar-frontoparietal pathways (Kotz and Schwartze, 2011), which also support beat-based timing (Penhune, Zatorre, and Evans, 1998). This explanation may also account for the greater ease in performing externally- guided than self-initiated movements, the latter of which also

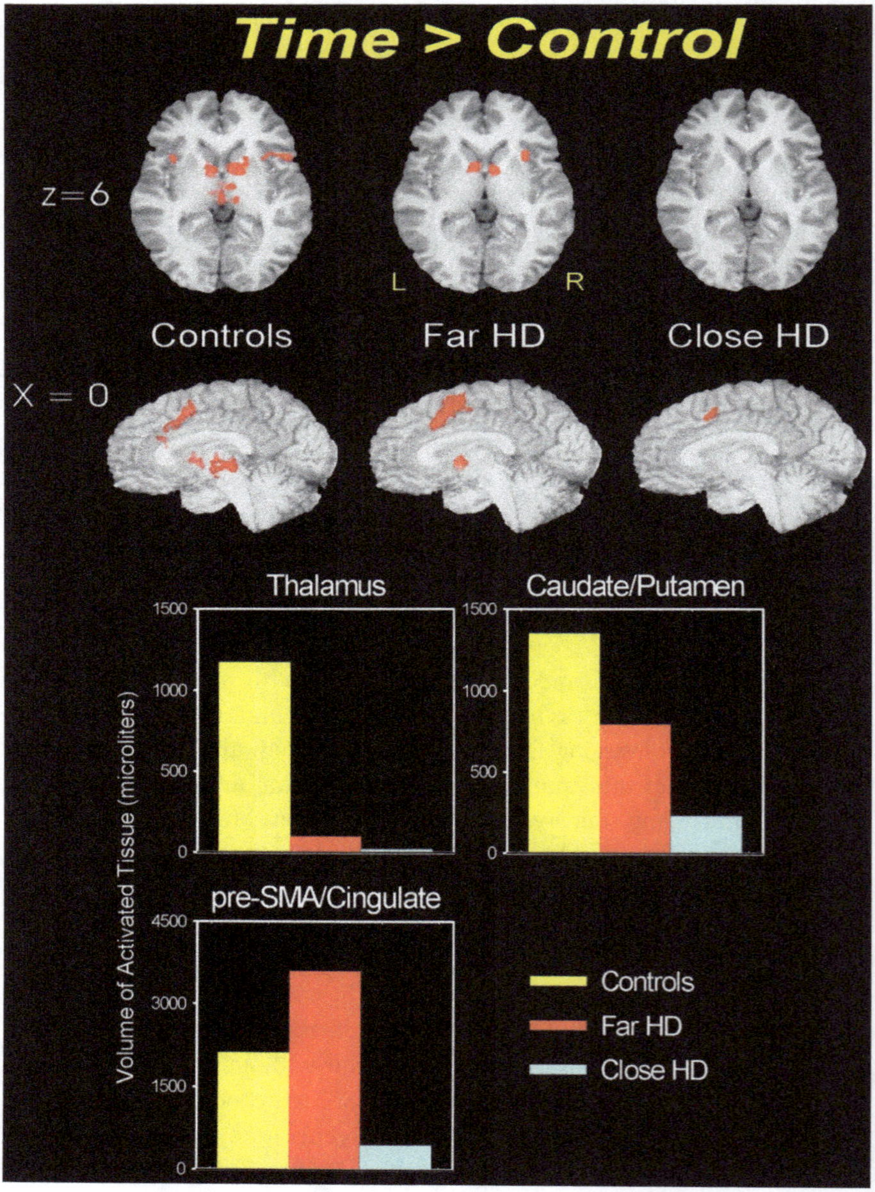

FIGURE 8.6 *Regions showing abnormal activation volume during time discrimination in prodromal* HD. *Top: Activation foci* ($p < 0.01$) *were derived from the timing minus control task comparison for the control, FAR, and CLOSE groups. L = left hemisphere; R = right hemisphere; z coordinate = mm superior to the anterior commissure-posterior commissure line. Bottom: Volume of activation in the thalamus, caudate/putamen, and the preSMA/cingulate. Figure adapted from Paulsen and colleagues* (2004).

emphasize motor circuit functioning (Cunnington et al., 1995; Jahanshahi et al., 1995; Wu et al., 2011). An alternative explanation is that predictably-timed external cues trigger the execution of a motor plan in premotor cortex, which then suppresses pathological activity in the STN, thereby facilitating movement (Amirnovin et al., 2004; Kuhn et al., 2004). Recent support for this proposal comes from the finding of reduced beta oscillations and increased direction tuning in the STN of individuals with PD when a motor plan was triggered by predictable, but not unpredictable cues (Sarma et al., 2012). Future research is needed to evaluate these hypotheses and to determine whether other compensatory routes to action might also be engaged depending upon the clinical phenotype (e.g., freezing of gait versus no freezing; Snijders et al., 2011). Altogether, this body of work underscores the clinical importance of internal timing disturbances in PD and suggests that timing is closely linked to motor planning disturbances, which are found both in PD and HD (Georgiou et al., 1994, 1995; Harrington and Haaland, 1991; Stout et al., 2011; Wu et al., 2011).

Temporal processing disturbances are also of clinical importance in prodromal HD, where timing ability is a strong marker of disease burden (Rowe et al., 2010; Stout et al., 2011). However, there are also subtle changes in a myriad of other functions including attention, working memory, processing speed, emotion processing, motor speed, and olfaction (Duff et al., 2010; Johnson et al., 2007; Paulsen et al., 2008; Pirogovsky et al., 2007; Say et al., 2011; Stout et al., 2011). With the development of treatments that putatively delay the onset of clinical signs and slow the progression of pathological changes, there has been a concerted effort to determine when changes on different measures can be reliably detected so as to evaluate their potential for outcomes in clinical trials. To this end, we used factor analysis to characterize cognitive domains underlying performances in prodromal HD on a large test battery that included the paced tapping task, and then evaluated their sensitivity in predicting time to diagnosis using a survival analysis (Harrington et al., 2012). Six factors were identified including speed-inhibition, working memory, motor planning, attention-information integration, sensory-perceptual processing, and declarative memory. Almost all tests clearly loaded on one of the six factors, except the paced-tapping task, which loaded on both the motor-planning factor (e.g., choice RT and planning tasks) and the sensory-perceptual factor (e.g., emotion and olfactory processing tasks). This finding was conceptually meaningful since timing proficiency is a facet of planning and perception. Importantly, only the motor-planning and sensory-perceptual factors uniquely predicted time to diagnosis, after controlling for measures of disease burden and motor symptoms. Although the neural basis of these results is unknown, tests of motor planning and timing both emphasize processing in frontostriatal circuits (Wu et al., 2011). In contrast, sensory-perceptual processes involved in emotion recognition, olfaction, and timing also probe for functioning in the ventral striatum and limbic systems (e.g., orbitofrontal, cingulate, and anterior insula; Fan et al., 2011; Harrington et al., 2011a; Walla, 2008). The insular cortex and cingulate motor area may be particularly important, as functioning in these regions is abnormal during motor timing in prodromal HD (Zimbelman et al., 2007). Thus, timing proficiency is an important marker of disease prognosis that may prove to be a useful outcome for clinical trials of HD treatments.

A more complete appreciation for the clinical utility of temporal processing measures in neurodegenerative disorders of the basal ganglia will likely come about as functional imaging investigations in PD and prodromal HD develop more fully. Despite current insights into the brain regions that are associated with timing disturbances in these diseases, much remains unknown about potential changes in interactions among brain regions. Regional activation, which is the conventional measure of brain dysfunction, is insensitive to communication among brain regions. Since temporal processing depends on the communication among brain regions, it will be important to study changes in the functional connectivity of brain circuits in diseases. Emerging research suggests that measures of functional connectivity are more sensitive to neurodegeneration (Rowe, 2010) and to the effects of dopamine therapy on timing in PD (Harrington et al., 2011b; Jahanshahi et al., 2010). These measures may also better elucidate potential compensatory routes to timing in basal ganglia disorders (Jahanshahi et al., 2010). In addition, there is a strong need for longitudinal studies that examine the rate of change in timing-related systems to help identify the best biomarkers of early brain-network dysfunction.

In the future, it will also be important to evaluate more complex forms of temporal processing in diseases, since they may be more sensitive to pathology. One area that has received only scant attention in functional imaging studies is intersensory timing. In naturalistic settings, temporal information from different senses is routinely combined into a single perceptual experience. We recently reported that striatum governs intersensory timing through its interactions with the cortex (Harrington et al., 2011a), which concurs with its role in multisensory integration (Nagy et al., 2006) and the integration of cortical oscillatory activity that comprises the clock signal (Matell and Meck, 2004). Another example is research into illusions of time, which are important because they reveal how the brain organizes and construes time depending of internal states (e.g., emotionally charged, high arousal; Dirnberger et al., 2012) or properties of stimuli (Harrington et al., 2011a; Wittmann et al., 2010b). This is relevant clinically because distortions in temporal illusions are beginning to elucidate mechanisms for timing in other diseases that alter dopamine system functioning such as schizophrenia (Carroll et al., 2008; Penney et al., 2005). These are merely two examples of many fascinating avenues for future research that are relevant to basal ganglia disorders. With the explosion of research into timing in healthy individuals and animal models in the last decade (Merchant et al., 2013), we are optimistic that this exciting work will soon translate into studies of neurodegenerative disorders.

Acknowledgments

This chapter was supported by funding from the Department of Veterans Affairs (1I01CX000146-01) and the NINDS (NS040068). The authors would like to thank Christopher Fong, Jason Reed, and Gabriel Castillo for their technical assistance.

References

Akkal, D., Dum, R.P., and Strick, P.L. 2007. "Supplementary Motor Area and Presupplementary Motor Area: Targets of Basal Ganglia and Cerebellar Output" *J Neurosci*, 27, 10659–73.

Allman, M.J. and Meck, W.H. 2012. "Pathophysiological Distortions in Time Perception and Timed Performance" *Brain*, 135, 656–77.

Amirnovin, R. et al. 2004. "Visually Guided Movements Suppress Subthalamic Oscillations in Parkinson's Disease Patients" *J Neurosci*, 24, 11302–6.

Artieda, J. et al. 1992. "Temporal Discrimination is Abnormal in Parkinson's Disease" *Brain*, 115, 199–210.

Aylward, E.H. et al. 2012. "Striatal Volume Contributes to the Prediction of Onset of Huntington Disease in Incident Cases" *Biol.Psychiatry*, 71, 822–8.

Aylward, E.H. et al. 2011. "Longitudinal Change in Regional Brain Volumes in Prodromal Huntington Disease" *J.Neurol.Neurosurg.Psychiatry*, 82, 405–10.

Balci, F. et al. 2008. "Pharmacological Manipulations of Interval Timing Using the Peak Procedure in Male C3H Mice" *Psychopharmacology* (*Berl*), 201, 67–80.

Balci, F. et al. 2012. "Epistasis Effects of Dopamine Genes on Interval Timing and Reward Magnitude in Humans" *Neuropsychologia*.

Baudrexel, S. et al. 2011. "Resting State fMRI Reveals Increased Subthalamic Nucleus-Motor Cortex Connectivity in Parkinson's Disease" *Neuroimage.*, 55, 1728–38.

Bechtel, N. et al. 2010. "Tapping Linked to Function and Structure in Premanifest and Symptomatic Huntington Disease" *Neurology*, 75, 2150–60.

Bergman, H. et al. 1998. "Physiological Aspects of Information Processing in the Basal Ganglia of Normal and Parkinsonian Primates" *Trends in Neuroscience*, 21, 32–8.

Beste, C. et al. 2007. "Time Processing in Huntington's Disease: A Group-Control Study" *PLoS.ONE.*, 2, e1263.

Bostan, A.C., Dum, R.P., and Strick, P.L. 2010. "The Basal Ganglia Communicate with the Cerebellum" *Proc.Natl.Acad.Sci.U.S.A.*, 107, 8452–6.

Brinkman, R.R. et al. 1997. "The Likelihood of Being Affected with Huntington Disease by a Particular Age, for a Specific CAG Size" *Am.J.Hum.Genet.*, 60, 1202–10.

Bronfeld, M. and Bar-Gad, I. 2011. "Loss of Specificity in Basal Ganglia Related Movement Disorders" *Front Syst.Neurosci.*, 5, 38.

Bueti, D. 2011. "The Sensory Representation of Time" *Front Integr.Neurosci.*, 5, 34.

Bueti, D. and Macaluso, E. 2011. "Physiological Correlates of Subjective Time: Evidence for the Temporal Accumulator Hypothesis" *Neuroimage.*, 57, 1251–63.

Bueti, D., van Dongen, E.V., and Walsh, V. 2008. "The Role of Superior Temporal Cortex in Auditory Timing" *PLoS.ONE.*, 3, e2481.

Buhusi, C.V. and Meck, W.H. 2002. "Differential Effects of Methamphetamine and Haloperidol on the Control of an Internal Clock" *Behav.Neurosci.*, 116, 291–7.

Carbon, M., Edwards, C., and Eidelberg, D. 2003. "Functional Brain Imaging in Parkinson's Disease" *Adv.Neurol*, 91, 175–81.

Carroll, C.A. et al. 2008. "Temporal Processing Dysfunction in Schizophrenia" *Brain and Cognition*, 67, 150–61.

Celone, K.A. et al. 2006. "Alterations in Memory Networks in Mild Cognitive Impairment and Alzheimer's Disease: An Independent Component Analysis" *J.Neurosci.*, 26, 10222–31.

Cerasa, A. et al. 2006. "Functional Changes in the Activity of Cerebellum and Frontostriatal Regions during Externally and Internally Timed Movement in Parkinson's Disease" *Brain Res Bull.*, 71, 259–69.

Clower, D.M., Dum, R.P., and Strick, P.L. 2005. "Basal Ganglia and Cerebellar Inputs to 'AIP'" *Cereb.Cortex*, 15, 913–20.

Cools, R. 2006. "Dopaminergic Modulation of Cognitive Function-Implications for l-DOPA Treatment in Parkinson's Disease" *Neurosci Biobehav.Rev.*, 30, 1–23.

Costa, R.M. et al. 2006. "Rapid Alterations in Corticostriatal Ensemble Coordination during Acute Dopamine-Dependent Motor Dysfunction" *Neuron*, 52, 359–69.

Coull, J.T., Cheng, R.K., and Meck, W.H. 2011. "Neuroanatomical and Neurochemical Substrates of Timing" *Neuropsychopharmacology*, 36, 3–25.

Coull, J.T., Nazarian, B., and Vidal, F. 2008. "Timing, Storage, and Comparison of Stimulus Duration Engage Discrete Anatomical Components of a Perceptual Timing Network" *J Cogn Neurosci*, 20, 2185–97.

Coull, J.T. et al. 2004. "Functional Anatomy of the Attentional Modulation of Time Estimation" *Science*, 303, 1506–8.

Cunnington, R. et al. 1995. "Movement-Related Potentials in Parkinson's Disease. Presence and Predictability of Temporal and Spatial Cues" *Brain*, 118 (Pt 4), 935–50.

Dirnberger, G. et al. 2012. "Give It Time: Neural Evidence for Distorted Time Perception and Enhanced Memory Encoding in Emotional Situations" *Neuroimage*, 63, 591–9.

Duff, K. et al. 2010. "Mild Cognitive Impairment in Prediagnosed Huntington Disease" *Neurology*.

Dumas, E.M. et al. 2012. "Early Changes in White Matter Pathways of the Sensorimotor Cortex in Premanifest Huntington's Disease" *Hum.Brain Mapp.*, 33, 203–12.

Dusek, P. et al. 2012. "Abnormal Activity in the Precuneus during Time Perception in Parkinson's Disease: An fMRI Study" *PLoS.ONE.*, 7, e29635.

Elsinger, C.L. et al. 2003. "Neural Basis for Impaired Time Reproduction in Parkinson's Disease: An fMRI Study" *J Int.Neuropsychol.Soc.*, 9, 1088–98.

Fan, J. et al. 2011. "Involvement of the Anterior Cingulate and Frontoinsular Cortices in Rapid Processing of Salient Facial Emotional Information" *Neuroimage*, 54, 2539–46.

Freeman, J.S. et al. 1996. "Abnormalities of Motor Timing in Huntington's Disease" *Parkinsonism and Related Disorders*, 2, 81–93.

Gatev, P., Darbin, O., and Wichmann, T. 2006. "Oscillations in the Basal Ganglia under Normal Conditions and in Movement Disorders" *Mov Disord.*, 21, 1566–77.

Georgiou, N. et al. 1993. "An Evaluation of the Role of Internal Cues in the Pathogenesis of Parkinsonian Hypokinesia" *Brain*, 116 (Pt 6), 1575–87.

Georgiou, N. et al. 1994. "Reduction in External Cues and Movement Sequencing in Parkinson's Disease" *J.Neurol.Neurosurg.Psychiatry*, 57, 368–70.

Georgiou, N. et al. 1995. "Reliance on Advance Information and Movement Sequencing in Huntington's Disease" *Mov Disord.*, 10, 472–81.

Gibbon, J. 1977. "Scalar Expectancy Theory and Weber's Law in Animal Timing" *Psychological Review*, 84, 279–325.

Gibbon, J., Church, R.M., and Meck, W.H. 1984. "Scalar Timing in Memory" *Annals of the New York Academy of Sciences*, 423, 52–77.

Grahn, J.A. 2009. "The Role of the Basal Ganglia in Beat Perception: Neuroimaging and Neuropsychological Investigations" *Ann.N.Y.Acad.Sci.*, 1169, 35–45.

Grahn, J.A. and Brett, M. 2009. "Impairment of Beat-based Rhythm Discrimination in Parkinson's Disease" *Cortex*, 45, 54–61.

Grahn, J.A. and Rowe, J.B. 2009. "Feeling the Beat: Premotor and Striatal Interactions in Musicians and Nonmusicians during Beat Perception" *J.Neurosci.*, 29, 7540–8.

Hammond, C., Bergman, H., and Brown, P. 2007. "Pathological Synchronization in Parkinson's Disease: Networks, Models and Treatments" *Trends Neurosci.*, 30, 357–64.

Harrington, D.L. and Haaland, K.Y. 1991. "Sequencing in Parkinson's Disease: Abnormalities in Programming and Controlling Movement" *Brain*, 114, 99–115.

Harrington, D.L., Haaland, K.Y., and Hermanowicz, N. 1998. "Temporal Processing in the Basal Ganglia" *Neuropsychology*, 12, 3–12.
Harrington, D.L., Haaland, K.Y., and Knight, R.T. 1998. "Cortical Networks Underlying Mechanisms of Time Perception" *J Neurosci*, 18, 1085–95.
Harrington, D.L. et al. 2004a. "Does the Representation of Time Depend on the Cerebellum?: Effect of Cerebellar Stroke" *Brain*, 127, 561–74.
Harrington, D.L. et al. 2004b. "Neural Representation of Interval Encoding and Decision Making" *Cognitive Brain Research*, 21, 193–205.
Harrington, D.L. et al. 2010. "Neural Modulation of Temporal Encoding, Maintenance, and Decision Processes" *Cereb.Cortex*, 20, 1274–85.
Harrington, D.L. et al. 2011a. "Neural Underpinnings of Distortions in the Experience of Time across Senses" *Front Integr.Neurosci.*, 5, 32.
Harrington, D.L. et al. 2011b. "Neurobehavioral Mechanisms of Temporal Processing Deficits in Parkinson's Disease" *PLoS.ONE.*, 6, e17461.
Harrington, D.L. et al. 2012. "Cognitive Domains that Predict Time to Diagnosis in Prodromal Huntington Disease" *J.Neurol.Neurosurg.Psychiatry*, 83, 612–9.
Hinton, S.C. et al. 2007. "Motor Timing Variability Increases in Preclinical Huntington's Disease Patients as Estimated Onset of Motor Symptoms Approaches" *J.Int. Neuropsychol.Soc.*, 13, 539–43.
Hohn, S. et al. 2011. "Behavioral and in Vivo Electrophysiological Evidence for Presymptomatic Alteration of Prefrontostriatal Processing in the Transgenic Rat Model for Huntington Disease" *J.Neurosci.*, 31, 8986–97.
Hoshi, E. et al. 2005. "The Cerebellum Communicates with the Basal Ganglia" *Nat. Neurosci*, 8, 1491–3.
Ivry, R.B. and Keele, S.W. 1989. "Timing Functions of the Cerebellum" *J Cogn Neurosci*, 1, 136–52.
Jahanshahi, M. et al. 1995. "Self-initiated Versus Externally Triggered Movements: 1. An Investigation Using Measurement of Regional Cerebral Blood Flow with PET and Movement-Related Potentials in Normal and Parkinson's Disease Subjects" *Brain*, 118, 913–33.
Jahanshahi, M. et al. 2010. "Dopaminergic Modulation of Striato-Frontal Connectivity during Motor Timing in Parkinson's Disease" *Brain*, 133, 727–45.
Johnson, S.A. et al. 2007. "Beyond Disgust: Impaired Recognition of Negative Emotions Prior to Diagnosis in Huntington's Disease" *Brain*, 130, 1732–44.
Jones, C.R. et al. 2008. "Basal Ganglia, Dopamine and Temporal Processing: Performance on Three Timing Tasks on and off Medication in Parkinson's Disease" *Brain and Cognition*.
Jones, C.R. et al. 2011. "Modeling Accuracy and Variability of Motor Timing in Treated and Untreated Parkinson's Disease and Healthy Controls" *Front Integr.Neurosci.*, 5, 81.

Koch, G., Oliveri, M., and Caltagirone, C. 2009. "Neural Networks Engaged in Milliseconds and Seconds Time Processing: Evidence from Transcranial Magnetic Stimulation and Patients with Cortical or Subcortical Dysfunction" *Philos. Trans.R.Soc.Lond B Biol.Sci.*, 364, 1907–18.

Koch, G. et al. 2004a. "Subthalamic Deep Brain Stimulation Improves Time Perception in Parkinson's Disease" *NeuroReport*, 15, 1071–3.

Koch, G. et al. 2004b "High-Frequency rTMS Improves Time Perception in Parkinson Disease" *Neurology*, 63, 2405–6.

Koch, G. et al. 2008. "Impaired Reproduction of Second but Not Millisecond Time Intervals in Parkinson's Disease" *Neuropsychologia*, 46, 1305–13.

Kosillo, P. and Smith, A.T. 2010. "The Role of the Human Anterior Insular Cortex in Time Processing" *Brain Struct.Funct.*, 214, 623–8.

Kotz, S.A. and Schwartze, M. 2011. "Differential Input of the Supplementary Motor Area to a Dedicated Temporal Processing Network: Functional and Clinical Implications" *Front Integr.Neurosci.*, 5, 86.

Kuhn, A.A. et al. 2004. "Event-Related Beta Desynchronization in Human Subthalamic Nucleus Correlates with Motor Performance" *Brain*, 127, 735–46.

Kwak, Y. et al. 2010. "Altered Resting State Cortico-Striatal Connectivity in Mild to Moderate Stage Parkinson's Disease" *Front Syst.Neurosci.*, 4, 143.

Lake, J.I. and Meck, W.H. 2012. "Differential Effects of Amphetamine and Haloperidol on Temporal Reproduction: Dopaminergic Regulation of Attention and Clock Speed" *Neuropsychologia*.

Lee, S.J. et al. 2012. "The Effects of Visual and Auditory Cues on Freezing of Gait in Patients with Parkinson Disease" *Am.J.Phys.Med.Rehabil.*, 91, 2–11.

Lewis, P.A. and Miall, R.C. 2003. "Brain Activation Patterns during Measurement of Sub- and Supra-second Intervals" *Neuropsychologia*, 41, 1583–92.

Lewis, S.J. et al. 2003. "Using Executive Heterogeneity to Explore the Nature of Working Memory Deficits in Parkinson's Disease" *Neuropsychologia*, 41, 645–54.

Lustig, C., Matell, M.S., and Meck, W.H. 2005. "Not 'Just' a Coincidence: Frontal-Striatal Interactions in Working Memory and Interval Timing" *Memory*, 13, 441–8.

Majsak, M.J. et al. 1998. "The Reaching Movements of Patients with Parkinson's Disease under Self-determined Maximal Speed and Visually Cued Conditions" *Brain*, 121 (Pt 4), 755–66.

Malapani, C., Deweer, B., and Gibbon, J. 2002. "Separating Storage from Retrieval Dysfunction of Temporal Memory in Parkinson's Disease" *J Cogn Neurosci*, 14, 311–22.

Maricq, A.V. and Church, R.M. 1983. "The Differential Effects of Haloperidol and Methamphetamine on Time Estimation in the Rat" *Psychopharmacology*, 79, 10–5.

Matell, M.S. and Meck, W.H. 2004. "Cortico-Striatal Circuits and Interval Timing: Coincidence Detection of Oscillatory Processes" *Cognitive Brain Research*, 21, 139–70.

Matell, M.S., Meck, W.H., and Nicolelis, M.A. 2003. "Interval Timing and the Encoding of Signal Duration by Ensembles of Cortical and Striatal Neurons" *Behavioral Neurosciences*, 117, 760–73.

Matell, M.S. et al. 2011. "A Heterogeneous Population Code for Elapsed Time in Rat Medial Agranular Cortex" *Behav.Neurosci.*, 125, 54–73.

McIntosh, G.C. et al. 1997. "Rhythmic Auditory-Motor Facilitation of Gait Patterns in Patients with Parkinson's Disease" *J Neurol Neurosurg.Psychiatry*, 62, 22–6.

Meck, W.H. 1996. "Neuropharmacology of Timing and Time Perception" *Cognitive Brain Research*, 3, 227–42.

Meck, W.H. 2006a. "Frontal Cortex Lesions Eliminate the Clock Speed Effect of Dopaminergic Drugs on Interval Timing" *Brain Res.*, 1108, 157–67.

Meck, W.H. 2006b. "Neuroanatomical Localization of an Internal Clock: A Functional Link between Mesolimbic, Nigrostriatal, and Mesocortical Dopaminergic Systems" *Brain Res*, 1109, 93–107.

Meck, W.H. et al. 2012. "Gene-Dose Dependent Effects of Methamphetamine on Interval Timing in Dopamine-Transporter Knockout Mice" *Neuropharmacology*, 62, 1221–29.

Melgire, M. et al. 2005. "Auditory/visual Duration Bisection in Patients with Left or Right Medial-Temporal Lobe Resection" *Brain and Cognition*, 58, 119–24.

Merchant, H., Harrington, D.L., and Meck, W.H. 2013. "Neural Basis of the Perception and Estimation of Time" *Annual Review of Neuroscience*, 36.

Merchant, H. et al. 2008. "Interval Timing and Parkinson's Disease: Heterogeneity in Temporal Performance" *Exp.Brain Res*, 184, 233–48.

Nagy, A. et al. 2006. "Multisensory Integration in the Basal Ganglia" *Eur.J Neurosci*, 24, 917–24.

Nieuwboer, A. et al. 2009. "Upper Limb Movement Interruptions are Correlated to Freezing of Gait in Parkinson's Disease" *Eur.J.Neurosci.*, 29, 1422–30.

Nopoulos, P.C. et al. 2010. "Cerebral Cortex Structure in Prodromal Huntington Disease" *Neurobiol.Dis.*, 40, 544–54.

O'Boyle, D.J., Freeman, J.S., and Cody, F.W.J. 1996. "The Accuracy and Precision of Timing of Self-paced, Repetitive Movements in Subjects with Parkinson's Disease" *Brain*, 119, 51–70.

Pastor, M.A. et al. 1992a. "Performance of Repetitive Wrist Movements in Parkinson's Disease" *Brain*, 115, 875–91.

Pastor, M.A. et al. 1992b. "Time Estimation and Reproduction is Abnormal in Parkinson's Disease" *Brain*, 115, 225.

Paulsen, J.S. et al. 2004. "fMRI Biomarker of Early Neuronal Dysfunction in Presymptomatic Huntington's Disease" *AJNR Am.J Neuroradiol.*, 25, 1715–21.

Paulsen, J.S. et al. 2008. "Detection of Huntington's Disease Decades before Diagnosis: The Predict-HD Study" *J Neurol Neurosurg.Psychiatry*, 79, 874–80.

Paulsen, J.S. et al. 2010. "Striatal and White Matter Predictors of Estimated Diagnosis for Huntington Disease" *Brain Res.Bull.*, 82, 201–7.

Penhune, V.B., Zatorre, R.J., and Evans, A. 1998. "Cerebellar Contributions to Motor Timing: A PET Study of Auditory and Visual Rhythm Reproduction" *J Cogn Neurosci*, 10, 752–65.

Penney, T.B. et al. 2005. "Interval-Timing Deficits in Individuals at High Risk for Schizophrenia" *Brain and Cognition*, 58, 109–18.

Perbal, S. et al. 2005. "Effects of Internal Clock and Memory Disorders on Duration Reproductions and Duration Productions in Patients with Parkinson's Disease" *Brain and Cognition*, 58, 35–48.

Pessiglione, M. et al. 2005. "An Effect of Dopamine Depletion on Decision-making: The Temporal Coupling of Deliberation and Execution" *J Cogn Neurosci*, 17, 1886–96.

Pirogovsky, E. et al. 2007. "Impairments in Source Memory for Olfactory and Visual Stimuli in Preclinical and Clinical Stages of Huntington's Disease" *J Clin.Exp. Neuropsychol.*, 29, 395–404.

Pouthas, V. et al. 2005. "Neural Network Involved in Time Perception: An fMRI Study Comparing Long and Short Interval Estimation" *Human Brain Mapping*, 25, 433–41.

Rammsayer, T.H. 1993. "On Dopaminergic Modulation of Temporal Information Processing" *Biol.Psychol.*, 36, 209–22.

Rammsayer, T. and Classen, W. 1997. "Impaired Temporal Discrimination in Parkinson's Disease: Temporal Processing of Brief Durations as an Indicator of Degeneration of Dopaminergic Neurons in the Basal Ganglia" *Int.J.Neurosci.*, 91, 45–55.

Ranen, N.G. et al. 1995. "Anticipation and Instability of IT-15 (CAG)n Repeats in Parent-Offspring Pairs with Huntington Disease" *Am.J.Hum.Genet.*, 57, 593–602.

Rao, S.M. et al. 1997. "Distributed Neural Systems Underlying the Timing of Movements" *J Neurosci*, 17, 5528–35.

Rao, S.M., Mayer, A.R., and Harrington, D.L. 2001. "The Evolution of Brain Activation during Temporal Processing" *Nature Neuroscience*, 4, 317–23.

Riesen, J.M. and Schnider, A. 2001. "Time Estimation in Parkinson's Disease: Normal Long Duration Estimation Despite Impaired Short Duration Discrimination" *J.Neurol.*, 248, 27–35.

Rowe, J.B. 2010. "Connectivity Analysis is Essential to Understand Neurological Disorders" *Front Syst.Neurosci.*, 4.

Rowe, K.C. et al. 2010. "Self-paced Timing Detects and Tracks Change in Prodromal Huntington Disease" *Neuropsychology*, 24, 435–42.

Sarma, S.V. et al. 2012. "The Effects of Cues on Neurons in the Basal Ganglia in Parkinson's Disease" *Front Integr.Neurosci.*, 6, 40.

Say, M.J. et al. 2011. "Visuomotor Integration Deficits Precede Clinical Onset in Huntington's Disease" *Neuropsychologia*, 49, 264–70.

Schwartze, M. et al. 2011. "The Impact of Basal Ganglia Lesions on Sensorimotor Synchronization, Spontaneous Motor Tempo, and the Detection of Tempo Changes" *Behav.Brain Res.*, 216, 685–91.

Smith, J.G. et al. 2007. "The Effect of Parkinson's Disease on Time Estimation as a Function of Stimulus Duration Range and Modality" *Brain and Cognition*, 64, 130–43.

Snijders, A.H. et al. 2011. "Gait-Related Cerebral Alterations in Patients with Parkinson's Disease with Freezing of Gait" *Brain*, 134, 59–72.

Spencer, R.M. and Ivry, R.B. 2005. "Comparison of Patients with Parkinson's Disease or Cerebellar Lesions in the Production of Periodic Movements Involving Event-based or Emergent Timing" *Brain and Cognition*, 58, 84–93.

Spildooren, J. et al. 2012. "Turning and Unilateral Cueing in Parkinson's Disease Patients with and without Freezing of Gait" *Neuroscience*, 207, 298–306.

Stoffers, D. et al. 2008. "Increased Cortico-Cortical Functional Connectivity in Early-Stage Parkinson's Disease: An MEG study" *Neuroimage*, 41, 212–22.

Stout, J.C. et al. 2011. "Neurocognitive Signs in Prodromal Huntington Disease" *Neuropsychology*, 25, 1–14.

Taniwaki, T. et al. 2003. "Reappraisal of the Motor Role of Basal Ganglia: A Functional Magnetic Resonance Image Study" *J.Neurosci.*, 23, 3432–38.

Thaut, M.H. et al. 1996. "Rhythmic Auditory Stimulation in Gait Training for Parkinson's Disease Patients" *Mov Disord.*, 11, 193–200.

Thompson, J.C. et al. 2010. "Automaticity and Attention in Huntington's Disease: When Two Hands are Not Better Than One" *Neuropsychologia*, 48, 171–8.

Tobin, A.J. and Signer, E.R. 2000. "Huntington's Disease: The Challenge for Cell Biologists" *Trends Cell Biol.*, 10, 531–6.

Vercruysse, S. et al. 2012. "Abnormalities and Cue Dependence of Rhythmical Upper-Limb Movements in Parkinson Patients with Freezing of Gait" *Neurorehabil.Neural Repair*, 26, 636–45.

Walla, P. 2008. "Olfaction and Its Dynamic Influence on Word and Face Processing: Cross-Modal Integration" *Prog.Neurobiol.*, 84, 192–209.

Wearden, A.J. 1999. "'Beyond the Fields We Know…': Exploring and Developing Scalar Timing Theory" *Behavioural Processes*, 45, 3–21.

Wearden, J.H. et al. 2008. "Stimulus Timing by People with Parkinson's Disease" *Brain and Cognition*, 67, 264–79.

Wiener, M. et al. 2010. "Fast Forward: Supramarginal Gyrus Stimulation Alters Time Measurement" *J.Cogn Neurosci.*, 22, 23–31.

Wiener, M. et al. 2012. "Parietal Influence on Temporal Encoding Indexed by Simultaneous Transcranial Magnetic Stimulation and Electroencephalography" *J.Neurosci.*, 32, 12258–67.

Wild-Wall, N. et al. 2008. "Time Estimation in Healthy Ageing and Neurodegenerative Basal Ganglia Disorders" *Neurosci Lett.*

Williams-Gray, C.H. et al. 2009. "The Distinct Cognitive Syndromes of Parkinson's Disease: 5 year Follow-up of the CamPaIGN Cohort" *Brain*, 132, 2958–69.

Wing, A.M., Keele, S.W., and Margolin, D.I. 1984. "Motor Disorder and the Timing of Repetitive Movements" In J.Gibbon and L. Allen (Eds.), *Ann NY Acad Sci: Timing Time Perc* (pp. 183–92).

Wittmann, M. et al. 2010a. "Accumulation of Neural Activity in the Posterior Insula Encodes the Passage of Time" *Neuropsychologia*, 48, 3110–20.

Wittmann, M. et al. 2010b. "The Neural Substrates of Subjective Time Dilation" *Front Hum.Neurosci.*, 4, 2.

Wu, T. and Hallett, M. 2005. "A Functional MRI Study of Automatic Movements in Patients with Parkinson's Disease" *Brain*, 128, 2250–9.

Wu, T., Chan, P., and Hallett, M. 2010. "Effective Connectivity of Neural Networks in Automatic Movements in Parkinson's Disease" *Neuroimage*, 49, 2581–7.

Wu, T. et al. 2010. "Neural Correlates of Bimanual Anti-phase and In-phase Movements in Parkinson's Disease" *Brain*, 133, 2394–409.

Wu, T. et al. 2011. "Effective Connectivity of Brain Networks during Self-initiated Movement in Parkinson's Disease" *Neuroimage.*, 55, 204–15.

Yu, H. et al. 2007. "Role of Hyperactive Cerebellum and Motor Cortex in Parkinson's Disease" *Neuroimage*, 35, 222–33.

Zimbelman, J.L. et al. 2007. "fMRI Detection of Early Neural Dysfunction in Preclinical Huntington's Disease" *J Int.Neuropsychol.Soc.*, 13, 758–69.

CHAPTER 9

Timing in the Cerebellum and Cerebellar Disorders

*Rebecca M.C. Spencer**

1 Introduction

Over the past century, the function of the cerebellum was largely assumed to be that of movement coordination. This assumption, along with its geographic segregation from the cortex, led the cerebellum to be overlooked in many early neuroimaging studies. Yet the cerebellum contains roughly 10% of the brain's mass (Hutchinson et al., 2003; Swanson, 1995). Perhaps more striking, the cerebellum contains 65–75% of the total neurons of the brain (Herculano-Houzel, 2010). In other words, for every one cortical neuron there are three to four cerebellar neurons. Moreover, neurons of the cerebellum are uniquely arranged into arrays of neural loops. These loops repeat millions of times throughout the cerebellum, and are arranged into lobules. Those lobules are, in turn, composed of folia, structures reminiscent of the fractalian structure of a tree or cauliflower (Voogd and Glickstein, 1998). Thus, the cerebellum is architecturally unique and capable of performing a redundant computation. It has been proposed that the computation performed by the cerebellum is that of timing (Braitenberg, 1983; see review in Spencer and Ivry, 2013).

2 Timing in Cerebellar Ataxia

In humans, focal cerebellar damage is rare. While a hemorrhagic stroke to the cerebellum is particularly rare, ischemic stroke, the more common cerebellar stroke, nonetheless accounts for only 1–7% of all strokes (Bogousslavsky, Van Melle, and Regli, 1988; Macdonell, Kalnins, and Donnan, 1987; Vemmos et al., 2000). Other focal cerebellar lesions, due to tumors and tumor resection, are also uncommon particularly in adults.

Cerebellar atrophy that typically results in widespread cerebellar damage can arise in a number of ways. Genetic degenerative diseases, primarily spinocerebellar ataxia (SCA), that affect the cerebellum are heterogeneous. SCA is a

* Department of Psychology, Neuroscience and Behavior Program, University of Massachusetts, Amherst, USA.

DOI 10.1163/9789004230699_010

group of autosomal dominant neurodegenerative disorders, many of which result in cerebellar degeneration. Prevalence estimates of SCA vary from .3 to 3 per 100,000 (van de Warrenburg et al., 2002). Moreover, of the approximately 30 different SCA subtypes (Schols et al., 2004), only a handful have a primarily cerebellar foci (e.g., SCA 6, 8, 15/16; Durr, 2010; Marelli et al., 2011). Of the degenerative cerebellar disorders, sporadic idiopathic cerebellar atrophy may have the greatest prevalence, estimated to have a prevalence rate of 8.4 per 100,000 (Muzaimi et al., 2004). Sporadic cases of cerebellar agenesis have also been reported (e.g., Glickstein, 1994; Macchi and Bentivoglio, 1977; Nowak, Timmann, and Hermsdorfer, 2007; Velioglu, Kuzeyli, and Ozmenoglu, 1998). Surprisingly, in many of these cases, motor performance is largely normal, suggesting possible compensatory mechanisms (for a review see: Glickstein, 1994; Macchi and Bentivoglio, 1977).

Seminal neurologists, Joseph Babinski and Gordon Holmes, are both credited for the earliest reports of the deficits associated with cerebellar damage in humans (Holmes, 1939, 1917; Trouillas et al., 1997). Babinski and Holmes described core symptoms associated with damage to the cerebellum – such as slurred speech, unsteady gait, and incoordination – and collectively referred to these as 'cerebellar syndrome'. Holmes, studying survivors of gunshot wounds to the back of the head in World War I, came to use the term 'ataxic' to specifically describe the movements of individuals with cerebellar syndrome. Ataxia remains the commonly used term to describe the movements associated with cerebellar disorders. Ataxia is broadly characterized by postural deficits, impairments in upper and lower limb movements, speech impairments, and oculomotor dysfunction (Trouillas et al., 1997). However, the array of symptomatology varies according to localization and extent of damage (Dichgans and Diener, 1984).

There is evidence that the montage of symptoms associated with cerebellar ataxia is due to impaired timing. Most prominently, ataxic movements are uncoordinated. It has been suggested that the lack of coordination between muscles is due to increased time and temporal variability of muscle initiation for individual sub-movements (Day et al., 1998). Upper and lower limb ataxia is also marked by intention tremor, dysmetria, and dysdiadochokinesis on a clinical exam. Unlike the resting tremor that visibly marks Parkinson's disease, intention tremor is not observed at rest; rather, it is most obvious at the endpoints of visually-guided movements. Intention tremor contributes to dysmetria. Dysmetria is the tendency to overshoot or undershoot a target such as the physician's finger. This, too, may be the result of impaired timing, in this case there is a breakdown in the timing between agonist and antagonist muscle contractions composing the movement (Flament and Hore, 1986). Impaired production of rapid alternating movements, termed dysdiadochokinesis, has likewise been attributed to impaired timing of sub-movements (Conrad and Brooks, 1974).

Empirical tests of timing in cerebellar patients also support the functional contribution of the cerebellum to movement timing. A repetitive tapping paradigm has been widely used to isolate motor timing in the laboratory setting. In such a task, participants are presented with a metronome that sets the goal pace, for instance a 500ms interval marked by brief metronome tones. At the beginning of a trial, the participant's goal is to tap at this goal pace (synchronization phase). After the metronome disengages (typically following 5–10 paced intervals), the participant's goal is to continue moving at that pace until the trial's end (continuation phase; e.g., after 15–20 unpaced taps). Importantly, individuals with cerebellar damage are impaired in repetitive tapping with a range of effectors (Ivry, Keele, and Diener, 1988). Specifically, temporal variability is greater for those with cerebellar damage relative to control groups (typically matched in age, education, handedness, and gender).

Moreover, a corpus of studies suggests that motor timing deficits in cerebellar patients may underlie movement inaccuracies. In a study of overarm ball throwing, inaccuracies of throws produced by individuals with cerebellar lesions were related to increased variability in the timing of the ball release within the rotation of the arm (Hore, Timmann, and Watts, 2002; Timmann, Richter, Bestmann, Kalveram, and Konczak, 2000; Timmann, Watts, and Hore, 1999). Likewise, ataxic speech may be greatly accounted for by a deficit in motor timing. The duration of syllable repetitions is slowed and irregular in ataxic speech. Speech slowing may be similar to that observed in individuals with Parkinson's disease (Ackermann et al., 1997b; Canter, 1963). However, cerebellar lesions also yield and increase the temporal variability of articulatory timing (Ackermann and Hertrich, 2000; Schalling, Hammarberg, and Hartelius, 2007).

These motor timing deficits are consistent with historical views of the cerebellum as being a motor structure, particularly for movement coordination. However, cerebellar patients are also impaired in non-motor, temporal perception tasks. A duration discrimination task provides an ideal probe of temporal perception with little motor demands. In a duration discrimination task, participants are presented with two intervals and asked to judge whether the second is shorter or longer than the first. By parametrically varying the difference between the intervals, a discrimination threshold and variability in discrimination performance can be obtained. While thresholds are generally similar between cerebellar patients and matched controls, variability in performance is greater for those with cerebellar lesions (Ivry and Keele, 1989) suggesting that the internal clock is 'noisy' (Ivry and Spencer, 2004a). Notably, differences are only observed for temporal discriminations. When the task requires discrimination of the amplitude (or loudness) of the tones (Ivry and Keele, 1989) or frequency (Casini and Ivry, 1999), performance does not differ between individuals with cerebellar lesions and controls.

The impairment in temporal perception may also underlie the deficit in speech perception observed in individuals with cerebellar ataxia. Distinguishing temporal phonetic features is critical to discriminating speech successfully. In English, to distinguish between '*rabid*' and '*rapid*' relies on precise temporal coding. Using similar words from the German language, such as '*boden*' and '*boten*', Ackermann and colleagues (1997a) found impaired discrimination of speech sounds in individuals with bilateral cerebellar degeneration, and this deficit was specifically associated with the temporal cues innate in the speech stimuli.

A set of recent studies further illustrates that cerebellar lesions also lead to impaired perceptual timing of interception, a skill necessary to catch a ball or to catch a closing elevator. In one study, individuals with cerebellar ataxias and matched control participants played a computer game in which a moving target must be intercepted by a 'cannon ball'. As such, the task required integration of an estimate of the speed of the moving target as well as the speed of the cannon ball once it was released (through a finger press). Participants with cerebellar ataxia performed significantly worse at the task than healthy controls (Bares et al., 2007; Bares et al., 2010a,b) and participants with Parkinson's disease (Bares et al., 2010a) supporting a role of the cerebellum in this form of perceptual time estimation.

In addition to temporal production and perception tasks, the cerebellum contributes uniquely to temporal aspects of associative learning. While early studies focused on the contribution of hippocampal and cortical areas (Gormezano et al., 1962; Oakley and Russell, 1972), only cerebellar lesions eliminate learning of the conditioned response (McCormick and Thompson, 1984). This is often demonstrated with an eye blink classical conditioning paradigm. In a typical eye blink conditioning task, the individual experiences trials in which an air puff to the eye, which causes an unconditioned eye blink, is paired with a tone. Over time, the individual is conditioned to respond to the tone alone with an eye blink. When a delay is implemented between the tone and the air puff, precise timing of the eye blink in response to the tone is necessary in order for the response to be adaptive (i.e., eye closed at the time at which an air puff might arrive). In animal and human cerebellar lesion studies of eye blink conditioning, the subjects are able to learn the conditioned response following a cerebellar lesion. Importantly, however, the adaptive timing of the response is lacking (Gerwig et al., 2003; Perrett, Ruiz, and Mauk, 1993;

Woodruff-Pak, Papka, and Ivry, 1996). In other words, the cerebellar lesioned animal will appropriately learn the response (a blink), but this response is ill-timed relative to the conditioned stimulus and, thus, not adaptive.

Collectively, these data demonstrate that temporal impairments may underlie ataxic movements. Perhaps more importantly, cerebellar lesions result in impaired timing in tasks explicitly measuring temporal perception and production (for reviews see Ivry and Spencer, 2004b; Ivry, 1996).

3 Timing: Localized or Distributed?

While cerebellar lesions result in timing impairment, this alone does not necessarily implicate the cerebellum in functioning as a single neural clock (i.e., a specialized dedicated timekeeper). Lesions in other subcortical or cortical regions may yield similar timing deficits that would support timing as a distributed function incorporating a network of brain areas (i.e., a distributed dedicated timekeeper). A third alternative is that multiple brain areas may have the ability to time depending on the relevant modality (i.e., intrinsic timing; for a review see Ivry and Schlerf, 2008).

A wealth of individual difference studies using repetitive tapping tasks, collectively favor a neural clock that functions across a range of durations. This individual difference work is carried out under the assumption that if there is a single neural clock, then a person with a "good clock" should perform well across a range of timing tasks and, likewise, a person with a rather "poor clock" should perform relatively poorly across tasks. In other words, one should see highly significant correlations in temporal performance across timed tasks. Indeed, temporal variability for a repetitive finger tapping task significantly correlates with variability for tapping with a range of other effectors. For instance, finger tapping (flexion/extension at the metacarpophalangeal joint) at a rate of 400ms per tap correlates with tapping with the arm (flexion/extension of the glenohumeral-shoulder-joint) and tapping the jaw (excursions of the temporomandibular joint) at the same rate (Franz, Zelaznik, and Smith, 1992). Likewise, temporal variability in the repetitive finger tapping task correlates with temporal variability for a repetitive foot tapping task (flexion/extension at the talocrural-ankle-joint; Keele et al., 1985).

Moreover, individual difference studies suggest that the capacity of the neural clock is not limited to motor tasks. Performance on repetitive tapping tasks is also predictive of performance on non-motor timing tasks (Keele et al., 1985). The duration discrimination task (described above) has been used to demonstrate this point. Importantly, performance on repetitive tapping tasks is predictive of performance on the duration discrimination task, suggesting a shared timing process across motor and perceptual timing tasks.

Thus, there is certainly reason to believe that temporal performance draws on a single neural resource. Taken in conjunction with the evidence provided in Section 2, evidence points to the cerebellum as being this neural clock. While this proposed function of the cerebellum as a timekeeper has gained favor, recent work in our lab and others has focused on limitations of the cerebellar clock.

4 Limitations on Cerebellar Timing

4.1 *Event versus Emergent Timing*

Given the rather consistent pattern of correlations that our lab and others have observed amongst timing tasks (Franz, Zelaznik, and Smith, 1992; Keele and Hawkins, 1982; Keele et al., 1985), we were surprised to find a lack of correlation between temporal variability on a repetitive finger tapping task and a repetitive continuous circle drawing task. As an aside, continuous cycling (either circling or 'wagging' of an effector) was of interest given that research on motor coordination, another hypothesized function of the cerebellum, was largely based on continuous bimanual movements (e.g., Carson et al., 1997; Haken, Kelso, and Bunz, 1985; Kelso, 1984; Semjen, Summers, and Cattaert, 1995). Timing of continuous movements is also particularly unique as temporal variability is remarkably low for healthy young adults. While the coefficient of variation, a measure of variability normalized to movement time, is typically 5–8% of the mean movement time for finger tapping (Keele et al., 1985; Theoret, Haque, and Pascual-Leone, 2001), it is as low as 3–5% for circle drawing (Spencer et al., 2003; Spencer and Zelaznik, 2003; Zelaznik, Spencer, and Ivry, 2002). Most importantly, these measures of temporal variability do not correlate across these two tasks (Robertson et al., 1999; Zelaznik, Spencer, and Ivry, 2002).

In a continuous circle drawing task, participants are given a circle template to equate movement amplitude across participants. A visual target at the top of the circle template assists participants in aligning movements to the metronome; participants are to pass through this target coincident with the metronome beep during the paced portion of a trial, just as they are to touch the table coincident with the metronome for the repetitive tapping task (Robertson et al., 1999; Zelaznik, Spencer, and Ivry, 2002). Nonetheless, when one performs this task, it is quite obvious that this movement goal is much less salient for the continuous circle drawing task than it is for the repetitive finger tapping task.

This led us to propose the Event Timing Hypothesis: Movements or intervals that are marked by salient features would share a timing process, while movements lacking such a structure would not be amenable to such a process (Ivry et al., 2002). To test this hypothesis, we compared variability on the continuous circle drawing task to a discrete circle drawing task. In the discrete circle drawing task, participants traced the circle template for every other movement interval and paused at the top of the trajectory for the intervening intervals. Thus, these tasks were equated for movement speed, amplitude, and biomechanical demands. In spite of this, there were no correlations between continuous and discrete circle drawing. Consistent with the Event Timing Hypothesis, however, performance on the discrete circle drawing did predict performance on the repetitive tapping task (Zelaznik, Spencer, and Ivry, 2002).

At this point, we know little about how continuous movements are timed. It has been suggested that timing for such tasks may be emergent. The idea of emergent timing is derived from Dynamical Systems Theory applied to studies of motor control (Turvey, 1977). Accordingly, timing for continuous tasks may emerge from the dynamics of the limbs. Alternatively, other properties of the movement, rather than timing being under the direct control of a neural clock, may be under direct neural control and result in well-timed movements. For instance, in a continuous circle drawing task, neural processes controlling movement velocity or jerk (Hogan and Flash, 1987) may result in precise temporal consistency without the direct control of time. Supporting the concept of an array of unique emergent processes, we found that temporal processes underlying continuous circling and continuous line drawing are distinct, and these processes are also independent of that used for discrete movements such as finger tapping (Spencer and Zelaznik, 2003). In other words, there seems to be no single 'continuous movement timer'. To illustrate this, we applied a slope analysis to the variability of timing across these tasks. The slope analysis, the examination of the relation between the timed interval and timing variance, assumes two independent sources of variability to movement timing: A duration-independent source associated with variability in implementing a timed movement command and a duration-dependent source associated with the central clock-like process (Wing and Kristofferson, 1973). When movement time is plotted against temporal variability for a given task a linear relationship is observed, particularly for millisecond to second range tasks (Gibbon, 1977; Gibbon and Church, 1990). Importantly, if two tasks share a neural timing process, they should have the same variability-interval relationship, or shared slope (Ivry and Corcos, 1993; Ivry and Hazeltine, 1995). Consistent with this, we observed the same variability-interval slopes for two line drawing tasks. Likewise, the slope was the same for two continuous circle drawing tasks. However the slopes were unique for continuous circle drawing relative to continuous line drawing. Thus, these results suggest a degree of independence for timing processes underlying continuous movement timing, but within any given task (e.g., variants of a continuous circling task), a shared timing source is used (Spencer and Zelaznik, 2003).

Even under indirect control, however, it is speculated that continuous movements may rely on a neural 'clock' to align early movement cycles to the metronome-instructed temporal goal. According to this Transformation Hypothesis, early cycles of a repetitive movement may utilize an explicit timing process to set the pace of the movement before a non-clock process (e.g., velocity-based control) may take over. Supporting this, Zelaznik and colleagues (2005) again compared temporal performance for repetitive finger tapping and repetitive continuous circle drawing but focused analyses on the initial, early cycles of the trials. Variability in the first cycle of these tasks was significantly correlated. Yet, consistent with work reviewed above, temporal variability for subsequent cycles did not correlate. This suggests that tasks with a precise temporal goal may initially require the event-based temporal representation, or internal clock, even though other non-clock timing processes may take over once the appropriate rate is adopted consistent with the Transformation Hypothesis (Zelaznik et al., 2005).

Individual difference studies into event timing in healthy individuals have provided insight as to how the cerebellar 'clock' measures time for movements and percepts. That is, the cerebellum may serve as a timer when salient features are available by which to start and stop the neural clock (Ivry et al., 2002). If such were the case, cerebellar lesions would yield impaired timing on repetitive movement tasks and duration perception tasks that require the timing between discrete events (e.g., metronome beeps, table touches when finger tapping), deficits that have been consistently reported. However, if the cerebellum provides only discrete/eventful timing measures, this yields a rather counterintuitive prediction, that is that individuals with cerebellar lesions may be unimpaired in the timing of continuous movements. We tested this in a series of studies of individuals with cerebellar ataxia.

In one such study, we compared temporal performance on repetitive finger tapping (on a table top; i.e., 'eventful'), continuous circle drawing, and discrete circle drawing (Spencer et al., 2003). The goal movement time for all tasks was 800ms per cycle. Participants were individuals with focal, unilateral lesions of the cerebellum due to stroke or tumor resection. With focal unilateral cerebellar lesions, impairments are observed unilaterally, in the ipsilesional limb (the cerebellum receives crossed projections from the cortex, and therefore, control is ipsi- rather than contra-lateral; Franz, Ivry, and Helmuth, 1996; Ivry, Keele, and Diener, 1988). Thus, this group allows for within-subject comparisons of the limb under control of the lesioned cerebellum to the unimpaired limb. Similar to previous studies (e.g., Ivry, Keele, and Diener, 1988), we found that temporal variability was greater with the ipsilesional, impaired limb relative to the contralesional, unimpaired limb for the table tapping task. Likewise, variability was greater for the ipsilesional limb relative to the contralesional limb for the discrete circle drawing task. Importantly, consistent with the predicted role of the cerebellum for timing of events, there was no difference in performance across limbs for the continuous movement task. Strikingly, the coefficient of variation was remarkably low, only about 3% for both the unimpaired and impaired limb in the cerebellar lesion group for the continuous circle drawing task (Figure 9.1; Spencer et al., 2003).

In a second experiment, we compared temporal variability for the repetitive finger tapping task under four conditions: (1) tapping on a table top, a variant of the task most commonly used in previous studies, at a rate of 1000ms per cycle; (2) a discrete air tapping task, also at a rate of 1000ms per cycle, in which taps were made with the finger held away from any table surface and movements were instructed to be 'discrete and staccato-like'; (3) a continuous air tapping task, at a rate of 1000ms per cycle, in which taps were, again, made without surface contact and participants were instructed to "move smooth and

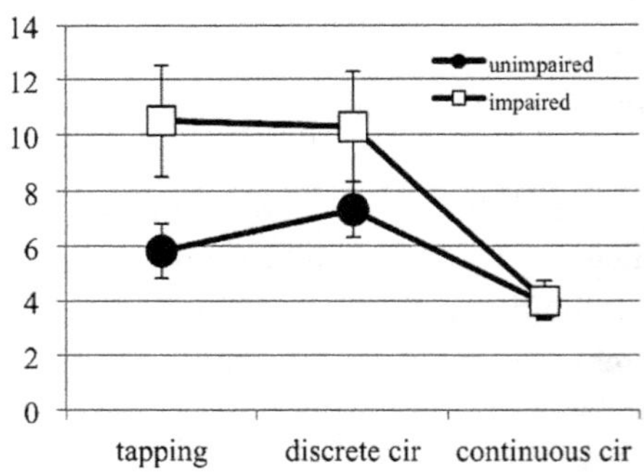

FIGURE 9.1 *Coefficient of variation (standard deviation divided by mean movement time) for repetitive finger tapping, discrete circle drawing and continuous circle drawing. Movements were performed by individuals with unilateral cerebellar lesions with both their impaired and unimpaired limbs. Error bars represent standard error.*

ADAPTED FROM SPENCER ET AL., 2003.

continuously"; and (4) a continuous air tapping task at a rate of 500ms per cycle (Spencer et al., 2003). Note that the last task, the fast continuous task, was included to equate for the faster velocity of the movement component of each cycle in the discrete tasks relative to the continuous tasks. That is, when tapping discretely, the velocity of the finger must be faster than in the continuous task in order to include a brief pause within the trajectory, thereby making the movement discrete. As such, one might argue that impairments associated with cerebellar lesions in timing of discrete movements, such as table tapping, are simply due to the speed of the movement, a deficit that may not be surprising to observe in conjunction with any movement disorder. Importantly, results were consistent with the Event Timing Theory and the hypothesized role of the cerebellum in timing movements with salient events. That is, for individuals with focal cerebellar lesions, variability was greater with the ipsilesional, impaired limb relative to the contralesional, unimpaired limb for the discrete tasks, table tapping, and discrete air tapping. There was no difference in performance across limbs for either the fast (500ms/cycle) or slow (1000ms/cycle) continuous air tapping tasks.

We found a similar pattern of results when comparing a group of individuals with bilateral cerebellar lesions (degeneration due to SCA or sporadic cerebellar ataxia) to healthy matched controls. Here, performance with the dominant hand was compared for individuals with bilateral cerebellar ataxia (SCA3, SCA6, or sporadic cerebellar ataxia) and a group of healthy age- and education-matched control participants. Individuals with bilateral cerebellar lesions exhibited greater variability than the control group for the discrete tasks. There was no difference in performance across groups for the continuous movement tasks (Spencer et al., 2003).

Finally, neuroimaging evidence supports the Event Timing Theory of cerebellar function. We examined the Blood-oxygen-level dependent (BOLD) response (i.e., neural activation) using functional magnetic resonance imaging (fMRI) when healthy young adults performed the continuous and discrete air tapping tasks described above. While activation in lobule VI of the cerebellum was observed for both conditions, additional activity (located in the superior vermis) was selectively activated for the discrete movement. This is consistent with the hypothesis that the cerebellum plays a unique role in discrete movement timing (Spencer et al., 2007).

Together, this work demonstrates that the cerebellar clock function is restricted. Specifically, the manner in which the cerebellum computes time, by representing intervals demarcated by salient events, by nature restricts this role to a subset of movement and perceptual tasks that are amenable to such a timekeeper.

4.2 *Interval Range of the Cerebellar Timing: Long versus Short Interval Timing*

Importantly, tasks discussed to this point involved intervals in the millisecond range, mostly less than 2000ms. For intervals greater than this, the contribution of the cerebellum to timing may decrease, with increasing weight of the non-temporal processes contributing greatest to performance variability. For instance, Mangels, Ivry, and Shimizu (1998) compared performance on a short, 400ms, interval task with performance on a long, 4 s, interval task in individuals with cerebellar lesions and age- and education- matched healthy control participants. The task was a duration perception task, ruling out the contribution of motor deficits to short versus fast movements. They found that individuals with cerebellar lesions were impaired, relative to healthy control participants, in perception of both the short and long intervals. To examine whether the neocortex further supports this timing function, via attention or working memory mechanisms, a group of individuals with focal lesions to the prefrontal cortex were also included. Individuals with lesions of the prefrontal cortex were impaired, relative to healthy controls, on the long interval perception tasks. There was no difference in performance for these two groups on the short, 400ms, perception task. The authors suggest that, for longer intervals, cortical contributions are necessary to maintain the standard in working memory and to attend to the task over the long interval. However, it may be somewhat surprising that the deficit in performance was not reduced for the longer, 4 s, interval for those with cerebellar lesions. One explanation for this finding is that, for longer interval timing, a sub-interval timing strategy may be used. That is, when faced with the task of timing a 4-s interval, one may internally time four 1-s intervals. As such, the cerebellar contribution to millisecond timing would be evident in this way.

In a meticulous analysis of temporal perception and production across a range of suprasecond intervals (2–12 s), a recent study found further evidence that the cerebellar contribution is greatest for short intervals although the deficit is still present for longer intervals (Gooch et al., 2010). In the duration production task under consideration, participants were required to reproduce the duration of a 2, 4, 6, 8, 10 or 12 s stimulus (a red square on a computer screen). The authors report a significant group (cerebellar patients versus controls) by interval interaction, reflecting highly variable temporal performance for the cerebellar patients relative to controls at the shortest intervals (2 s and, less so, 4 s) with little group differences observed at the longer intervals. Likewise, a temporal estimation task, requiring participants to produce intervals of 2, 4, 5, 6, 10 or 12 s from memory, revealed consistently worse performance for individuals with cerebellar lesions compared to controls across all intervals, although the deficit was again greatest at the shorter durations. Notably, when the patient group was divided into those with middle-to-superior cerebellar lesions and those with damage outside of middle-to-superior cerebellum, only those with middle-to-superior damage differed from controls. This suggests that the cerebellar clock may be localized to superior cerebellum, an issue we will return to below.

Individuals with Parkinson's disease are also reported to be impaired in temporal perception and production tasks. Movements of individuals with Parkinson's disease are visibly slow, a deficit termed bradykinesia. As such, one might predict that individuals with Parkinson's disease would be slow on a repetitive tapping task. Counter-intuitively, however, these individuals actually tend to speed up movements (Harrington et al., 2004; Ivry and Keele, 1989; O'Boyle, Freeman, and Cody, 1996). Whether temporal variability is increased in individuals with Parkinson's disease, as would be expected for an impaired neural clock, is unclear. Considering intervals in the millisecond range (less than 2 s), some studies report increased temporal variability in Parkinson's disease (Harrington, Haaland, and Hermanowicz, 1998) relative to matched controls, while others find no difference (Ivry and Keele, 1989; Spencer and Ivry, 2005). While Parkinson's disease patients are often examined as a model of basal ganglia dysfunction, isolation of the disease to basal ganglia is not pure. Importantly, Aparicio and colleagues (2005) found that individuals with focal lesions of the basal ganglia, most in the putamen, some extending to the caudate, internal capsule or globus pallidus, due to stroke, were unimpaired on a repetitive tapping task. Variability of taps, with a goal pace of 400ms, did not differ for basal ganglia lesion patients relative to healthy controls nor was there a difference in variability between the contralesional, impaired limb and the ipsilesional, unimpaired limb. While this work remains to be replicated, it calls into question the conclusion of some that the basal ganglia is an internal clock for millisecond range intervals.

It is worth considering that the basal ganglia may contribute to timing of suprasecond intervals, those intervals greater than the 200–2000ms that seems to be the expertise of the cerebellum. Supporting this, Matell and colleagues (2003) illustrated the ability of the striatum to time intervals of 10 and 40 s. By recording striatal neurons in rats trained to time press a lever 10 or 40 s after a stimulus onset, it was clear that these neurons distinguished between the two intervals just as response rates were distinct for the two trial types. More importantly, a relationship between the time of peak firing and performance was observed (Matell, Meck, and Nicolelis, 2003).

The selective role of basal ganglia for suprasecond intervals and not subsecond intervals was distinguished in a recent study in humans using deep brain stimulation (DBS) to the subthalamic nucleus (STN; Wojtecki et al., 2011). DBS is used to treat symptoms of Parkinson's disease. When the stimulation was off, these participants were impaired relative to matched controls; specifically, 15-s intervals were underestimated in length. Low frequency (10 Hz) stimulation resulted in further underestimation when reproducing the 15-s interval. Conversely, high frequency (130 Hz) stimulation reduced the amount of underestimation, resulting in performance similar to that of controls. However, neither high nor low frequency DBS to the same region influenced duration discrimination in the millisecond range (800–1600ms), supporting a distinction in the temporal range that the basal ganglia may expertly time.

The corollary to the consideration of a separate, perhaps basal ganglia-based, timing process for longer intervals, is to consider whether there is a lower limit to the cerebellar timing function. Can the cerebellum time intervals on the range of 50 or 100ms? Mounting evidence suggests that timing for such ultra-short intervals is also distinct. In fact, ultra-short intervals may be unique from other intervals in that timing for ultra-short intervals might not draw on the function of a single neural clock but rather the timing process may be modality dependent, drawing on cortical areas associated with that modality. Termed the State-Dependent Network model, Karmarkar and Bounomano (2007) proposed that networks of neurons within a given cortical region can accurately detect time. The basis for the State-Dependent Network model is that time is implicitly represented in the synaptic properties or state of a neural network (Buonomano and Mauk, 1994; Karmarkar and Buonomano, 2007; Yamazaki and Tanaka, 2005). By simulating a series of state-dependent networks, Karmarkar and Buonomano (2007) showed that, when presented with a brief (e.g., 100ms) interval marked by two tones, the first tone of the pair will generate activity in the network. This activity then changes predictably over time and includes fast and slow inhibitory post-synaptic potentials and short-term synaptic plasticity in the connections between nodes in the network. When the second tone of the interval arrives, the dynamical state of the network has been changed as a result of the first tone, so there is a different pattern of activity in the network in response to the second tone even if the second tone is identical to the first tone in duration, pitch, and frequency. The temporal interval defined by the two tones can be interpreted from the final state of the network. Importantly, if the stimulus (100ms interval marked by two tones) is preceded by any event, the state of the network is changed. As such, if standard and test intervals are presented in two different contexts, the dynamics of the state-dependent network would differ and the ability to encode duration would fail. This creates a 'reset problem', which has been used to demonstrate support for the State-Dependent Network (SDN) hypothesis. In the Reset Task designed to test this problem, similar to the duration perception task described above, human participants were presented with an interval of approximately 100ms marked by two tones and were to respond as to whether the interval was short or long relative to a 100ms standard interval (Figure 9.2A; Spencer, Karmarkar and Ivry, 2009). Performance was similar if the test interval was preceded by a distractor interval also of 100ms. This result is consistent with the SDN hypothesis given that a fixed distractor interval would allow for building a stable representation of this fixed state based on trial-by-trial feedback. More importantly, when the distractor interval was of varying length (50–150ms), discrimination thresholds were greatly increased relative to the no-distractor condition. This is consistent with the prediction of the SDN hypothesis. That is, that the variable length distractor results in the network being in a different state each time the target interval is presented and a stable representation cannot be extracted (Karmarkar and Buonomano, 2007).

Recently, we revisited this effect, considering attentional issues that may also produce the context effects observed in the Reset Task (Spencer, Karmarkar, and Ivry, 2009). We predicted that using very brief intervals (i.e., 100ms and 100ms +/– 50ms), which may trigger automatic attentional capture (Posner, 1978), imposed unique sources of variability. In other words, attention would be drawn to the distractor interval by attentional capture thereby impairing attention when presented with the target interval. To test this, we replicated the Reset Task using both conditions from the original Karmarkar and Buonomano study (2007; approximately 100ms for both the distractor and target intervals; 100–100 task) and a novel condition with a 300ms distractor and 100ms target interval (300–100 task). First, with respect to the 100–100 task, our results replicated those of Karmarkar and Buonomano (2007) in that thresholds were elevated in both the fixed and variable distractor conditions, relative to the no-distractor conditions, and thresholds were greatly increased in the presence of a variable distractor interval (Figure 9.2B; Spencer, Karmarkar and Ivry, 2009). This was not the case in the 300–100 task. While there was a small but significant main effect of a distractor, performance with a variable distractor was not worse than a fixed distractor in this condition. This result limits the range of the SDN model. That is, the original model assumes that any distractor interval (100- or 300-ms) should alter the state of the network when presented with the target interval. Rather, these results are consistent with our hypothesis that unpredictable ultra-short intervals (<150ms) may disrupt timing of the target interval. Importantly for the purposes of the present discussion, this work supports a distinction in timing for ultra-short (50–150ms) and longer millisecond range (200–2000ms) timing. Furthermore, we report that variants of the Reset Task with a 300ms target interval and either a 100ms (100–300 task) or 300ms

(300–300 task) distractor interval also failed to find the increased threshold in the variable distractor condition relative to the fixed condition or no-distractor conditions. These results indicate how attentional processes can influence the perception of brief intervals, as well as point to important constraints of SDN models. As such, we posit that the SDN model, which proposes that timing may be computed in cortical regions supporting the task modality, may support ultra-short interval timing (<200ms) but not millisecond range timing (200–2000ms; Spencer, Karmarkar, and Ivry, 2009).

5 Localization of Timing within the Cerebellum

Given the evidence supporting the function of the cerebellum in millisecond range timing, we turn to consideration of where such a function might be located within the cerebellum. Architecturally, the cerebellum is composed

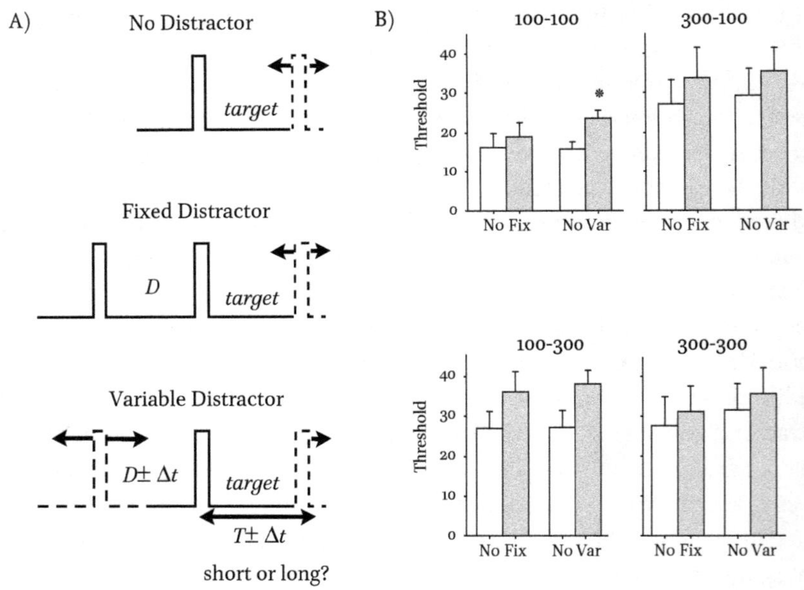

FIGURE 9.2 *A) The Reset Task described by Karmarkar and Buonomano (2007). Participants judged the length of the target (T) relative to a previously presented standard interval when the target was presented without a distractor (top), with a fixed interval distractor (D; middle), or a variable length distractor (bottom). B) Consistent with Karmarkar and Buonomano (2007), we found an increased threshold with a variable distractor condition for the 100–100 condition. This was not the case in the 100–300, 300–300, or 300–100 conditions.*

ADAPTED FROM SPENCER, KARMARKAR, AND IVRY, 2009.

of repeating loops. These loops underlie three longitudinal divisions – the vermis, the paravermis, and the hemispheres. Each of these divisions is, in turn, composed of multiple lobules. This elaborate arrangement may either suggest a highly specialized single map (Apps and Hawkes, 2009) or multiple functional units (Grodd et al., 2001). Mounting evidence suggests a specialized role of superior cerebellum in timing. As described above, timing impairments are exaggerated in individuals with superior cerebellar lesions relative to those with inferior lesions (Gooch et al., 2010). Like Gooch and colleagues (2010), Harrington and colleagues (2004) studied timing in individuals with superior versus inferior cerebellar lesions and found that timing impairments were associated with superior lesions. In this study, Harrington and colleagues compared two groups of cerebellar lesion patients: those with lesions to inferior cerebellum (approximately lobules VII–X) and those with superior cerebellar lesions (approximately lobules I–VI). Those individuals with superior, but not inferior cerebellar lesions, were more variable relative to healthy controls on the duration perception task with standards of 300 and 600ms, although this reached only trend-level for significance ($p = .07$; see Ivry and Spencer, 2004a). Moreover, those with superior cerebellar lesions were more variable than the control group on a repetitive finger tapping task at both 300 and 600ms. Specifically, total variability was modeled to dissociate the variability due to the central timekeeping process (i.e., 'clock' variability) from variability due to noisy motor implementation (i.e., motor variability; Wing and Kristofferson, 1973). The deficit observed in individuals with superior cerebellar damage was associated with the former, supporting an impaired clock process in superior cerebellum.

Functional imaging studies, by and large, find anterior cerebellum is active in conjunction with timed movement and perception tasks. For instance, using positron emission tomography (PET), Penhune and colleagues (1998) found that performance on the repetitive finger tapping task produces activation in lobule V beyond that seen in a control condition in which participants simply listened to the metronome. Likewise, neuroimaging of perceptual timing tasks also yields cerebellar activation. However, the localization within the cerebellum for these tasks is less clear. For example, one study of visual timing pointed to a critical region around the border of left Crus I/II, activation (Lewis and Miall, 2003). Other studies have pointed to a region within lobule VI (Aso et al., 2010) or superior vermis (Tesche and Karhu, 2000; Xu et al., 2006).

The anterior cerebellum has a known somatotopic arrangement, akin to the homunculus of the motor cortex. Using functional neuroimaging, mapping of the activation in this region suggests an inverted body map with the foot superior to the hand and the hand superior to the lips and facial representations

(Grodd et al., 2001; Habas, Axelrad, and Cabanis, 2004; Nitschke et al., 1996; Rijntjes et al., 1999). It has been proposed that underlying this body map is a series of interval representations (Ivry, 1996). In other words, the cerebellum may have an effector by interval arrangement akin to the ocular dominance columns in visual cortex. This model would be consistent with the observed unilateral impairment associated with lateralized cerebellar lesions.

In addition to the somatotopic organization of the anterior lobe, somatotopy has been observed in the inferior aspect of the posterior lobe of the cerebellum (lobules VIIB and VIII; Habas, Axelrad, and Cabanis, 2004; Rijntjes et al., 1999) and, most recently, we have reported evidence of a somatotopic arrangement in lateral cerebellum, lobules VI and VII (Schlerf, Verstynen, Ivry, and Spencer, 2010). In that study, we used fMRI to map movements of the fingers and toes that were either simple (i.e., flex/extend all fingers/toes simultaneously) or complex (i.e., flex/extend individual digits sequentially). Movements were repetitive at a rate of approximately 500ms. For all movements, ispilateral activation was observed in anterior cerebellum and this activation was somatotopically organized, that is, foot movement responses were distinct from, and inferior to, hand movement responses. Interestingly, when movements were complex, activation was also observed in lobules VI and VII. This lateral cerebellar activation was particularly noteworthy for two reasons. First, unlike the unilateral activity observed in the anterior cerebellum, activation in lateral cerebellum in conjunction with complex movements was bilateral. Second, the activation was somatotopically arranged here as well. This is particularly interesting in lateral cerebellum given that this region has previously been associated with the non-motor functions of the cerebellum (Akshoomoff and Courchesne, 1992; Desmond, Chen, and Shieh, 2005). Such a role is supported by connectivity studies, which illustrate that lateral cerebellum is functionally connected to prefrontal cortex (Kelly and Strick, 2003; Krienen and Buckner, 2009; Middleton and Strick, 2001). Thus, finding somatotopic organization here would be akin to finding somatotopic organization in the frontal cortex. To make sense of this arrangement, from these observations, we propose that the lateral cerebellum may support the anterior cerebellar function under conditions with high cognitive demands – such as movements with high complexity (Schlerf, Verstynen, Ivry, and Spencer, 2010) or working memory when a task requires response mapping (Spencer and Ivry, 2009).

6 Conclusions

It should be clear from this Chapter that the cerebellum plays a critical role in timing and, for this reason, timing deficits are central to the impairments associated with cerebellar lesions and disorders. However, this chapter also emphasizes a critical distinction in temporal processing for ultra-short intervals, millisecond, and supra-second intervals. Timing of ultra-short intervals (approximately those less than 200ms) may utilize SDN in relevant cortical regions. Supra-second intervals (approximately those greater than 2 s) may utilize basal ganglia processes with an increase in weight from cortical areas, such as the frontal cortex, for cognitive demands such as working memory. Central to our review is the cerebellum's role for millisecond range intervals (approximately 200–2000ms). While such a distinction may seemingly diminish the functional capacity of the cerebellum, we posit that a deficit in millisecond range timing has substantial ramifications as evidenced by the ataxic movements and impaired perceptions of individuals with cerebellar lesions and disorders.

References

Ackermann, Hermann et al. 1997a. "Categorical Speech Perception in Cerebellar Disorders" *Brain and Language* 60: 323–31.

Ackermann, Hermann et al. 1997b. "Kinematic Analysis of Articulatory Movements in Central Motor Disorders" *Movement Disorders* 12(6): 1019–27.

Ackermann, Hermann, and Ingo Hertrich. 2000. "The Contribution of the Cerebellum to Speech Processing" *Journal of Neurolinguistics* 13: 95–116.

Akshoomoff, Natacha A., and Eric Courchesne. 1992. "A New Role for the Cerebellum in Cognitive Operations" *Behavioral Neuroscience* 106: 731–8.

Aparicio, Paul D., Joern Diedrichsen, and Richard B. Ivry. 2005. "Effects of Focal Basal Ganglia Lesions on Timing and Force Control" *Brain and Cognition* 58: 62–74.

Apps, Richard, and Richard Hawkes. 2009. "Cerebellar Cortical Organization: A One-Map Hypothesis" *Nature Reviews Neuroscience* 10(9): 670–81.

Aso, Kenji et al. 2010. "Cerebro-Cerebellar Interactions Underlying Temporal Information Processing" *Journal of Cognitive Neuroscience* 22(12): 2913–25.

Bares, Martin et al. 2007. "Impaired Predictive Motor Timing in Patients with Cerebellar Disorders" *Experimental Brain Research* 180(2): 355–65.

Bares, Martin et al. 2010a. "Predictive Motor Timing Performance Dissociates between Early Diseases of the Cerebellum and Parkinson's Disease" *Cerebellum* 9(1): 124–35.

Bares, Martin et al. 2010b. "The Neural Substrate of Predictive Motor Timing in Spinocerebellar Ataxia" *Cerebellum* 9(1): 124–35.

Bogousslavsky, Julien, Guy Van Melle, and Franco Regli. 1988. "The Lausanne Stroke Registry: Analysis of 1,000 Consecutive Patients with First Stroke" *Stroke* 19(9): 1083–92.

Braitenberg, Valentino. 1983. "The Cerebellum Revisited" *Journal of Theoretical Neurobiology* 2: 237–41.

Buonomano, Dean V., and Michael D. Mauk. 1994. "Temporal Discrimination and the Timing of Motor Responses" *Neural Computation* 6(1): 38–55.

Canter, Gerald. 1963. "Speech Characteristics of Patients with Parkinson's Disease: Intensity, Pitch and Duration" *Journal of Speech and Hearing Disorders* 28(3): 221–9.

Carson, Richard G. et al. 1997. "The Dynamics of Bimanual Circle Drawing" *Quarterly Journal of Experimental Psycholology Section A: Human Experimental Psychology* 50(3): 664–83.

Casini, Laurence, and Richard B. Ivry. 1999. "Effects of Divided Attention on Temporal Processing in Patients with Lesions of the Cerebellum or Frontal Lobe" *Neuropsychology* 13(1): 10–21.

Conrad, B., and V.B. Brooks. 1974. "Effect of Dentate Cooling on Rapid Alternating Arm Movements" *Journal of Neurophysiology* 37: 792–804.

Day, B.L. et al. 1998. "Influence of Vision on Upper Limb Reaching Movements in Patients with Cerebellar Ataxia" *Brain* 121(2): 357–72.

Desmond, John E., S.H. Annabel Chen, and Perry B. Shieh. 2005. "Cerebellar Transcranial Magnetic Stimulation Impairs Verbal Working Memory" *Annals of Neurology* 58(4): 553–60.

Dichgans, J., and H.C. Diener. "Clinical Evidence for Functional Compartmentalization of the Cerebellum" In *Cerebellar Functions*, edited by J.R. Bloedel and J. Dichgans, 126–47. Berlin-Heidelberg: Springer Verlag, 1984.

Durr, Alexandra. 2010. "Autosomal Dominant Cerebellar Ataxias: Polyglutamine Expansions and Beyond" *Lancet Neurology* 9(9): 885–94.

Flament, D., and J. Hore. 1986. "Movement and Electromyographic Disorders Associated with Cerebellar Dysmetria" *Journal of Neurophysiology* 55(6): 1221–33.

Franz, Elizabeth A., Richard B. Ivry, and Laura L. Helmuth. 1996. "Reduced Timing Variability in Patients with Unilateral Cerebellar Lesions during Bimanual Movements" *Journal of Cognitive Neuroscience* 8: 107–18.

Franz, Elizabeth A., Howard N. Zelaznik, and Anne Smith. 1992. "Evidence of Common Timing Processes in the Control of Manual, Orofacial, and Speech Movements" *Journal of Motor Behavior* 24(3): 281–7.

Gerwig, Marcus et al. 2003. "Comparison of Eyeblink Conditioning in Patients with Superior and Posterior Inferior Cerebellar Lesions" *Brain* 126: 71–94.

Gibbon, John. 1977. "Scalar Expectency Theory and Weber's Law in Animal Timing" *Psychological Review* 84: 279–325.

Gibbon, John, and Russell M. Church. 1990. "Representation of Time" *Cognition* 37(1–2): 23–54.

Glickstein, Mitchell. 1994. "Cerebellar Agenesis" *Brain* 117(5): 1209–12.

Gooch, Cynthia M. et al. 2010. "Interval Timing Disruptions in Subjects with Cerebellar Lesions" *Neuropsychologia* 48(4): 1022–31.

Gormezano, I. et al. 1962. "Nictitating Membrane: Classical Conditioning and Extinction in the Albino Rabbit" *Science* 138: 33–4.

Grodd, Wolfgang et al. 2001. "Sensorimotor Mapping of the Human Cerebellum: Fmri Evidence of Somatotopic Organization" *Human Brain Mapping* 13: 55–73.

Habas, Chrisophe, H. Axelrad, and E.-A. Cabanis. 2004. "The Cerebellar Second Homunculus Remains Silent during Passive Bimanual Movements" *NeuroReport* 15: 1571–4.

Haken, Hermann, J.A. Scott Kelso, and H. Bunz. 1985. "A Theoretical Model of Phase Transition in Human Hand Movements" *Biological Cybernetics* 51: 347–56.

Harrington, Deborah L., Kathleen Y. Haaland, and Neal Hermanowicz. 1998. "Temporal Processing in the Basal Ganglia" *Neuropsychology* 12(1): 3–12.

Harrington, Deborah L. et al. 2004. "Does the Representation of Time Depend on the Cerebellum? Effect of Cerebellar Stroke" *Brain* 127(3): 1–14.

Herculano-Houzel, Suzana. 2010. "Coordinated Scaling of Cortical and Cerebellar Numbers of Neurons" *Frontiers in Neuroanatomy* 4: 12.

Hogan, Neville, and Tamar Flash. 1987. "Moving Gracefully: Quantitative Theories of Motor Coordination" *Trends in Neuroscience* 10: 170–4.

Holmes, Gordon. 1939. "The Cerebellum of Man" *Brain* 62: 1–30.

———. 1917. "The Symptoms of Acute Cerebellar Injuries Due to Gunshot Injuries" *Brain* 40: 461–535.

Hore, Jon, Dagmar Timmann, and Sherry Watts. 2002. "Disorders in Timing and Force of Finger Opening in Overarm Throws Made by Cerebellar Subjects" *Annals of the New York Academy of Sciences* 978: 1–15.

Hutchinson, Siobhan et al. 2003. "Cerebellar Volume of Musicians" *Cerebral Cortex* 13(9): 943–9.

Ivry, R., and R.M.C. Spencer. 2004a. "Evaluating the Role of the Cerebellum in Temporal Processing: Beware of the Null Hypothesis" *Brain* 127(8): E13.

———. 2004b. "The Neural Representation of Time" *Current Opinion in Neurobiology* 14: 225–32.

Ivry, Richard B. 1996. "The Representation of Temporal Information in Perception and Motor Control" *Current Opinion in Neurobiology* 6: 851–7.

Ivry, Richard B., and Daniel M. Corcos. "Slicing the Variability Pie: Component Analysis of Coordination and Motor Dysfunction" In *Variability and Motor Control*, edited by K.M. Newell and D.M. Corcos, 415–47. Urbana, IL: Human Kinetics, 1993.

Ivry, Richard B., and R. Elliot Hazeltine. 1995. "Perception and Production of Temporal Intervals across a Range of Durations: Evidence for a Common Timing Mechanism" *Journal of Experimental Psychology: Human Perception and Performance* 21(1): 3–18.

Ivry, Richard B., Steve W Keele, and H.C. Diener. 1988. "Dissociation of the Lateral and Medial Cerebellum in Movement Timing and Movement Execution" *Experimental Brain Research* 73: 167–80.

Ivry, Richard B., and Steve W. Keele. 1989. "Timing Functions of the Cerebellum" *Journal of Cognitive Neuroscience* 1: 136–52.

Ivry, Richard B., and John E. Schlerf. 2008. "Dedicated and Intrinsic Models of Time Perception" *Trends in Cognitive Sciences* 12(7): 273–80.

Ivry, Richard B., Rebecca M.C. Spencer, Howard N. Zelaznik, and Joern Diedrichsen. 2002. "The Cerebellum and Event Timing" *Annals of the New York Academy of Sciences* 978: 302–17.

Karmarkar, Uma R., and Dean V. Buonomano. 2007. "Timing in the Absence of Clocks: Encoding Time in Neural Network States" *Neuron* 53(3): 427–38.

Keele, Steve W., and Harold L. Hawkins. 1982. "Explorations of Individual Differences Relevant to High Level Skill" *Journal of Motor Behavior* 14(1): 3–23.

Keele, Steve W. et al. 1985. "Do Perception and Motor Production Share Common Timing Mechanisms: A Correctional Analysis" *Acta Psychologica* 60(2–3): 173–91.

Kelly, Roberta M., and Peter L. Strick. 2003. "Cerebellar Loops with Motor Cortex and Prefrontal Cortex of a Nonhuman Primate" *Journal of Neuroscience* 23(23): 8432–44.

Kelso, J.A. Scott. 1984. "Phase Transitions and Critical Behavior in Human Bimanual Coordination" *Americal Journal of Physiology* 15: 1000–4.

Krienen, Fenna M., and Randy L. Buckner. 2009. "Segregated Fronto-Cerebellar Circuits Revealed by Intrinsic Functional Connectivity" *Cerebral Cortex* 19(10): 2485–97.

Lewis, Penelope A., and R. Chris Miall. 2003. "Brain Activation Patterns during Measurement of Sub- and Supra-second Intervals" *Neuropsychologia* 41: 1583–92.

Macchi, G., and M. Bentivoglio. "Agenesis or Hypoplasia of Cerebellar Structures" In *Handbook of Clinical Neurology*, edited by P.J. Vinken and G.W. RBruyn, 367–93. Amsterdam: North-Holland, 1977.

Macdonell, Richard A., Renate M. Kalnins, and Geoffrey A. Donnan. 1987. "Cerebellar Infarction: Natural History, Prognosis, and Pathology" *Stroke* 18(5): 849–55.

Mangels, Jennifer A., Richard B. Ivry, and Naomi Shimizu. 1998. "Dissociable Contributions of Prefrontal and Neocerebellar Cortex to Time Perception" *Cognitive Brain Research* 7: 15–39.

Marelli, Cecilia et al. 2011. "Sca15 Due to Large Itpr1 Deletions in a Cohort of 333 White Families with Dominant Ataxia" *Archives of Neurology* 68(5): 637–43.

Matell, Matthew S., Warren H. Meck, and Miguel A.L. Nicolelis. 2003. "Interval Timing and the Encoding of Signal Duration by Ensembles of Cortical and Striatal Neurons" *Behavioral Neuroscience* 117(4): 760–73.

McCormick, David A., and Richard F. Thompson. 1984. "Cerebellum: Essential Involvement in the Classically Conditioned Eyelid Response" *Science* 223(4633): 296–9.

Middleton, Frank A., and Peter L. Strick. 2001. "Cerebellar Projections to the Prefrontal Cortex of the Primate" *Journal of Neuroscience* 21(2): 700–12.

Muzaimi, M.B. et al. 2004. "Population Based Study of Late Onset Cerebellar Ataxia in South East Wales" *Journal of Neurology, Neurosurgery, and Psychiatry* 75(8): 1129–34.

Nitschke, Matthias F. et al. 1996. "Somatotopic Motor Representations in the Human Anterior Cerebellum: A High-Resolution Functional mri Study" *Brain* 119: 1023–9.

Nowak, Dennis A., Dagmar Timmann, and Joachim Hermsdorfer. 2007. "Dexterity in Cerebellar Agenesis" *Neuropsychologia* 45(4): 696–703.

O'Boyle, Donald J., John S. Freeman, and Fredrick W.J. Cody. 1996. "The Accuracy and Precision of Timing of Self-paced, Repetitive Movements in Subjects with Parkinson's Disease" *Brain* 119: 51–70.

Oakley, David A., and I. Steele Russell. 1972. "Neocortical Lesions and Pavlovian Conditioning" *Physiology and Behavior* 8(5): 915–26.

Penhune, Virginia B., Robert J. Zatorre, and Alan C. Evans. 1998. "Cerebellar Contributions to Motor Timing: A Pet Study of Auditory and Visual Rhythm Reproduction" *Journal of Cognitive Neuroscience* 10(6): 752–65.

Perrett, Stephan P., Blenda P. Ruiz, and Michael D. Mauk. 1993. "Cerebellar Cortex Lesions Disrupt Learning-Dependent Timing of Conditioned Eyelid Responses" *Journal of Neuroscience* 13(4): 1708–18.

Posner, Michael I. *Chronometric Explorations of Mind*. Hillsdale, NJ: Lawerence Erlbaum Associates, 1978.

Rijntjes, Michel et al. 1999. "Multiple Somatotopic Representations in the Human Cerebellum" *NeuroReport* 10: 3653–8.

Robertson, Shannon D. et al. 1999. "Correlations for Timing Consistency among Tapping and Drawing Tasks: Evidence against a Single Timing Process for Motor Control" *Journal of Experimental Psychology: Human Perception and Performance* 25: 1316–30.

Schalling, Elika, Britta Hammarberg, and Lena Hartelius. 2007. "Perceptual and Acoustic Analysis of Speech in Individuals with Spinocerebellar Ataxia (Sca)" *Logopedics Phoniatrics Vocology* 32(1): 31–46.

Schlerf, John E. et al. 2010. "Evidence of Somatotopy in the Lateral Cerebellar Hemisphere Associated with Complex Actions" *Journal of Neurophysiology* 103(6): 3330–6.

Schols, Ludger et al. 2004. "Autosomal Dominant Cerebellar Ataxias: Clinical Features, Genetics, and Pathogenesis" *Lancet Neurology* 3(5): 291–304.

Semjen, Andras, Jeffrey J. Summers, and Daniel Cattaert. 1995. "Hand Coordination in Bimanual Circle Drawing" *Journal of Experimental Psychology: Human Perception and Performance* 21(5): 1139–57.

Spencer, Rebecca M.C., and Richard B. Ivry. "Cerebellum and Timing" In *Handbook of the Cerebellum and Cerebellar Disorders*, edited by Mario Manto, Jeremy

D. Schmahmann, Ferdinando Rossi, Donna L. Gruol and Norkyuki Koibuchi. New York: Springer, 2013.

———. 2005. "Comparison of Patients with Parkinson's Disease or Cerebellar Lesions in the Production of Periodic Movements Involving Event-based or Emergent Timing" *Brain and Cognition* 58: 84–93.

———. 2009. "Sequence Learning Is Preserved in Individuals with Cerebellar Degeneration When the Movements are Directly Cued" *Journal of Cognitive Neuroscience* 21(7): 1302–10.

Spencer, Rebecca M.C., Uma R. Karmarkar, and Richard B. Ivry. 2009. "Evaluating Dedicated and Intrinsic Models of Temporal Encoding by Varying Context" *Philosophical Transactions of the Royal Society of London, Series B* 364(1525): 1853–63.

Spencer, Rebecca M.C. et al. 2007. "Cerebellar Activation during Discrete and Not Continuous Timed Movements: An Fmri Study" *NeuroImage* 36: 378–87.

Spencer, Rebecca M.C. et al. 2003. "Disrupted Timing of Discontinuous Movements by Cerebellar Lesions" *Science* 300: 1437–9.

Spencer, Rebecca M., and Howard N. Zelaznik. 2003. "Weber (Slope) Analyses of Timing Variability in Tapping and Drawing Tasks" *Journal of Motor Behavior* 35(4): 371–81.

Swanson, Larry W. 1995. "Mapping the Human Brain: Past, Present, and Future" *Trends in Neurosciences* 18(11): 471–4.

Tesche, Claudia D., and Jari J.T Karhu. 2000. "Anticipatory Cerebellar Responses during Somatosensory Omission in Man" *Human Brain Mapping* 9: 119–42.

Theoret, Hugo, Jasmine Haque, and Alvaro Pascual-Leone. 2001. "Increased Variability of Paced Finger Tapping Accuracy Following Repetitive Magnetic Stimulation of the Cerebellum in Humans" *Neuroscience Letters* 306: 29–32.

Timmann, Dagmar et al. 2000. "Predictive Control of Muscle Responses to Arm Perturbations in Cerebellar Patients" *Journal of Neurology, Neurosurgery, and Psychiatry* 69(3): 345–52.

Timmann, Dagmar, Sherry Watts, and Jon Hore. 1999. "Failure of Cerebellar Patients to Time Finger Opening Precisely Causes Ball High-Low Inaccuracy in Overarm Throws" *Journal of Neurophysiology* 2: 103–14.

Trouillas, P. et al. 1997. "International Coopertative Ataxia Rating Scale for Pharmacological Assessment of the Cerebellar Syndrome" *Journal of Neurological Sciences* 145: 205–11.

Turvey, Michael T. "Preliminaries to a Theory of Action with Reference to Vision" In *Perceiving, Acting, and Knowing*, edited by Robert Bransford, 211–265. Hillsdale, NJ: Lawerence Erlbaum, 1977.

van de Warrenburg, B.P. et al. 2002. "Spinocerebellar Ataxias in the Netherlands: Prevalence and Age at Onset Variance Analysis" *Neurology* 58(5): 702–8.

Velioglu, Sibel K., Kayhan Kuzeyli, and Mehmet Ozmenoglu. 1998. "Cerebellar Agenesis: A Case Report with Clinical and Mr Imaging Findings and a Review of the Literature" *European Journal of Neurology* 5(5): 503–6.

Vemmos, K.N. et al. 2000. "The Athens Stroke Registry: Results of a Five-year Hospital-based Study" *Cerebrovascular Diseases* 10(2): 133–41.

Voogd, Jan, and Mitchell Glickstein. 1998. "The Anatomy of the Cerebellum" *Trends in cognitive sciences* 2(9): 307–13.

Wing, Alan M., and A. Kristofferson. 1973. "Response Delays and the Timing of Discrete Motor Responses" *Perception and Psychophysics* 14: 5–12.

Wojtecki, Lars et al. 2011. "Modulation of Human Time Processing by Subthalamic Deep Brain Stimulation" *PLoS One* 6(9): e24589.

Woodruff-Pak, Diane S., Michelle Papka, and Richard B. Ivry. 1996. "Cerebellar Involvement in Eyeblink Classical Conditioning in Humans" *Neuropsychology* 10: 443–58.

Xu, Duo et al. 2006. "Role of the Olivo-Cerebellar System in Timing" *Journal of Neuroscience* 26(22): 5990–5.

Yamazaki, Tadashi, and Shigeru Tanaka. 2005. "Neural Modeling of an Internal Clock" *Neural Computation* 17(5): 1032–58.

Zelaznik, Howard N., Rebecca M.C. Spencer, and Richard B. Ivry. 2002. "Dissociation of Explicit and Implicit Timing in Repetitive Tapping and Drawing Movements" *Journal of Experimental Psychology: Human Perception and Performance* 28: 575–88.

Zelaznik, Howard N. et al. 2005. "Timing Variability in Circle Drawing and Tapping: Probing the Relationship between Event and Emergent Timing" *Journal of Motor Behavior* 37(5): 395–403.

CHAPTER 10

Striatal and Frontal Pathology: Parkinson's Disease and Patients with Lesions of the Basal Ganglia and Frontal Cortex

Catherine R.G. Jones and Marjan Jahanshahi***

1 Introduction

Interval timing is the timing of perceptual or motor events in the millisecond and seconds-range. Everyday examples of perceptual timing are keeping track of elapsed time so that you can estimate that your kettle has boiled, or judging that your favorite television program is due to start. In the laboratory, perceptual timing has been investigated using a wide variety of tasks, including duration discrimination, time estimation, time production and time reproduction. Motor timing is employed in everyday life when we find ourselves clapping in time with music or dancing to the beat. In the laboratory, the more prosaic activity of tapping in time with a regularly paced tone and continuing to do so after cessation of the tone (synchronization-continuation task; S-C) has been used. It has been proposed that at the core of our ability to efficiently time movements and events is a brain-based 'internal clock' that enables precise calculations (see Merchant et al., 2013 for a recent review). However, successful temporal processing requires more than just a working clock. Ancillary cognitive processes are an integral part of a complex neural architecture that gives rise to accurate motor and perceptual timing. As an example, any task in which two intervals are compared requires an on-line representation of the previously presented interval, alongside dynamic updating of the current interval. Therefore, simultaneous maintenance and encoding are required, placing demands on attention and working memory.

This conceptualization of timing is supported by the most dominant model of interval timing, the scalar expectancy theory (SET; Gibbon, 1977; Gibbon et al., 1984). SET conceives that our representations of time are supported in a three-stage process consisting of clock, memory, and decision-making

* School of Psychology, Cardiff University, UK.
** Sobell Department of Motor Neuroscience and Movement Disorders, UCL Institute of Neurology, London, UK.

DOI 10.1163/9789004230699_011

components (see Figure 10.1). The clock stage is comprised of a pacemaker that emits pulses, with the pacemaker connected to an accumulator. At the onset and offset of an interval that is to be timed, pulses are gated from the pacemaker to the accumulator via a switch, which is operated by a timing signal. The accumulator therefore holds a representation of the current time value, which can be transferred to working memory. The reference memory component provides a more permanent store for important durations (i.e., a standard/ reference duration). The decision process is provided by a comparator, which compares the current time in working memory to the stored time in reference memory. The 'scalar' part of the model comes from applying the observation that human (and animal) timing conforms to the scalar property, whereby the standard deviation of timed responses increases with the mean of the interval being timed. There are challenges to the SET model of interval timing, particularly regarding the biological plausibility of the model (e.g., Matell and Meck, 2000, 2004). However, the model is a salient demonstration of the widespread belief that clock processes are supported by a range of cognitive operations

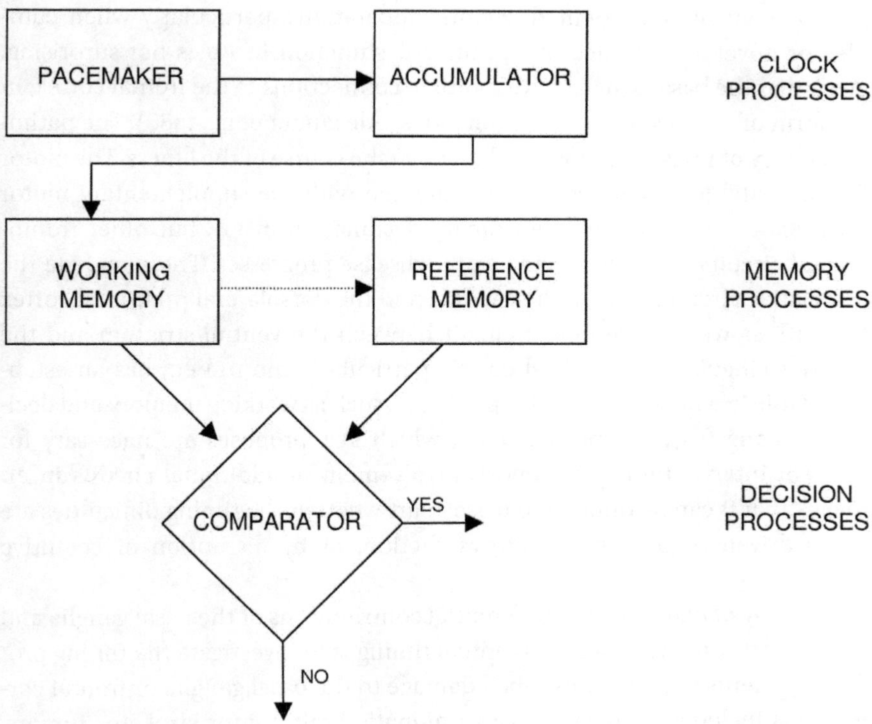

FIGURE 10.1 *Scalar expectancy theory (Gibbon, 1977; Gibbon et al., 1984).*

(attention, memory, decision making), which may have their own distinct neural correlates.

The focus of clinical investigation into interval timing has been on identifying the structure(s) that are integral to the clock process. A large body of literature has investigated the role of the basal ganglia, particularly the temporal profile of individuals with Parkinson's disease (PD). PD is the result of a loss of dopamine producing neurons in the substantia nigra pars compacta (SNc), a pathology that leads to akinesia (lack and poverty of movement), bradykinesia (slowed movement initiation and execution), rigidity (increased muscle tone), and tremor at rest. However, aside from the cardinal motor difficulties, cognitive and affective difficulties are also prevalent. These include depression, anxiety, apathy, hallucinations and delusions, sleep disturbance and cognitive decline (e.g., Chaudhuri and Schapira, 2009). Particularly, impaired executive function is observed from early on in the disease (for a review see Dirnberger and Jahanshahi, 2013), with more widespread cognitive decline emerging as the disease progresses. Executive functions are primarily mediated by the prefrontal cortex and include a range of cognitive processes, including planning, task switching, and inhibition. These are integrated and coordinated to enable achievement of goal-oriented cognitive operations, particularly when complex or novel. Occurrence of executive dysfunction in PD is not surprising, given that the basal ganglia have multiple connections to the frontal cortex, in the form of a series of frontostriatal loops (Alexander et al., 1986). The pathophysiology of PD is complex and changes in the course of the illness. The motor frontostriatal loop, connecting the putamen with the supplementary motor area (SMA), is the site of the primary dysfunction in PD, but other frontostriatal circuits become affected as the disease progresses. These include the associative circuit between the caudate and the dorsolateral prefrontal cortex (DLPFC), as well as the limbic circuit between the ventral striatum and the anterior cingulate. The frontal cortex, particularly the DLPFC, has an established role in a range of cognitive processes, such as working memory and decision making (e.g., D'Esposito, 2007), which SET proposes are necessary for efficient interval timing. Therefore, involvement of additional circuits in PD means that it can be difficult to tease apart whether the timing difficulties are being driven by a core timing dysfunction, or by disruption of cognitive processes.

One way of exploring the differential contributions of the basal ganglia and frontal cortex to motor and perceptual timing is to investigate the timing profile of patients with circumscribed damage to the basal ganglia or frontal cortex. This includes individuals with traumatic brain injury, stroke or tumors. The capacity of the human brain for the development of compensatory neural circuits means that caution must be exercised when investigating the effects of acquired brain injury. Further, studies are limited by the difficulty of finding patients with comparable pathology, as well as with lesions that are solely limited to the region of interest. However, such patients provide an unparalleled insight into the function of the basal ganglia and frontal cortex in temporal processing. Therefore, the present chapter draws on evidence from patients with circumscribed lesions and patients with PD to better understand the roles of the basal ganglia and frontal cortex in interval timing. There is a great deal of heterogeneity among patients with acquired lesions. We have limited our evaluation to those with a discrete lesion, e.g., caused by tumor, stroke or head injury. Patients with degenerative disease (e.g., dementia) have not been included. For the frontal cortex, studies that investigate a range of lesions sites have been included if the reporting of the data enables conclusions to be drawn about the specific contribution of the frontal cortex. In complement to the current chapter, Chapter 8 provides an overview of timing in neurodegenerative disorders of the basal ganglia, particularly PD and Huntington's disease. Chapter 11 also includes a summary of functional imaging research in PD, which is not covered in the current chapter. We start with an overview of the tasks that have been employed to assess timing, before reviewing the empirical evidence.

2 Measures of Perceptual and Motor Timing

Perceptual timing tasks are far more varied than motor timing tasks, and refer to any subjective judgment of time that is not defined by movement. Sometimes a motor response in a perceptual task may be tied to the temporal decision (e.g., pressing a button when a certain period of time is perceived to have elapsed), thus the categorization of tasks is not clear cut. Indeed, studies have shown a significant correlation between performance on motor and perceptual timing tasks (e.g., Keele et al., 1985; Merchant et al., 2008), which is compatible with there being a shared neural substrate. Further, although this chapter focuses on the perceptual vs. motor distinction there are other ways of categorizing temporal tasks, including the length of interval used (e.g., Lewis and Miall, 2003) and whether timing is an explicit demand of the task or implicit and emergent (e.g., intercepting a moving target; see Coull and Nobre, 2008).

Perceptual timing is commonly measured in the visual or auditory modality, using presentation of simple auditory (e.g., pure tone) or visual (e.g., small square) stimuli to define temporal intervals. Another variation is whether the intervals are be 'filled' (i.e., the stimulus is present for the duration of the interval) or 'unfilled' (i.e., the onset and offset of the interval is bounded by two stimuli, but the actual interval is empty). Filled intervals can sometimes take the form of an active and additional task, for example, counting or reading aloud random numbers. All perceptual tasks share certain cognitive demands including sustained attention to the temporal interval(s) as well as decision-making/evaluative processes. However, the reliance on other cognitive processes varies with task demands. The duration discrimination task presents two stimuli sequentially and requires the participant to discriminate which is longer, or whether the second stimuli is longer or shorter than the first. The appeal of this task is that the perceptual temporal decision is not confounded by movement. At the cognitive level, the demand on working memory is high as both intervals need to be simultaneously held on-line and compared.

The time estimation task and time production task share similarities, as both require units of time to be accurately applied to an interval. In the time estimation task the participant has to estimate the length of a presented interval to the nearest second. In contrast, the time production task requires the participant to indicate (e.g., press a button) when a specified interval of time has elapsed. Therefore, in both of these tasks the participant has to map their understanding of the common units of time onto their internal sense of time passing. The time reproduction task presents an interval of time but rather than compare it to another interval, the participant has to reproduce the duration of the interval as accurately as possible using a response key. Similar to the duration discrimination task, this places considerable demand on working memory. For the time estimation, production, and reproduction tasks, sometimes participants are instructed to count out intervals, either at a self-paced and self-preferred rate, at a self-paced but specified rate (e.g., 1 s), or at an externally paced rate (e.g., provided by a metronome). These modifications mean that participants are both required to accurately perceive a discrete interval and time a short continuous sequence that is intrinsically tied to motor production. Thus, many purported perceptual timing tasks have an implicit motor timing element. Chronometric counting is distinct from interval timing, albeit with possible overlap (e.g., Hinton et al., 2004). Therefore, some studies require that random numbers are read aloud to inhibit counting. However, all of these tasks are ostensibly dual task paradigms, which is differentially demanding for patients with neurological damage compared to healthy controls (e.g., Brown and Marsden, 1991; Perbal et al., 2005). Although requiring participants not to count or read random numbers may present as an obvious solution, it then becomes difficult to control the mental activities that participants engage in during the task.

Three classic tasks from the animal timing literature, the peak interval procedure, the temporal generalization task and the temporal bisection task have been adapted to investigate perceptual timing in patient studies. These three tasks share a commonality in demanding that a learnt interval be stored in short term memory and recalled when required. In the peak interval procedure, the participants are trained on a visually presented target duration that they will need to reproduce. In the testing phase, the visual stimulus is presented for a longer period. Participants have to press a button when they think the target duration has elapsed but they are instructed to make multiple guesses; pressing the button before the estimated duration has elapsed and continuing until they judge it has passed. In many ways analogous to a time reproduction task, this procedure enables a response curve to be plotted, showing a peak at the time where responses are most frequent. With time plotted on the x-axis, a curve with a peak shifted to the right indicates relative overestimation, whereas a peak shifted to the left indicates underestimation. In the temporal generalization task participants are repeatedly exposed to a target duration, which becomes learnt. During the testing phase participants have to indicate whether the durations they are presented with are the same or different to the learnt target. The proportion of 'yes' responses can be plotted to create a temporal generalization gradient, illustrating the probability of a response as a function of signal duration. With duration plotted on the x-axis, a rightward shift in the peak of the generalization function would indicate overestimation of the target duration, whereas a leftward shift would suggest underestimation. The temporal bisection task requires participants to learn a 'short' and a 'long' target duration. In the testing phase, participants are presented with a range of durations including the learnt targets. Each duration has to be classified as more similar to the short or long target. The resultant sigmoid curve plots the probability of making a 'long' response as a function of stimulus duration. With durations plotted on the x-axis, a leftward shift in the curve indicates a relative overestimation of time. The bisection point (or point of subjective equality; PSE) is the duration at which both responses occur with equal probability.

Motor timing is assessed using a repetitive tapping task, often called the S-C task. First, the participant taps in time with a regularly paced cue, typically an auditory pure tone. Following a criterion number of taps the pacing cue is switched off and the participant has to maintain the same rhythm unaided. Thus, there are two discrete phases to the task, a synchronization phase and a continuation phase, which are measured separately. The accuracy of the tapping rate, usually measured by the mean inter-response interval, is important in determining whether tapping is unusually slow or fast. Variability has classically been measured using the Wing and Kristofferson (1973a,b) model, which decomposes tapping variability into 'clock' and 'motor' components, although the model is not without caveats (see Jones et al., 2011). The S-C task is considered to be more 'automatic' and less cognitively demanding than perceptual tasks, particularly when short inter-tap intervals are used (Lewis and Miall, 2003).

3 Timing in Patients with Parkinson's Disease

3.1 *Perceptual Timing*

Perceptual timing tasks in patients with PD are summarized in Table 10.1. Of the 37 tasks, there are significant differences between the PD and control group on 23 tasks (62%). Data have indicated underestimation on time estimation tasks (Lange et al., 1995; Pastor et al., 1992a) and overestimation on time production tasks (Lange et al., 1995; Jones et al., 2008) in patients with PD. This pattern of results is compatible with a slowed internal clock (although see Wojtecki et al., 2011). The slowed clock hypothesis is particularly compelling as it neatly reflects the bradykinetic presentation of patients with PD. However, many other studies fail to find evidence of impairment on these two tasks (Perbal et al., 2005; Riesen and Schnider, 2001; Wearden et al., 2008; Wild-Wall et al., 2008). Further, patterns of impairment on other timing tasks are not amenable to the slowed clock hypothesis. For example, in the time reproduction task the internal clock is being used to both estimate and reproduce; this means the speed of the clock could not influence the findings, assuming the rate was regular.

The time reproduction task is the most discriminating task, and two thirds of studies reported some degree of impairment in PD (see Table 10.1). It is conceivable that the motor response, which is intrinsically tied to the temporal decision, may compound or drive the poor performance on the task. However, this interpretation does not explain the pattern of within-task overestimation on relatively short intervals compared to underestimation on relatively long intervals (see Koch et al., 2004; 2005). Patients with PD have also shown evidence of impairment on the duration discrimination task, although preserved performance is also reported (see Table 10.1). Harrington et al. (1998a) reported dissociation in PD between impaired duration discrimination and preserved frequency discrimination, which is compatible with a timing-specific deficit and not general cognitive dysfunction. However, the duration task, unlike the frequency task, requires attention be maintained for the duration of the presented stimuli. Therefore, the additional attentional demands of the duration task provide an alternative explanation for the pattern of results.

TABLE 10.1 *Studies of perceptual timing in patients with Parkinson's disease (PD) compared to healthy controls, organized by type of task.*

Study	Medication state	Modality	Standard duration (s)	Results
Time estimation				
Lange et al. (1995)	On/Off	–	10, 30, 60	Underestimation (Off)
Pastor et al. (1992a)	Off	V	3, 9, 27	Underestimation
Riesen and Schnider (2001)	On	V	12, 24, 48	No impairment
Wearden et al. (2008)	On/Off	A	0.77–1.18	No impairment
Time production				
Jones et al. (2008)	On	A	30, 60, 120	Overestimation
Lange et al. (1995)	On/Off	–	10, 30, 60	Overestimation (Off)
Perbal et al. (2005)	On	V	5, 14, 38	Counting: No impairment
				Random number: underestimation
Wild-Wall et al. (2008)	On	V	1.2	No impairment
Wojtecki et al. (2011)*	Off	V	5, 15	Underestimation (15 s)
Time reproduction				
Jones et al. (2008)	On	A	0.25, 0.5, 1, 2	↓ variability. Violation of Scalar property
Koch et al. (2004)*	Off	V	5, 15	Overestimate 5 s, Underestimate 15 s
Koch et al. (2005)	Off	V	5, 15	Left-hemi PD: Overestimate 5 s, underestimate 15 s
				Right-hemi PD: Overestimate 5 s
Koch et al. (2008)	On/Off	V	Short: 0.4, 0.45, 0.5, 0.55, 0.6	Standard: Underestimation (Long)
			Long: 1.6, 1.8, 2, 2.2, 2.4	1 hour delay: No impairment
Merchant et al. (2008)	On/Off	A	0.35, 0.45, 0.55, 0.65, 0.85, 1	↑ variability
Pastor et al. (1992a)**	Off	V	2, 3, 4.5, 6, 9	Overestimation

TABLE 10.1 *Studies of perceptual timing in patients with Parkinson's disease* (PD) *compared to healthy controls, organized by type of task* (cont.)

Study	Medication state	Modality	Standard duration (s)	Results
Perbal et al. (2005)	On	V	5, 14, 38	Counting: No impairment Random number: ↑ variability
Torta et al. (2010)*	On/Off	–	Seconds-range, varied for each participant	UF: No impairment F: Underestimation (Off)
Wojtecki et al. (2011)*	Off	A	5, 15	No impairment
Duration discrimination				
Guehl et al. (2008)*	Off	A	0.05	Impaired
Harrington et al. (1998a)	On	A	0.3, 0.6	Impaired
Harrington et al. (2011)†	On/Off	A&V	1.2, 1.8	Impaired
Hellström et al. (1997)	On	A	0.4, 0.8, 1.2, 1.6	No impairment at group level Gender differences reported
Ivry and Keele (1989)	On	A	0.4	No impairment
Rammsayer and Classen (1997)	On	A	0.05	Impaired
Riesen and Schnider (2001)	On	V	0.2, 1	Impaired
Wearden et al. (2008)	On/Off	A	0.4 with short delay (1.1s) between tones; 0.35 – 0.65 with long delay (2, 4, 8s)	Short delay: No impairment Long delay: Impairment
Wojtecki et al. (2011)*	Off	A	1.2	No impairment

Temporal bisection				
Merchant et al. (2008)	On/Off	A	0.35, 1	↑ variability (Off)
Smith et al. (2007)	On	A&V	Short (0.1, 0.5) & Long (1, 5)	Impaired (Long)
Wearden et al. (2008)	On/Off	A	0.2, 0.8	No impairment
Temporal generalization				
Wearden et al. (2008)	On/Off	A	0.4	No impairment
Peak interval procedure				
Malapani et al. (1998)	On/Off	V	8, 21	↓ accuracy for 21 s (Off) ↑ variability for 8 s (Off)

V, visual; A, auditory; A&V, auditory and visual separately; F, filled interval; UF, unfilled interval

* Patients had undergone STN DBS, data in the table are reported when stimulation was turned off.

** Pastor et al. (1992a) tested 5 variations of the time reproduction task. As all tasks produced the same result we present the data as one result in the table, with reference to the different manipulations in the main body of the article.

† Data collected during a neuroimaging (fMRI) experiment.

The temporal generalization and bisection tasks present a mixed picture, with no consistent evidence of impairment (see Table 10.1). However, the peak-interval procedure has produced more convincing results. Here patients show evidence of a 'migration' effect whereby relatively short intervals are overestimated and relatively long intervals are underestimated (Malapani et al., 1998; Malapani et al., 2002). As previously alluded to, this effect has been replicated in studies of time reproduction (Koch et al., 2004; 2005). Systematic investigation of the pattern of responding under different conditions has indicated that deficits in the storage and retrieval of the durations in memory contribute to the migration effect (Malapani et al., 2002). Further, the effect is abolished if the reproductions of the short and long intervals are separated by a delay of one hour (Koch et al., 2008). The profile of performance across tasks investigating the migration effect are, therefore, compatible with a cognitive explanation for the perceptual timing difficulties seen in PD.

Other studies also present data that are compatible with a cognitive explanation. For example, Guehl et al. (2008) employed a duration discrimination task where the standard interval of 50ms was defined by two clicks. Although participants with PD were impaired on this task they were unimpaired on a version where trains of clicks paced at 50ms intervals were used and participants had to decide which of two trains had one long interval (>50ms) in the middle. Therefore, although the durations being judged were identical across the two tasks, the task context was different. The authors suggested that performance is impaired on the standard duration discrimination task because the onset of the stimuli cannot be predicted as easily i.e., it requires a greater allocation of attentional resources. An attentional explanation was also put forward by Torta et al. (2010), who found that participants with PD were impaired when reproducing the amount of time it had taken them to complete a simple motor task but unimpaired when the interval to reproduce was unfilled. The dissociation was argued to be driven by the demands of the concurrent motor task, which left less attentional resources for the timing task. Finally, Wearden et al. (2008) assessed patients with PD on five different perceptual timing tasks. The only task that they were impaired on was a duration discrimination task that required the standard interval (350–650ms) to be held in memory for 2–8 s. Therefore, impaired memory function may also be a contributing factor to performance.

As was mentioned in the Introduction of this chapter, PD is principally a movement disorder but cognitive deficits are a feature of the disease. These include executive dysfunction and mild cognitive impairment in the early stages, and dementia in the later stages (Dirnberger and Jahanshahi, 2013; Kehagia, Barker, and Robbins, 2013). If cognitive difficulties contribute to the temporal profile seen in PD then tasks with additional cognitive load should be more likely to elicit impairment. In support of this, the majority (5 of 6; 83%) of tasks that required random numbers to be read out during timing reported a deficit (Koch et al., 2004, 2005; Malapani et al., 1998; Riesen and Schnider, 2001). A cognitive explanation would also predict greater impairment at longer intervals, where demands on memory and attention would be more extensive. Indeed, Jones et al. (2008) used exploratory factor analysis to establish that the time production of seconds-range intervals and a measure of attention (Paced Auditory Serial Addition Test) shared common variance and were distinct from time reproduction and warned reaction time tasks involving millisecond-range interval timing. If the tasks in Table 10.1 are divided by the arbitrary cut-off of 1 s (i.e., standard intervals of 1–1000ms vs. standard intervals >1000ms), 6 of 15 (40%) studies using the shorter intervals reported poor performance in the PD group, compared to 16 of 26 (62%) using the longer interval range. The proportion of those performing poorly at longer intervals is increased further if the cutoff is set at 5 s (71%; 12 of 17). This pattern across studies indicates that longer intervals are more challenging for patients with PD. However, the presence of deficits in tasks with very short intervals (e.g., Guehl et al., 2008; Rammsayer and Classen, 1997) as well as preserved performance in studies where the intervals are substantial (e.g., Perbal et al., 2005; Riesen and Schnider, 2001; Wojtecki et al., 2011) needs to be considered. Further, neuroimaging research has found basal ganglia activation in both millisecond and seconds-range temporal processing (Jahanshahi et al., 2006).

Teasing apart the specific factors that contribute to temporal dysfunction in PD can be aided by investigating correlates of poor performance. Pastor et al. (1992a) found that time reproduction was associated with measures of natural tapping rate, reaction time, and movement time. This suggests that perceptual timing deficits relate specifically to bradykinesia in PD. However, although some studies have found correlations between perceptual timing and disease severity (Pastor et al., 1992a; Wearden et al., 2008) this association is not always found (Rammsayer and Classen, 1997; Smith et al., 2007). In terms of temporal deficits being driven by non-motor deficits, Rammsayer and Classen (1997) reported a significant correlation in PD between increased self-reported depression and greater impairment on a duration discrimination task. Further, poor performance on the time production task (Wild-Wall et al., 2008) and time reproduction task (Perbal et al., 2005) was associated with higher levels of cognitive decline. In contrast, Merchant et al. (2008) found that performance on a range of cognitive tasks (working memory, go/no-go reaction time, and verbal learning) did not discriminate those with PD who performed well or poorly on a range of timing tasks. However, their timing tasks all used intervals

≤1 s, and it may be that general cognitive ability is an additional source of variance only for the more cognitively demanding seconds-range intervals.

3.2 *Motor Timing*

Motor timing studies in PD are summarized in Table 10.2. The influential Wing and Kristofferson (1973a,b) model was designed to characterize tapping in the continuation phase. As such, the synchronization phase has not been analyzed in most studies. However, both healthy participants and those with PD perform better at synchronization than continuation tapping (e.g., Jones et al., 2011) and there is no convincing evidence that the pattern of impairment in PD differs significantly between the two phases (e.g., Jones et al., 2011; Joundi et al., 2012; Pastor et al., 1992b; Wojtecki et al., 2011).

Given that PD is a movement disorder it is perhaps not surprising that motor timing deficits are reported more consistently than perceptual timing deficits. Of the 13 studies reported in Table 10.2, 10 (77%) report group differences in variability or accuracy. However, closer inspection reveals a more complex picture. For accuracy, tapping rate on the S-C task in PD has been shown to be faster (Harrington et al., 1998a; Ivry and Keele, 1989; Jones et al., 2011; O'Boyle et al., 1996), slower (Pastor et al., 1992b), and unimpaired (Duchek, Balota, and Ferraro, 1994; Joundi et al., 2012; Spencer and Ivry, 2005; Wojtecki et al., 2011). One interpretation for these inconsistent findings is that performance depends on the duration being timed. It has been proposed that individuals with PD can demonstrate 'hastening' and festination at the shorter intervals (Claassen et al., 2013; Jones et al., 2011). Festination is a clinical phenomenon often observed in PD and is the tendency to speed up when performing a repetitive movement. A review of the motor timing literature (see Jones et al., 2011; Jones and Jahanshahi, 2014) indicates that the interval range of 400–600ms is the threshold at which performance switches from impaired to unimpaired in PD. In contrast, there does not seem to be a threshold effect for variability, although having tested patients on a range of intervals (250, 500, 1000, and 2000ms), Jones et al. (2011) found that variability for both patients with PD and healthy controls was lowest at 500ms.

As the clinical motor features of PD are compatible with difficulties in motor timing, it is surprising when studies fail to find an association between clinically assessed motor dysfunction and performance on the S-C task (O'Boyle et al., 1996; Spencer and Ivry, 2005). However, positive associations have been reported by some (Harrington et al., 1998a). To our knowledge, there are no studies that have assessed whether there are cognitive or affective correlates of motor timing performance in PD, which should be addressed in future research.

TABLE 10.2 *Studies of the synchronization-continuation (S-C) task in patients with Parkinson's disease (PD) compared to healthy controls*

Study	Medication state	Duration (ms)	Synchronzation: variability	Synchronization: accuracy	Continuation: variability	Continuation: accuracy
Cerasa et al. (2006)†	Off	750	↑	PD = Controls	PD = Control	PD = Controls
Duchek et al. (1994)	On	550	–	–	↓ (MV) PD = Control (SD)	PD = Controls
Elsinger et al. (2003)†	On & Off	600	↑	PD = Control	↑	Faster
Harrington et al. (1998a)	On	300, 600	–	–	↑ (CV, TV)	Faster
Ivry and Keele (1989)	On	550	–	–	PD = Control	Faster
Jahanshahi et al. (2010)†	On & Off	1000	PD = Control	PD = Control	PD = Control	PD = Control
Jones et al. (2011)	On & Off; De novo	250, 500, 1000, 2000	No consistent pattern across durations	Faster at 250ms	No consistent pattern across durations	Faster at 250ms
Joundi et al. (2012)*	On	500, 2000	↑ (CoV)	PD = Control	↑ (CoV, CV, MV)	PD = Control
Merchant et al. (2008)	On & Off	350, 450, 550, 650, 850, 1000	–	–	↑ (SD)	Not reported
O'Boyle et al. (1996)	On & Off	550	–	–	↑ (CV, MV, TV) (only CV sig. when On)	Faster (only sig. when On)

TABLE 10.2 *Studies of the synchronization-continuation (S-C) task in patients with Parkinson's disease (PD) compared to healthy controls* (cont.)

Study	Medication state	Duration (ms)	Synchronzation: variability	Synchronization: accuracy	Continuation: variability	Continuation: accuracy
Pastor et al. (1992b)	On & Off	400, 500, 666, 1000, 2000	–	Slower at 400 and 500ms	↑ (CV, MV, TV)	Slower at 400 and 500ms
Spencer and Ivry (2005)	On & Off	550	–	–	PD = Control	PD = Control
Wojtecki et al. (2011)*	Off	800	PD = Control	PD = Control	PD = Control	PD = Control

CoV, coefficient of variation; SD, standard deviation. Using the Wing and Kristofferson (1973a,b) model: CV, clock variance; MV, motor variance; TV, total variance. The Wing and Kristofferson model is only applied in the continuation phase.
* Patients had undergone STN DBS, data reported in the table are when stimulation was turned off.
† Data collected during a neuroimaging (PET or fMRI) experiment.

3.3 *Summary*

There is clear evidence of impairment in both motor and perceptual timing in the milliseconds and seconds-range in PD, although also a considerable number of studies do not report deficits. Dopaminergic medication can benefit timing performance in PD. This was the case for 57% (8 of 14) of the perceptual timing tasks (see Table 10.1) and 40% (2 of 5) of the motor timing tasks (see Table 10.2) that reported impairment 'off' medication. Interpreting studies that do not report a positive effect of medication is difficult given issues such as individual differences in response to medication, the varying types of medication used, and study-specific differences such as the length of medication withdrawal. Historically, perceptual timing dysfunction has been interpreted as evidence of a slowed internal clock. However, the pattern of deficits across tasks is often not compatible with this explanation. Further, evidence suggests that compromised frontal activity in PD, likely affecting ancillary cognitive processes, may be contributing to or driving the difficulties in perceptual timing. Of course, it is possible that poor temporal processing in PD is driven by a combination of a core timing deficit *and* cognitive impairment. One piece of evidence that provides a compelling argument for a core timing deficit is work by Grahn and Brett (2009). Here a rhythm discrimination task established that patients with PD were as proficient as controls at discrimination when the rhythm did not contain a beat but showed poorer performance when a beat was present. Rhythms with a beat structure are easier to discriminate than rhythms without a beat, so a global cognitive or perceptual impairment could not be driving the timing deficit. What is clear is that the data from studies of PD support the hypothesis that the integrity of the basal ganglia and frontal cortex are essential to motor and perceptual timing.

4 Timing in Patients with Lesions of the Basal Ganglia

Patients with focal lesions to the basal ganglia provide an important approach for assessing the contribution of the basal ganglia to temporal processing. The primary difference between patients with basal ganglia lesions and patients with PD is that the latter have a degenerative disorder that involves areas and systems of the brain beyond the basal ganglia. Therefore, patients with acquired lesions to the basal ganglia allow a more constrained interpretation of the contribution of the basal ganglia to timing function. Ultimately, these data enable us to consider whether the timing dysfunction observed in PD is driven by top-down cognitive factors rather than a critical impairment to an internal timekeeper.

Aparicio et al. (2005) assessed 6 patients with right-sided lesions of the basal ganglia and found no evidence of impairment on the S-C task with an inter-tap interval of 400ms. A similar study was conducted more recently by Schwartze et al. (2011) who adopted an innovative adaptive version of the S-C task. Here, 10 patients with unilateral lesions of the basal ganglia (8 left- and 2 right-sided) were assessed tapping at their most comfortable rate and on an adaptive S-C task, which included a change in tempo (acceleration or deceleration over 3 intervals) at the end of the synchronization phase. Compared to healthy controls, the patient group showed more variable tapping when required to tap at their most comfortable rate and showed a more heterogeneous distribution of individual rates. This suggests difficulty with internally generated temporal pacing following basal ganglia damage, perhaps not surprising given that a primary projection site of the basal ganglia, the pre-supplementary motor area, has a critical role in initiating freely chosen movement (e.g., Halsband et al., 1993). For the S-C task, there was no fundamental difference in tapping accuracy between groups, and differences in variability were significant in a *t*-test but not in a more comprehensive ANOVA. These findings are therefore superficially compatible with Aparicio and colleagues. However, Schwartze et al.'s innovative adaptive paradigm uncovered a novel finding. They found that patients were less able to correct their tapping rate to the accelerated or decelerated rhythm. Analysis of the profile of error correction revealed that the patients were having difficulty with period, rather than phase, correction. Period correction is thought to reflect an internal timekeeper and is assumed to rely on cognitive control. As such, the authors speculate that impairment to attention-dependent aspects of temporal processing lead to deterioration in performance in individuals with basal ganglia lesions when the rhythm changes. Reflecting this interpretation, the motoric disturbance was accompanied by reduced sensitivity for accurately identifying whether the tempo had changed, and in which direction.

The previous studies have only used patients with unilateral lesions, which do not necessarily lead to the expression of behavioural dysfunction. Indeed, when required to tap as fast as possible, Aparicio et al.'s (2005) patients did not show significantly reduced tapping speed. Whether the absence of clear behavioural markers is due to compensation by the unaffected hemisphere or reflects a relatively minor insult is unclear. Regardless, it makes interpreting intact temporal performance difficult. Coslett et al. (2010) addressed this issue by assessing two patients with bilateral damage to the basal ganglia on a comprehensive battery of perceptual and motor timing tasks. Importantly, both patients showed significant evidence of motor-related dysfunction, including dystonic posturing and akinesia; whereas their cognitive profiles did not indicate any areas of impairment. In contrast to the data presented by Schwartze et al. (2011), both patients performed well on measures of perceptual timing. These included time estimation, reproduction, and production tasks in the seconds-range (2–12 s) and a duration discrimination task of both millisecond (300, 600ms) and seconds-range (2, 8 s) intervals. However, when required to perform the S-C task with an inter-stimulus interval of 400ms, one patient showed abnormally short responses, while the other showed significantly elevated variability. This finding appears in contrast to Aparicio et al. and Schwartze et al., neither of whom found evidence of impairment with standard S-C tapping. However, the participants in Schwartze et al.'s study tapped with their least affected (i.e., ipsilesional) hand. Aparicio et al.'s participants were assessed on both hands and the pattern of data indicated aspects of performance, particularly motor variability, were more compromised in the contralesional hand. However, this effect did not reach statistical significance.

The pattern of data across studies highlights the difficulty in separating the non-specific aspects of motor preparation and execution from the specific temporal contribution. Schwartze et al.'s approach of manipulating the temporal component of the task whilst the motor demands remain constant is an elegant way of dissociating the processes that contribute to motor timing. Schwartze et al.'s patients demonstrate a difficulty in attending to time, which is a deficit that should have compromised performance on Coslett et al.'s perceptual timing tasks, particularly as most were in the seconds-range. However, Schwartze et al.'s S-C task required the monitoring of rhythm changes, introducing a perceptual timing element to a motor timing task that is typically considered to be processed automatically and with limited cognitive engagement (Lewis and Miall, 2003). This 'double hit' may have created a demanding task that was better placed to elicit subtle difficulties in attending to time.

In conclusion, the majority of the motor timing data do not support a link between basal ganglia pathology and motor timing dysfunction. However, the finding (Coslett et al., 2010) of motor timing impairment in patients with bilateral lesions, in contrast to patients with unilateral lesions, suggests that the basal ganglia are important for motor timing but that both hemispheres need to be compromised. Pervasive cognitive deficits are absent in patients with lesions to the basal ganglia, in contrast to individuals with PD, which provides greater clarity when interpreting results. Impairment in the detection and response to rhythm change (Schwartze et al., 2011) is compatible with the monitoring of time being compromised and merits further investigation. However, Coslett et al.'s (2010) multi-task assessment of perceptual timing found no evidence of impairment in their two patients with bilateral lesions. Clearly, there is not enough data on perceptual timing in individual with basal ganglia lesions to draw any firm conclusions and further research is needed.

5 Timing in Patients with Lesions to the Frontal Cortex

5.1 *Perceptual Timing*

Studies reporting perceptual timing in patients with lesions to the frontal cortex are summarized in Table 10.3. As with patients with basal ganglia lesions, these patients allow a more specific investigation of the contribution of a particular brain region than is afforded by studying patients with a degenerative disorder. The clearest finding from surveying the data in Table 10.3 is that perceptual timing impairments in frontal patients are more common than in patients with PD and are present in 88% (14 of 16) of tasks across 11 studies. Notably, these impairments are seen for both millisecond and seconds-range intervals. Taking a closer look at the specific pattern of findings is important for understanding the nature of the deficit(s) that are leading to the profile of impairment.

The frontal cortex performs a wide range of cognitive operations and the locus of the lesion site is an important consideration for interpretation. The orbitofrontal (OFC) is associated with impulsive behavior (e.g., Torregrossa et al., 2008), which is intrinsically linked to insensitivity to reward and punishment (Berlin et al., 2004). Delay aversion is a critical component of impulsivity and may impact on any task where a response has to be withheld until a specific time. In keeping with this, Berlin et al. found that patients with lesions to the OFC underestimated the second-range intervals that they were required to produce. They also overestimated the duration of seconds-range intervals that were presented to them. The increased time estimation and decreased time production were positively correlated with the 'attention' subscale of the Barrett Impulsiveness Scale, a self-report impulsivity questionnaire. Whether lesions to the OFC cause a speeded internal clock, which contributes to impulsive behaviour (i.e., the subjective experience of time passing slows, which exacerbates intolerance to reward delay), or if the impulsivity itself drives the temporal distortion (e.g., difficulty in waiting for the interval to end makes it seem longer) is an important question. Of course, as Berlin et al. discuss, it is also possible that OFC patients have a speeded clock *and* an intolerance of delayed gratification. In contrast to those with OFC lesions, patients with damage to the prefrontal cortex (PFC), including the DLPFC, did not show atypical timing behaviour. This is perhaps surprising given the role of the DLPFC in attention (e.g., Katsuki and Constantinidis, 2012).

Well-reported case studies can give a unique insight into the qualitative temporal distortions experienced by those with focal lesions. Binkofski and Block (1996) detail a patient with a tumor in the left superior PFC, including the DLPFC, who described experiencing time as passing more quickly than

TABLE 10.3 *Studies of perceptual timing in patients with lesions of the frontal lobe compared to healthy controls, organized by type of task.*

Study	Lesion location	Laterality	Modality	Standard duration (s)	Results
Duration discrimination					
Casini and Ivry (1999)	PFC	1 R, 4 L	A	0.4	Impaired
Harrington et al. (1998b)	Anterior or posterior cortex	18 R, 19 L	A	0.3, 0.6	Impaired – R hemisphere patients only (anterior and posterior location)
Ivry and Keele (1989)	All included posterior frontal lobe	6 R, 6 L	A	0.4	Unimpaired
Mangels et al. (1998)	PFC	2 R, 5 L	A	0.4, 4	Impaired (4 s; 0.4 s with long ISI -4 s)
Temporal bisection					
Nichelli et al. (1995)	Frontal	6 R, 2 L, 3 B	A	Short: 0.1, 0.9 Long: 8, 32	Impaired accuracy Impaired precision (Long)
Time estimation					
Berlin et al. (2004)	PFC & OFC	PFC: 6 R, 3 L, 1 B OFC: 4 R, 4 L, 1 B	V	10, 30, 60, 90	OFC: Overestimation (total and 90 s)
Coslett et al. (2010)	Anterior or posterior cortex	16 R, 15 L	AV	2, 4, 6, 8, 10, 12	↑ variability
Koch et al. (2002)	DLPFC	1 R	V	5, 10, 30, 60, 90	Underestimation (90 s)
Mimura et al. (2000)	Frontal	2 R, 2 L, 4 B	V	10, 30, 60, 90, 120	↓ accuracy, overestimation particularly for shorter intervals

TABLE 10.3 *Studies of perceptual timing in patients with lesions of the frontal lobe compared to healthy controls, organized by type of task.*

Study	Lesion location	Laterality	Modality	Standard duration (s)	Results
Mimura et al. (2000)	Frontal	2 R, 2 L, 4 B	–	10, 30, 60, 90, 120	↓ accuracy, tendency for overestimation
Time production					
Berlin et al. (2004)	PFC & OFC	PFC: 6 R, 3 L, 1 B OFC: 4 R, 4 L, 1 B	V	10, 30, 60, 90	OFC: Underestimation (90 s)
Binkofski and Block (1996)	PFC	1 L	–	60	Overestimation
Coslett et al. (2010)	Anterior or posterior cortex	16 R, 15 L	AV	2, 4, 6, 8, 10, 12	Unimpaired
Mimura et al. (2000)	Frontal	2 R, 2 L, 4 B	V	10, 30, 60, 90, 120	↑ variability
Time reproduction					
Coslett et al. (2010)	Anterior or posterior cortex	16 R, 15 L	AV	2, 4, 6, 8, 10, 12	↑ variability
Kagerer et al. (2002)	Precentral, postcentral, subcortical	Precentral: R 5, L 5 Postcentral: R 5, L 10 Subcortical: L 9	A&V	1, 1.5, 2, 2.5, 3, 3.5, 4, 4.5, 5, 5.5	Right precentral patients underestimated at intervals >3.5 s

PFC, prefrontal cortex; OFC, orbitofrontal cortex; R, right; L, left; B, bilateral; ISI, inter-stimulus interval; A, auditory; V, visual; AV, audiovisual; A&V, auditory and visual separately.

usual, with events (e.g., television programs) appearing to pass in 'accelerated motion'. Additionally, when driving, the velocity of his car seemed intolerably fast, with external objects rushing towards him at an excessive rate. Similarly, a patient with a lesion to the right DLPFC reported judging the durations of events in daily life as shorter than they actually were (Koch et al., 2002). This resulted in disruption to daily life, where the patient would leave work earlier than the scheduled time, convinced that the working day must have elapsed. However, data from the experimental measures of perceptual timing in both case reports presents apparently contradictory findings. Binkofski and Block's patient showed gross overproduction when asked to indicate to the examiner when 60 s had elapsed. If his subjective experience of accelerated time was due to a speeded internal clock (i.e., a clock with a pacemaker running too quickly) then he would have demonstrated underproduction. Similarly, Koch et al.'s patient showed underestimation when asked to verbally label the length of a 90 s period of time. A speeded clock would lead to overestimation, as more clock 'ticks' would accumulate in the designated interval. Further, performance for Koch et al.'s patient did not differ from control participants for intervals below 90 s. Koch and colleague's paradigm included reading aloud random numbers paced at a variable rate, which although inhibits counting is likely to have been differentially challenging for a patient with a PFC lesion. In summary, both studies demonstrate patients with PFC lesions who show remarkable and intrusive perceptual timing distortions. However, a straightforward clock explanation (i.e., a clock running too fast) does not fit with the pattern of experimental data. Related to this, Mimura et al. (2000) reported individual differences in the tendency to show either slowed or speeded responding in eight patients with frontal lesions tested on measures of time reproduction and time production. Thus, evidence for a simple speeded clock explanation in patients with frontal lesions is unlikely, mirroring the limited evidence for a slowed clock in PD. Of course, given the extent of the frontal cortex, it could be argued that clock speed is moderated by a specific region of the frontal cortex. Meta-analysis across patient studies, and neuroimaging and transcranial magnetic stimulation (TMS) in healthy populations could all be used to address this possibility.

Further investigation of the role of the PFC in timing was conducted by Casini and Ivry (1999), who tested duration discrimination (400ms standard) and frequency discrimination in patients with lesions to the PFC, specifically the DLPFC. Compared to the results for the two tasks performed separately, the patients with DLPFC lesions showed impaired performance when the tasks were performed together. In contrast, the dual task manipulation only compromised duration discrimination for patients with cerebellar lesions. It was suggested that the timing deficits observed in patients with PFC lesions, unlike those with cerebellar lesions, can be explained by general deficits such as attention. Mangels et al. (1998) found that patients with lesions to the right or left PFC were impaired on a duration discrimination task when the intervals were long (4000 compared to 400ms) or the interval between the two durations was long (4000 compared to 1000ms). This suggests that it was the monitoring demands, specifically the executive control of working memory, rather than the temporal demands that drove the deficit. The patients with PFC lesions were also less able to spontaneously use regularly paced auditory tones to help them time the 4000ms interval, although they could benefit when given explicit instructions on how to utilize the markers. Again, this suggests that cognitive difficulties contributed to the poor timing performance. Harrington et al. (1998b) controlled for general cognitive deficits by excluding participants who were impaired at frequency discrimination. Of the remaining participants, duration discrimination was only impaired in patients with lesions to the right hemisphere (including the DLPFC and parietal cortex, particularly the supramarginal gyrus). The timing performance of the patients with right hemisphere lesions correlated with a measure of attention switching and the authors proposed a right prefrontal-inferior parietal network that influences temporal processing through its role in working memory and attention. In a final investigation of duration discrimination in patients with frontal lesions, Ivry and Keele (1989) found no evidence of impairment. Therefore, duration discrimination tasks tend toward showing a deficit in patients with frontal lesions but with the pattern of deficits consistent with non-specific cognitive processes driving the impairment.

A cognitive explanation was also the conclusion arrived at by Mimura et al. (2000). For their frontal patients, poor performance on the Wisconsin Card Sorting Task, a measure of executive functioning, correlated with increased verbal overestimation of units of time. Conversely, poor performance on a measure of delayed memory correlated with increased underestimation. Consequently, Mimura et al. argued that working memory deficits drove the atypical perceptual timing profile in patients with frontal lesions. However, it is also important to note that Coslett et al. (2009) did not find any correlation between performance on three perceptual timing tasks and the extent of damage in the lateral frontal lobe. In contrast, there were correlations between the extent of damage in the parietal lobe and performance on the time reproduction task.

Using an assessment of time reproduction, Kagerer et al. (2002) replicated Harrington et al.'s (1998b) finding of compromised timing in patients with right hemisphere lesions only. Here, patients with lesions to the right and left hemispheres were unimpaired in the time reproduction of intervals <3000ms but patients with lesions to right precentral regions showed difficulty with intervals >3500ms. Further, studies using repetitive TMS to create a 'virtual lesion' in healthy participants, report that disruption to the right DLPFC impairs temporal processing (e.g., Jones et al., 2004). However, the right hemisphere was not implicated in Coslett et al.'s (2009) study on perceptual timing in 31 patients with lesions across anterior and posterior cortex.

The severity of the temporal deficits seen in patients with frontal lesions is worth commenting on. The overestimation of the patients with OFC lesions in Berlin et al. (2004) was on average 20 s for a 90 s interval, although the large standard error bars suggest a considerable spread of responses. Mimura et al. (2000) reported at least two patients with remarkable overestimation, estimating 90 s as being greater than 150 s. In contrast, another two patients estimated the same interval as being less than 50 s. However, the task included a dual task component, reading aloud random numbers, which may have distorted the estimations. When the participants counted the intervals at the self-paced rate of 1 s, performance was much improved, although one exceptionally poor performer remained. The patient assessed by Binkofski and Block (1996) produced average time production values of 286 s, when he was asked to produce 60 s; while Koch et al.'s (2002) patient underestimated a 90 s interval as 40 s. Deviations of this extent are not seen in patients with PD. For example, the PD patients described by Lange et al. (1995) estimated 60 s intervals as lasting about 40 s and produced time intervals of around 70 s when asked to produce 60 s intervals. These quantitative differences, as well as the tendency toward speeded vs. slowed responding, indicates that comparable pathology cannot be driving the deficits in patients with frontal lesions and those with PD. The most parsimonious explanation for the profile across patient groups is that aberrant cognitive processes are driving the extreme performance in the frontal patients, and that this cognitive disruption is less evident in the PD group and in patients with lesions to the basal ganglia.

5.2 *Motor Timing*

The classic S-C task has not been used very often to test patients with frontal lesions. Ivry and Keele (1989) assessed 12 patients with lesions that included damage to the posterior frontal lobe. The patients showed increased variability compared to controls during the continuation phase on a version of the task that required tapping every 550ms. Analysis using the Wing and Kristofferson (1973a,b) model attributed the variance to disruption in motor execution, rather than core clock processes. A larger study assessed 43 patients with a range of frontal lesions using a longer inter-tap interval of 1.5 s (Picton et al.,

2006). Increased variability on both phases of the task was found in patients with lesions to the right lateral frontal cortex, including some with lesions of the basal ganglia. In addition, patients with lesions to the superior medial frontal cortex showed competent performance in early trials but their variability became abnormally high as the task progressed. In contrast to Ivry and Keele, analysis using the Wing and Kristofferson model implicated clock processes in both types of dysfunction. More detailed investigation of the lesion locations led Picton et al. (2006) to attribute the role of the right lateral frontal cortex in monitoring time. Any temporal task naturally demands monitoring and the S-C task necessitates monitoring the clock, i.e., a counting of clock ticks, and also monitoring that the executed responses are accurate. Disruption to monitoring would lead to increased variability. This interpretation is compatible with the finding that the same patients were unable to effectively monitor foreperiods in a simple reaction time task (Stuss et al., 2005). In contrast, the deteriorating performance of the patients with lesions of the superior medial frontal cortex was related to the role of this region in sustaining performance and counteracting distraction.

Using a different type of task, Halsband et al. (1993) assessed patients with lesions to the primary motor cortex, premotor cortex or supplementary motor area (SMA) on a range of tasks that assessed rhythm reproduction from memory. These tasks used more complex rhythms than the simple isochronous S-C task. Lesions to both the premotor cortex and the SMA impacted upon rhythm reproduction, despite unimpaired performance on measures of rhythm discrimination and manual dexterity. The two patients with SMA lesions were assessed on their ability to tap in time with a complex rhythm and then in the absence of the pacing stimulus; thus the task was similar to the classic S-C task. Here, patients were only impaired when the pacing stimulus was absent. This suggested that it is the internal generation of remembered rhythms that is challenging, rather than timing with the external sensory guidance provided by cues. Notably, damage to the left hemisphere had more impact on performance than damage to the right hemisphere, with the effects being seen in both hands. The pattern of deficits suggests that these frontal premotor regions are necessary for the temporal aspects of motor programming to be successfully employed.

5.3 *Summary*

Localization of the key regions of the frontal cortex involved in temporal processing is limited in patient studies as lesion sites are varied. For example, the area of the frontal cortex that was most commonly compromised in Mangels et al.'s (1998) study was the DLPFC, but the lesions often extended to more ventral regions. Another limitation of this type of research is the plastic, compensatory changes that can occur following a lesion. The recruitment of alternative regions to enable performance on a task may mask or 'muddy' the results. Some studies have assessed patients with chronic lesions that are years old (e.g., 9–18 years in Mangels et al., 1998), whereas others have used patients with a more recent history (e.g., Coslett et al., 2009).

However, assessing across the perceptual and motor tasks it is clear that dysfunctional temporal processing is characteristic of patients with frontal lesions. Particularly, data are consistently interpreted as reflecting the impact of disrupted cognition, including working memory and monitoring, attention, and inhibition on performance.

6 Concluding Comments

Surveying the data, it is undisputable that patients with PD or lesions to the frontal cortex present with motor and perceptual timing deficits. The picture for patients with basal ganglia lesions is less clear, primarily because this patient group is understudied. However, the data suggest that when both the right and left basal ganglia are compromised motor timing deficits will emerge.

Seconds-range timing has been considered more cognitive (e.g., Lewis and Miall, 2003) and relying on the mesocortial pathway, while millisecond-range timing is argued to be the preserve of the nigrostriatal pathway (Rammsayer and Classen, 1997; Wiener et al., 2011). However, it is important to note that as PD progresses the extent of dysfunction extends beyond the frontostriatal motor loop. This includes connections from the ventral tegmental area in the midbrain to the frontal cortex (i.e., the mesocortical pathway). Recently, there has been discussion that executive dysfunction may influence, if not drive, the temporal processing deficits seen in PD (Parker et al., 2013). There is suggestion that perceptual timing requires the same resources, including those used in inhibition, updating, and shifting, that are employed during general executive processing (Brown et al., 2013). Therefore, damage to prefrontal regions that are engaged in executive functions should be sufficient to cause temporal disruption in PD. However, the temporal profile of patients with purely frontal damage is distinct from patients with PD. Most notably, temporal deficits, particularly perceptual, are more extreme in patients with frontal lesions. Further, although dopaminergic medication can ameliorate the motor symptoms of PD and predominantly has a positive effect on temporal processing, it does not consistently have a positive effect on executive performance (e.g., Pascual-Sedano et al., 2008) and can even worsen performance (e.g., Cools and

D'Esposito, 2011). Given the evidence of motor timing difficulties in patients with bilateral lesions to the basal ganglia (Coslett et al., 2010), it can be proposed that a core timing dysfunction in PD is exacerbated as the disease progresses and further frontal regions become compromised.

Merchant et al. (2013) have discussed how degeneracy offers a contributing explanation for why the profile of timing performance for a particular patient group can vary, both within and across studies. Degeneracy refers to the capacity of multiple neural systems to sustain the same behavioral function, such that damage to one neural system can be compensated by pre-existing neural systems that can support the behavior. Degeneracy can be observed in functional imaging studies when patients with PD recruit the cerebellum in the absence of adequate fronto-striatal activation and show behaviorally preserved performance (e.g., Jahanshahi et al., 2010). Further, it would explain why capturing temporal processing difficulties in patients with discrete lesions to the basal ganglia can prove elusive, particularly in cases where the lesion is unilateral and redundancy (i.e., the presence of identical mechanisms in both hemispheres) is also a factor. That a relatively intact frontal cortex, certainly in early stage PD, is not enough to compensate for temporal dysfunction suggests that the basal ganglia have a core role in interval timing.

The consideration of the contributory roles of the basal ganglia and associated cortical regions is sympathetic to current conceptualization of interval timing as being driven by both a core timing network and by areas that are activated in a context-dependent fashion (e.g., dependent on modality or cognitive load), rather than a single locus (see Merchant et al., 2013). The core timing network is proposed to engage cortical, thalamic, and basal ganglia structures, particularly the striatum and SMA (Merchant et al., 2013). This conceptualization fits well with the profile of timing deficits across the three patient groups explored in this chapter, although clearly patients with damage to the basal ganglia remain significantly understudied. A clear direction for future work would be to compare the three patient groups directly, assessing both cognitive and temporal performance, to enable detailed profiling of performance.

References

Alexander, G.E., M.R. DeLong, and P.L. Strick. 1986. "Parallel Organization of Functionally Segregated Circuits Linking Basal Ganglia and Cortex" *Annu Rev Neurosci* 9: 357–81.

Aparicio, P., J. Diedrichsen, and R.B. Ivry. 2005. "Effects of Focal Basal Ganglia Lesions on Timing and Force Control" *Brain Cogn* 58(1): 62–74.

Berlin, H.A., E.T. Rolls, and U. Kischka. 2004. "Impulsivity, Time Perception, Emotion and Reinforcement Sensitivity in Patients with Orbitofrontal Cortex Lesions" *Brain* 127(5): 1108–26.

Binkofski, F., and R.A. Block. 1996. "Accelerated Time Experience after Left Frontal Cortex Lesion" *Neurocase* 2: 485–93.

Brown, R.G., and C.D. Marsden. 1991. "Dual Task Performance and Processing Resources in Normal Subjects and Patients with Parkinson's Disease" *Brain* 114(1): 215–31.

Brown, S.W., S.A. Collier, and J.C. Night. 2013. "Timing and Executive Resources: Dual-Task Interference Patterns between Temporal Production and Shifting, Updating, and Inhibition Tasks" *J Exp Psychol Hum Percept Perform* 39(4): 947–63.

Casini, L., and R.B. Ivry. 1999. "Effects of Divided Attention on Temporal Processing in Patients with Lesions of the Cerebellum or Frontal Lobe" *Neuropsychology* 13(1): 10–21.

Cerasa, A. et al. 2006. "Functional Changes in the Activity of Cerebellum and Frontostriatal Regions during Externally and Internally Timed Movement in Parkinson's Disease" *Brain Res Bull* 71(1–3): 259–69.

Chaudhuri, K.R., and A.H. Schapira. 2009. "Non-Motor Symptoms of Parkinson's Disease: Dopaminergic Pathophysiology and Treatment" *Lancet Neurol* 8(5): 464–74.

Claassen, D.O. et al. 2013. "Deciphering the Impact of Cerebellar and Basal Ganglia Dysfunction in Accuracy and Variability of Motor Timing" *Neuropsychologia* 51: 267–74.

Cools, R., and M. D'Esposito. 2011. "Inverted-U-Shaped Dopamine Actions on Human Working Memory and Cognitive Control" *Biol Psychiatry* 69(12): e113–25.

Coslett, H.B. et al. 2009. "Cognitive Timing: Neuropsychology and Anatomic Basis" *Brain Res* 1254: 38–48.

Coslett, H.B., M. Wiener, and A. Chatterjee. 2010. "Dissociable Neural Systems for Timing: Evidence from Subjects with Basal Ganglia Lesions" *PLoS One* 5(4): e10324.

Coull, J., and A. Nobre. 2008. "Dissociating Explicit Timing from Temporal Expectation with Fmri" *Curr Opin Neurobiol* 18(2): 137–44.

D'Esposito, M. 2007. "From Cognitive to Neural Models of Working Memory." *Philos Trans R Soc Lond B Biol Sci* 362(1481): 761–72.

Dirnberger, G., and M. Jahanshahi. 2013. "Executive Dysfunction in Parkinson's Disease: A Review [In eng]" *J Neuropsychol* 7(2): 193–224.

Duchek, J.M. et al. 1994. "Component Analysis of a Rhythmic Finger Tapping Task in Individuals with Senile Dementia of the Alxheimer Type and in Individuals with Parkinson's Disease" *Neuropsychology* 8(2): 218–26.

Elsinger, C.L. et al. 2003. "Neural Basis for Impaired Time Reproduction in Parkinson's Disease: An Fmri Study" *J Int Neuropsychol Soc* 9(7): 1088–98.

Gibbon, J. 1977. "Scalar Expectancy Theory and Weber's Law in Animal Timing" *Psychol Rev* 84: 279–325.

Gibbon, J. et al. 1984. "Scalar Timing in Memory" *Ann N Y Acad Sci* 423: 52–77.

Grahn, J.A., and M. Brett. 2009. "Impairment of Beat-Based Rhythm Discrimination in Parkinson's Disease" *Cortex* 45(1): 54–61.

Guehl, D. et al. 2008. "Auditory Temporal Processing in Parkinson's Disease" *Neuropsychologia* 46(9): 2326–35.

Halsband, U. et al. 1993. "The Role of Premotor Cortex and the Supplementary Motor Area in the Temporal Control of Movement in Man" *Brain* 116(1): 243–66.

Harrington, D.L., K.Y. Haaland, and N. Hermanowicz. 1998a. "Temporal Processing in the Basal Ganglia" *Neuropsychology* 12(1): 3–12.

Harrington, D.L., K.Y. Haaland, and R.T. Knight. 1998b. "Cortical Networks Underlying Mechanisms of Time Perception" *J Neurosci* 18(3): 1085–95.

Hellström, A. et al. 1997. "Tone Duration Discrimination in Parkinson's Disease" *Neuropsychologia* 35(5): 737–40.

Hinton, S.C., Harrington, D.L., Binder, J.R., Durgerian, S., and Rao, S.M. 2004. "Neural Systems Supporting Timing and Chronometric Counting: an fMRI Study" *Cog Brain Res*, 21(2): 183–92.

Ivry, R.B., and S.W. Keele. 1989. "Timing Functions of the Cerebellum" *J Cogn Neurosci* 1(2): 136–52.

Jahanshahi, M. et al. 2006. "The Substantia Nigra Pars Compacta and Temporal Processing" *J Neurosci* 26(47): 12266–73.

Jahanshahi, M. et al. 2010. "Dopaminergic Modulation of Striato-Frontal Connectivity during Motor Timing in Parkinson's Disease" *Brain* 133(3): 727–45.

Jones, C.R.G. and Jahanshahi, M. 2014. "Contributions of the Basal Ganglia to Temporal Processing: Evidence from Parkinson's Disease" *Timing & Time Perception*, 2(1): 87–127.

Jones, C.R. et al. 2004. "The Right Dorsolateral Prefrontal Cortex is Essential in Time Reproduction: An Investigation with Repetitive Transcranial Magnetic Stimulation" *Exp Brain Res* 158(3): 366–72.

Jones, C.R. et al. 2008. "Basal Ganglia, Dopamine and Temporal Processing: Performance on Three Timing Tasks On and Off Medication in Parkinson's Disease" *Brain Cogn* 68(1): 30–41.

Jones, C.R.G. et al. 2011. "Modeling Accuracy and Variability of Motor Timing in Treated and Untreated Parkinson's Disease and Healthy Controls" *Front Integr Neurosci* 5(81).

Joundi, R.A. et al. 2012. "High-Frequency Stimulation of the Subthalamic Nucleus Selectively Decreases Central Variance of Rhythmic Finger Tapping in Parkinson's Disease" [In eng] *Neuropsychologia* 50(10): 2460–6.

Kagerer, F.A. et al. 2002. "Cortical Involvement in Temporal Reproduction: Evidence for Differential Roles of the Hemispheres" *Neuropsychologia* 40(3): 357–66.

Katsuki, F., and C. Constantinidis. 2012. "Unique and Shared Roles of the Posterior Parietal and Dorsolateral Prefrontal Cortex in Cognitive Functions" *Front Integr Neurosci* 6: 17.

Keele, S.W. et al. 1985. "Do Perception and Motor Production Share Common Timing Mechanisms: A Correctional Analysis" *Acta Psychol* 60(2–3): 173–91.

Kehagia, A.A., R.A. Barker, and T.W. Robbins. 2013. "Cognitive Impairment in Parkinson's Disease: The Dual Syndrome Hypothesis" *Neurodegener Dis* 11(2): 79–92.

Koch, G. et al. 2002. "Selective Deficit of Time Perception in a Patient with Right Prefrontal Cortex Lesion" *Neurology* 59(10): 1658–9.

Koch, G. et al. 2004. "Subthalamic Deep Brain Stimulation Improves Time Perception in Parkinson's Disease" *Neuroreport* 15(6): 1071–3.

Koch, G. et al. 2005. "Memory for Time Intervals is Impaired in Left Hemi-Parkinson Patients" *Neuropsychologia* 43(8): 1163–7.

Koch, G. et al. 2008. "Impaired Reproduction of Second but Not Millisecond Time Intervals in Parkinson's Disease" *Neuropsychologia* 46(5): 1305–13.

Lange, K.W. et al. 1995. "Subjective Time Estimation in Parkinson's Disease" *J Neural Transm Suppl* 46: 433–8.

Lewis, P.A., and R.C. Miall. 2003. "Distinct Systems for Automatic and Cognitively Controlled Time Measurement: Evidence from Neuroimaging" *Curr Opin Neurobiol* 13(2): 250–5.

Malapani, C. et al. 1998. "Coupled Temporal Memories in Parkinson's Disease: A Dopamine-Related Dysfunction" *J Cogn Neurosci* 10(3): 316–31.

Malapani, C., B. Deweer, and J. Gibbon. 2002. "Separating Storage from Retrieval Dysfunction of Temporal Memory in Parkinson's Disease" *J Cogn Neurosci* 14(2): 311–22.

Mangels, J.A., Ivry, R.B., and Shimizu, N. 1998. "Dissociable Contributions of the Prefrontal and Neocerebellar Cortex to Time Perception" *Cogn Brain Res* 7: 15–39.

Matell, M.S., and W.H. Meck. 2000. "Neuropsychological Mechanisms of Interval Timing Behavior" *Bioessays* 22(1): 94–103.

Matell, M.S., and W.H. Meck. 2004. "Cortico-Striatal Circuits and Interval Timing: Coincidence Detection of Oscillatory Processes" *Brain Res Cogn Brain Res* 21(2): 139–70.

Merchant, H. et al. 2008. "Interval Timing and Parkinson's Disease: Heterogeneity in Temporal Performance" *Exp Brain Res* 184(2): 233–48.

Merchant, H., D.L. Harrington, and W.H. Meck. 2013. "Neural Basis of the Perception and Estimation of Time" *Annu Rev Neurosci* 36: 313–36.

Mimura, M., M. Kinsbourne, and M. O'Connor. 2000. "Time Estimation by Patients with Frontal Lesions and by Korsakoff Amnesics" *J Int Neuropsychol Soc* 6(5): 517–28.

Nichelli, P. et al. 1995. "Duration Processing after Frontal Lobe Lesions" *Ann NY Acad Sci* 769: 183–90.

O'Boyle, D.J., J.S. Freeman, and F.W. Cody. 1996. "The Accuracy and Precision of Timing of Self-Paced, Repetitive Movements in Subjects with Parkinson's Disease" *Brain* 119(1): 51–70.

Parker, K.L. et al. 2013. "Executive Dysfunction in Parkinson's Disease and Timing Deficits" *Front Integr Neurosci* 7: 75.

Pascual-Sedano, B. et al. 2008. "Levodopa and Executive Performance in Parkinson's Disease: A Randomized Study" *J Int Neuropsychol Soc* 14(5): 832–41.

Pastor, M.A. et al. 1992a. "Time Estimation and Reproduction Is Abnormal in Parkinson's Disease" *Brain* 115(1): 211–25.

Pastor, M.A. et al. 1992b. "Performance of Repetitive Wrist Movements in Parkinson's Disease" *Brain* 115(3): 875–91.

Perbal, S. et al. 2005. "Effects of Internal Clock and Memory Disorders on Duration Reproductions and Duration Productions in Patients with Parkinson's Disease" *Brain Cogn* 58(1): 35–48.

Picton, T.W. et al. 2006. "Keeping Time: Effects of Focal Frontal Lesions" *Neuropsychologia* 44(7): 1195–209.

Rammsayer, T., and W. Classen. 1997. "Impaired Temporal Discrimination in Parkinson's Disease: Temporal Processing of Brief Durations as an Indicator of Degeneration of Dopaminergic Neurons in the Basal Ganglia" *Int J Neurosci* 91(1–2): 45–55.

Riesen, J.M., and A. Schnider. 2001. "Time Estimation in Parkinson's Disease: Normal Long Duration Estimation Despite Impaired Short Duration Discrimination" *J Neurol* 248(1): 27–35.

Schwartze, M. et al. 2011. "The Impact of Basal Ganglia Lesions on Sensorimotor Synchronization, Spontaneous Motor Tempo, and the Detection of Tempo Changes" *Behav Brain Res* 216(2): 685–91.

Smith, J.G. et al. 2007. "The Effect of Parkinson's Disease on Time Estimation as a Function of Stimulus Duration Range and Modality" *Brain Cogn* 64(2): 130–43.

Spencer, R.M., and R.B. Ivry. 2005. "Comparison of Patients with Parkinson's Disease or Cerebellar Lesions in the Production of Periodic Movements Involving Event-Based or Emergent Timing" *Brain Cogn* 58(1): 84–93.

Stuss, D.T. et al. 2005. "Multiple Frontal Systems Controlling Response Speed" *Neuropsychologia* 43(3): 396–417.

Torregrossa, M.M., J.J. Quinn, and J.R. Taylor. 2008. "Impulsivity, Compulsivity, and Habit: The Role of Orbitofrontal Cortex Revisited" *Biol Psychiatry* 63(3): 253–5.

Torta, D.M. et al. 2010. "Dissociation between Time Reproduction of Actions and of Intervals in Patients with Parkinson's Disease" *J Neurol* 257(8): 1356–61.

Wearden, J.H. et al. 2008. "Stimulus Timing by People with Parkinson's Disease" *Brain Cogn* 67(3): 264–79.

Wiener, M., F.W. Lohoff, and H.B. Coslett. 2011. "Double Dissociation of Dopamine Genes and Timing in Humans" *J Cogn Neurosci*, 23(10): 2811–21.

Wild-Wall, N. et al. 2008. "Time Estimation in Healthy Ageing and Neurodegenerative Basal Ganglia Disorders" *Neurosci Lett* 442: 34–8.

Wing, A.M., and Kristofferson., A.B. 1973a. "Response Delays and Timing of Discrete Motor Responses" *Percept Psychophys* 14: 5–12.

Wing, A.M., and Kristofferson., A.B. 1973 "Timing of Interresponse Intervals" *Percept Psychophys* 13: 455–60.

Wojtecki, L. et al. 2011. "Modulation of Human Time Processing by Subthalamic Deep Brain Stimulation" *PLoS One* 6(9): 12.

CHAPTER 11

Bayesian Models of Interval Timing and Distortions in Temporal Memory as a Function of Parkinson's Disease and Dopamine-Related Error Processing

Bon-Mi Gu, Anita J. Jurkowski**, Jessica I. Lake†, Chara Malapani††, and Warren H. Meck‡*

1 Introduction

The fidelity of timing and time perception in the milli-seconds to multi-seconds range is subject to distortion in humans and other animals (Droit-Volet and Meck, 2007; Eagleman, 2008; Matthews and Meck, 2014; Merchant et al., 2013a; Grondin, 2010). This failure to represent objective time in a veridical manner occurs in healthy individuals, including children, young adults, and aged participants as a function of temporal context (Balci et al., 2009; Lustig and Meck, 2001, 2005, 2011; van Wassenhove et al., 2008). Of particular interest for this chapter, however, are the pathophysiological distortions in time perception and timed performance that occur as a function of dopamine (DA) depletion in the basal ganglia associated with Parkinson's disease (PD; see Allman and Meck, 2012; Coull et al., 2012, 2013; Harrington et al., 2014; Höhn et al., 2011; Jahanshahi et al., 2006; Jones et al., 2008; Jones and Jahanshahi, 2011, 2014). Patients with PD have a loss of DA neurons in the substantia nigra pars reticulata as well as the substantia nigra pars compacta (SNPC) which projects to the striatum – a brain region shown to be of critical importance to interval timing (Jahanshahi et al., 2006; Koch et al., 2008, 2009; Parker et al., 2013; Rakitin and Stern, 2006; Smith et al., 2007). PD patients experience difficulties when producing temporal components of movement, and in programming and synchronizing motor responses. These problems are reflected in increased reaction time and movement time (Bloxham et al., 1987; Evarts et al., 1981; Jurkowski and Stacy, 2005; Jurkowski et al., 2002, 2005a,b), prolonged

* Department of Psychology, University of Michigan, Ann Arbor, MI.
** Department of Psychology, University of Missouri, Columbia, MO.
† Department of Psychology, University of California, Los Angeles, Los Angeles, CA.
†† Department of Psychiatry, Columbia University, New York, NY.
‡ Department of Psychology and Neuroscience, Duke University, Durham, NC.

DOI 10.1163/9789004230699_012

inter-onset latencies when performing sequential or simultaneous movements (Benecke et al., 1986; Roy et al., 1993), impaired ability to maintain a fixed rhythm in tapping tasks (Nakamura et al., 1978; O'Boyle et al., 1996; Stelmach and Worringham, 1988; Wing et al., 1984) and increased speech production time (Lieberman et al., 1992; Volkmann et al., 1992). Parameters such as reaction time, inter-onset latency, rhythm maintenance, and temporal organization of speech all require accurate timing, and dysfunctions are well correlated with the clinical phenomenon of bradykinesia in PD (Benecke et al., 1986; Hallett and Khoshbin, 1980; Nakamura et al., 1978; Schirmer, 2004; Volkmann et al., 1992). There is evidence that the timing deficits in PD patients may not be limited to the motor domain. Non-motor deficits include impaired temporal discrimination of pairs of stimuli in the somaesthetic, visual, and auditory modalities (Artieda et al., 1992), absence of interlimb synchronization in processing of bimodal simultaneous stimuli (Malapani et al., 1994), difficulty in recognizing the emotional, linguistic, and temporal features expressed in speech (Schirmer, 2004), and impairments in attention, working memory, and time estimation (Brown and Marsden, 1988; Lee et al., 2010; Pastor et al., 1992; Perbal et al., 2005; Pouthas and Perbal, 2004). Recent evidence suggests that a unified model of timing and time perception may be possible based on coordinated activity in a core striatal network that is interconnected with olivocerebellar circuits and the cerebral cortex (Harrington et al., 2014; Koch et al., 2009; Kotz and Schwartze, 2011; Meck, 2005; Merchant et al., 2013a; Teki et al., 2012; see Gibbon et al., 1997, and Malapani et al., 1998a, for an alternative proposal for striato-thalamo-cerebral interconnectivity). This interactive timekeeping mechanism is used in both production and perception of intervals in the millisecond range and seconds-to-minutes range with humans and other animals (Ivry and Hazeltine, 1995; Ivry and Keele, 1989; Jahanshahi et al., 2010a; Keele et al., 1985; Meck et al., 2008; Merchant et al., 2008b, 2013b). Thus, timing may be an independent/modular process employed whenever its specific computation is needed (Malapani and Gibbon, 2001; Merchant et al., 2013a; Van Rijn et al., 2014). In fact, accurate timing is required in many behaviors across species. For example, the skilled movements of musicians, athletes, and predators require precise timing of activity across different groups of muscles. It has been suggested that several motor or perceptual tasks that require accurate timing have access to the same internal timing mechanism, because normative performance varies in similar ways across tasks (Buhusi and Meck, 2005; Malapani and Fairhurst, 2002; Ivry and Hazeltine, 1995).

Neuropsychological research provides a complementary way to address the issue of whether a common timing mechanism is used in both production and perception and to investigate possible locations of the hypothesized internal timekeepers within neural systems (Coull et al., 2011; Meck et al., 2008;

Merchant et al., 2013a). This approach asks whether different patient groups will show either similarities or dissociations in performance on specific timing tasks as a function of the location of their neurological lesion. It has been shown that patients with lesions in the cerebellum are impaired in both motor and perceptual tasks requiring accurate timing, in "short" sub-second time ranges (Ivry et al., 1988; Ivry and Keele, 1989) as well as in "long" *supra*-second time ranges (Malapani et al., 1998b; Nichelli et al., 1996). Sub-second timing performance in perceptual and production timing tasks with comparable levels of difficulty was found to be unimpaired in PD patients (Ivry and Keele, 1989), leading to the hypothesis that timing in this range doesn't directly involve the basal ganglia (Ivry, 1996). However, more recent findings obtained in patients with diseases originating in the basal ganglia have shown that both time estimation and motor tasks requiring accurate timing are impaired in PD patients, in both sub-second and *supra*-second time ranges (Allman and Meck, 2012; Harrington et al., 1998a,b, 2011a,b; Jahanshahi et al., 2010a,b; Malapani et al., 1998b; Merchant et al., 2008a, 2013a; O'Boyle et al., 1996; Pastor et al., 1992).

2 Isolating the Internal Clock by Dividing Interval Timing into Separate Stages of Information Processing

Distortions in timing that are induced by pharmacological, lesion, or stimulation techniques are postulated to be diagnostic of the underlying psychological processes (Allman et al., 2014a,b; Coull et al., 2011, 2013; Meck, 2006a,b,c). These temporal processes have been divided into three major stages of information processing (i.e., clock, memory, and decision stages – see Allman and Meck, 2012; Bueti and Macaluso, 2011; Hinton and Meck, 1997; Matell and Meck, 2000; Pouthas and Perbal, 2004) in order to isolate the underlying timing mechanisms (Roberts, 1981, 1993) and understand the time course of temporal discrimination (Tarantino et al., 2010).

2.1 *Clock Stage*

Systemic administration of DA agonists and antagonists has been found to cause horizontal leftward (e.g., amphetamine) and rightward (haloperidol) shifts in psychometric functions relating signal duration to temporal judgments (Lake and Meck, 2013; Maricq and Church, 1983; Meck, 1983, Meck et al., 2012a). These horizontal shifts are proportional to the durations being timed and have shown a positive correlation with the efficacy of the drugs at the D2 receptor (e.g., Meck, 1986). This result is compatible with the hypothesis that increasing (or decreasing) the effective level of DA in the brain causes an increase (or decrease) in the speed of oscillatory/pacemaker processes used for timing (Coull et al., 2011; MacDonald and Meck, 2004, 2005, 2006; Matell and Meck, 2004): *if the clock is caused to run faster (or slower) than when a target duration was first learned, then participants will think that the end of the interval and its associated feedback is due earlier (or later) than it actually is, and consequently timing functions will be horizontally shifted to the left (or right).* These horizontal shifts have been interpreted as indicative of effects upon "clock speed" (Cheng et al., 2006; Matell et al., 2004, 2006; Williamson et al., 2008). As testing continues under the influence of the drug, however, these under- or over-estimations of signal duration begin to readjust such that by the end of testing, psychometric functions should have gradually shifted back to normal. If the drug is removed abruptly and testing continues, however, what one should see is an immediate rebound effect. That is to say, participants that have been tested with the DA agonists show a subsequent rightward shift upon removal of the DA agonist, while those tested with DA antagonists show a leftward shift. This is thought to be due to participants re-learning the target durations while under the influence of the drug (Buhusi, 2003; Meck, 1983, 1996). An example of how this might work is as follows: When participants have re-learned the target duration with a clock that runs faster (or slower) than normal, and are then returned to the "non-drug" baseline condition, the normal speed of the clock is slower (or faster) than it was during re-learning. Participants will then expect feedback later (or earlier) than the re-learned time, and the timing functions will be shifted to the right (or left), which is in the opposite direction to the shift induced by the initial administration of the drug (Oprisan and Buhusi, 2011).

2.2 *Memory Stage*

Horizontal displacement of psychometric functions may also result from distortions in the content of temporal memories that are stored for later retrieval (Coull et al., 2011; Meck, 1983, 1996). This distortion in the content of temporal memory has been described in terms of a multiplicative translation constant that is responsible for producing scalar transforms of sensory input taken from an internal clock (e.g., Meck, 2002a,b). An example of how this might work is as follows: the number of pacemaker pulses integrated by an accumulator during the presentation of a signal can provide a representation of the perceived duration for the current trial. When this clock reading is transferred to memory (presumably following feedback), the transfer may occur with some bias that we have referred to as a memory-storage or memory translation constant. In scalar timing theory (Gibbon et al., 1984; Wearden and Lejeune, 2008), this has formally been referred to as the k* parameter, which is a multiplicative constant describing how the reference memory works during encoding and retrieval (Meck, 1983, 2002a; 2006c; Meck and Angell, 1992; Meck and Church, 1987; Meck et al., 1986; Van Rijn et al., 2014). If the remembered time of feedback reliably differed from the obtained clock reading, then a participant would consistently expect the feedback to occur at a time other than the programmed time. It would be later than the programmed time if the memory storage constant was greater than 1.0. Alternatively, the participant would consistently expect feedback to occur earlier than the programmed time if its memory storage constant was less than 1.0 (Meck, 1983, 2002a). A distortion at this level, known as the "memory-speed" effect, causes the real time of an event to be consistently remembered as "shorter" or "longer" than it actually is. Because the physical time is always misremembered, no amount of training will eliminate the distortion. Unlike changes in clock speed, changes in memory-storage speed would not be self-correcting even in the case where participants are "surprised" by the mismatch between their clock readings and a value sampled from reference memory. This is because any "updating" of memory in this case would lead to the continued distortion of stored values for as long as the modification in memory-storage speed remains in effect (i.e., as long as K* is $>$ or $<$1.0). Pharmacological research has shown that the memory constant is related to the amount of acetylcholine (ACh) available in the brain, such that increasing (or decreasing) synaptic ACh decreases (or increases) the value of K* in a proportional manner to the timed interval (Meck, 1983, 1996, 2002a; Meck and Church, 1987; Oprisan and Buhusi, 2011).

2.3 *Decision Stage*

In virtually all interval-timing tasks, the decision to respond is based on the ratio of a currently accumulating duration to a remembered standard duration (Gibbon et al., 1984, 1997; Meck, 1983). If one observed decreasing variability with increasing target duration, the use of a *difference rule* rather than a *ratio rule* may be inferred, which would represent a violation of scalar variability across signal durations (Hinton and Rao, 2004; Rakitin et al., 1998; Wearden and Lejeune, 2008). Evidence regarding the neuropharmacology or anatomical locus of the decision stage is relatively lacking, but the interaction between the dorsal and ventral striatum has been implicated in the setting of response thresholds (Agostino et al., 2013; MacDonald et al., 2012). One piece of evidence concerning the neural processes that play a role in specific aspects of this decision process is the finding that participants systematically "search" for the target duration by systematically varying their temporal criterion on a trial-by-trial basis between a lower and upper boundary. This "temporal search" is revealed by an individual-trials analysis (Church et al., 1994; Rakitin et al., 1998) showing that successive peak times in the peak-interval procedure alternate between opposite sides of the median peak time more often that would be expected by chance (Meck, 1988; Meck et al., 1984). This finding suggests that duration discrimination mechanisms operate similar to a feedback-control system in order to minimize the amount of error in the estimation of the target duration and appear to depend upon hippocampal-striatal interactions (MacDonald et al., 2014; Meck, 1988; Meck et al., 2013; Yin and Meck, 2014; Yin and Troger, 2011). As such, a variety of feedback techniques have been used to affect this level of alternation and the decision stage in temporal generalization and reproduction tasks (e.g., Lustig and Meck, 2005; Wearden and Grindrod, 2003).

Recent work has demonstrated that human participants are sensitive to the probabilistic structure of the mean and variation of a distribution of event durations. This sensitivity can be modeled by a Bayesian-type decision process that participants use for the optimal estimation of duration and coincidence timing (Cicchini et al., 2012; Jazayeri and Shadlen, 2010; Miyazaki et al., 2005). These results highlight the statistical nature of psychophysical judgments as well as the efficiency and flexibility of decision-making processes involved in timing (Meck et al., 2012b). It has recently shown that the hyperbolic discounting of reward value emerges from the scalar property of interval timing (Cui, 2011). Moreover, there is an increasing awareness that decision-making is influenced by subjective time being reliably different from objective time and that these differences are modulated by sources and forms of variability in clock, memory, and decision stages (Gibbon et al., 1997; Gu and Meck, 2011; Meck et al., 2012b; Jazayeri and Shadlen, 2010; Ray and Bossaerts, 2011).

In summary, there are a number of important implications derived from the findings described above: (1) clock and memory stages of temporal processing are functionally distinguishable; (2) DA function appears to be importantly involved in the "clock" stage and perhaps in the "memory" stage as well – although ACh function appears to be more strongly implicated (Coull et al., 2011; Oprisan and Buhusi, 2011). However, findings in bilateral PD patients with substantial losses of DA cell bodies in the SNPC, show reliable "migration effects" when trained and tested with multiple target durations during the same session. This outcome was unanticipated based on the hypothesis that DA is solely involved in the clock stage – which would predict a proportional rightward shift for both target durations as clock speed slows down rather than migration towards each other (Malapani et al., 1998b; Meck and Benson, 2002). These PD findings also challenge current timing theories (e.g., scalar timing theory and striatal beat-frequency theory, SBF; Gibbon et al., 1984, 1997; Matell and Meck, 2004), which cannot account for all of the quantitative properties of the observed clock and memory effects without some additional assumptions about memory processes and decision-making (Gu and Meck, 2011; Lustig et al., 2005). We describe these effects as part of the review below.

3 Migration Effects Observed in Temporal Reproduction

The peak-interval (PI) timing procedure is a duration reproduction task in which human participants are instructed to reproduce durations by centering a sequence of responses around the target duration(s) that typically range from 6–21 s (Rakitin et al., 1998; Yin et al., 2015). The target duration is initially pre-trained by presenting the participant with a series of fixed-time (FT) trials (e.g., N = 10 trials) during which the signal (e.g., blue rectangle) briefly changes color (blue–> magenta) at the target duration. Participants were then asked to reproduce this duration from memory during peak trials by either centering a sequence of key presses (e.g., Lake and Meck, 2013; Lustig and Meck, 2005; Malapani et al., 1998b; Rakitin et al., 1998, 2005, 2006) or pressing a key just before and releasing it just after the target duration (Malapani et al., 2002) before terminating the trial and the continuous presentation of the blue rectangle. Trials are separated by an intertrial interval (ITI) during which a feedback histogram can be presented on the computer screen displaying the frequency of responses plotted on a relative time scale with the letter "T" marking the target duration and a bracket of ±15% displayed around this target duration – thus, providing information concerning the accuracy and precision of the responses produced on the immediately preceding trial. Additional messages informed the participants if they had responded too short or too long with respect to the target duration or whether one of their responses was a direct hit. Participants were instructed that they should refrain from counting or tapping so as to avoid sub-dividing or any other type of extraneous timekeeping. To further discourage counting, random digits were intermittently superimposed over the blue rectangle, and participants were instructed to read the digits aloud. Inter-digit times were determined randomly using two uniform distributions (see Malapani et al., 1998b; Rakitin et al., 1998, for additional details). In the following sessions, participants typically received 20 FT trials, 30 peak trials with ITI feedback, and 10 peak trials without ITI feedback during blocked sessions each with a different target duration over the course of two or more days. Data collection sessions were preceded by pre-training sessions during both days. Target duration pairs used were: 11 and 17 s (Hinton, 2003; Hinton and Meck, 2004; Meck and Malapani, 2004), 7 and 17 s (Lake and

Meck, 2013), 7 and 14 s (Gu et al., 2015; Lustig and Meck, 2005), 8 and 21 s (Malapani et al., 1998b), and 6 and 17 s (Malapani et al., 2002; Rakitin and Malapani, 2005; Rakitin et al., 2005, 2006).

Malapani et al. (1998b) showed that when PD patients given DA-replacement therapy (e.g., L-dopa + apomorphine) are trained to time using corrective feedback they are able to reproduce single or multiple target durations (e.g., 8 and/or 21 s visual signals trained individually or within the same session) with the same accuracy and precision as age-matched controls. However, when these same PD patients are tested in an OFF medication state (again with feedback) a very different phenomenon emerges. Under these conditions, the 8s target duration is over-produced and the 21s target duration is under-produced relative to the ON medication state when both durations are trained and tested within the same session, as illustrated in Figure 11.1. This distortion was characterized as a "migration" effect, as though there was a mutual attraction, or "coupling" in memory for these two target durations. It is important to note that the 21 s target duration was reproduced as "shorter" than veridical when there was a resident memory of an 8 s target duration trained previously, but was reproduced as "longer" than veridical when it was the only trained target duration tested in the OFF medication state, indicating a "slowing" of temporal processing (e.g., Fournet et al., 2000; Meck and Benson, 2002) as illustrated in Figure 11.2.

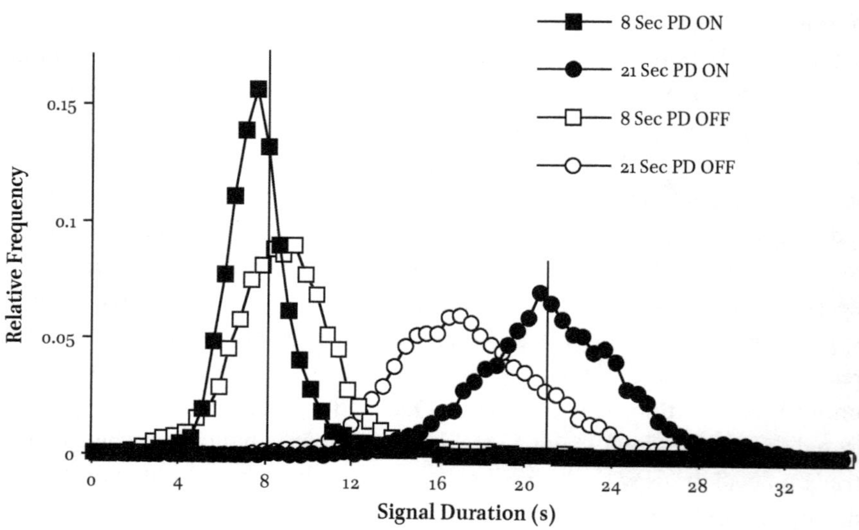

FIGURE 11.1 *Relative frequency distributions for the two target durations (8 and 21 s) showing accuracy and precision of duration reproduction in Parkinson's disease (PD) patients ON and OFF their levodopa medication.*

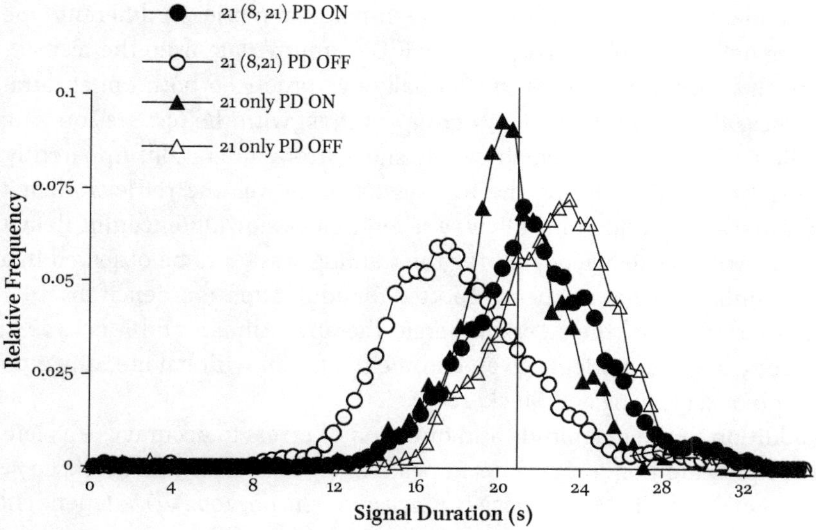

FIGURE 11.2 *Relative frequency distributions for a 21-s target duration plotted as a function of whether it was trained in conjunction with a 8-s target duration or by itself (21 only). The peak functions illustrate the accuracy and precision of duration reproduction in Parkinson's disease (PD) patients ON and OFF their dopaminergic medication as a function of the multiple or single target duration training.*

These findings have been followed up using an "encode/decode" paradigm (Malapani et al., 2002) that is able to reveal whether the dysfunctional coupling is part of the original learning and storage of the appropriate target durations (encoding) or introduced upon retrieval of these memories at test (decoding). Using this training (with feedback) –> test (without feedback) paradigm, it was demonstrated that the migration effect appears whenever PD patients are tested OFF medication, whether this is during training or test phases of the experiment. These results replicate and extend the earlier PD-migration effect first reported by Malapani et al. (1998b) showing that compared to the ON medication state, PD patients tested OFF medication produce responses that are "long" relative to the "short" target duration and responses that are "short" relative to the "long" target duration. The PD-migration effects reported by Malapani et al. (2002) were observed during both of the OFF training sessions, as seen in the OFF-OFF and OFF-ON groups' first-day performances, and in the test sessions, as seen in the OFF-OFF and ON-OFF groups' second-day performances. These results indicate that the PD-migration effect is not entirely dependent on the presence or absence of ITI feedback and is unlikely attributable to a selective problem with memory encoding. These findings led Malapani et al. (2002) to attribute migration in PD patients to a

DA-dependent dysfunction of retrieving temporal memories. A different type of timing deficit was observed in the OFF-ON groups' data from the test session. In this case, PD patients substantially overproduced both target durations. These data from the OFF-ON group contrast with the test session data from the OFF-OFF group, which clearly show strong migration. Apparently, restoring DA function during the test session improves the retrieval deficit (manifested as migration) and allows a second, clock slowing/encoding deficit associated with DA deficiency during the training session to be observed in a manner similar to what has been reported for adult attention-deficit disorder patients tested off of their dopaminergic therapy (Allman and Meck, 2012; Meck, 2003, 2005) or patients given chronic treatment with haloperidol, a DA receptor blocker (Lustig and Meck, 2005).

In addition to the magnitude and direction of errors in accuracy (e.g., leftward or rightward horizontal shifts in peak time – see Lustig and Meck, 2005; Malapani et al., 1998b; Meck, 2002a,b; Meck and Benson, 2002), DA-dependent distortions in storage and retrieval are also distinguishable by changes in the source and pattern of variability in treated and untreated PD patients (see Jones et al., 2011). Analysis of timing data precision showed that the PD-migration effects were associated with a violation of the scalar property of interval timing (Gibbon et al., 1997; Hinton and Rao, 2004; Rakitin et al., 1998). That is, the relative widths of the Gaussian-shaped PI response distributions for the two target durations were systematically different from each other when responses are plotted in proportion to the maximal level of responding on a relative time scale – such that the 21 s function was sharper than the 8 s function as illustrated in Figure 11.3. Timing functions for multiple target durations would normally be expected to superimpose when plotted on a relative time scale as a result of scalar sources of variance dominating non-scalar (e.g., Possion) sources (see Cheng and Meck, 2007; Gibbon, 1992; Rakitin et al., 1998). Moreover, the coefficient of variation (CV) expressed as the standard deviation of timed responses expressed as a proportion of the obtained peak time should be constant (Gibbon et al., 1997), but was observed to be different for the two target durations in all PD groups tested OFF medication, replicating the previous findings of Malapani et al. (1998b). The storage deficit associated with the rightward shifts for both target durations, however, continued to reflect the scalar property of interval timing; that is, the standard deviation of response times increased in proportion to the magnitude of the timing errors (Malapani et al., 2002).

In summary, DA deficiency leading to dysfunction in the basal ganglia results in two separable temporal memory deficits (see Malapani and Rakitin, 2003; Shea-Brown et al., 2006). Distortions that occur while storing multiple

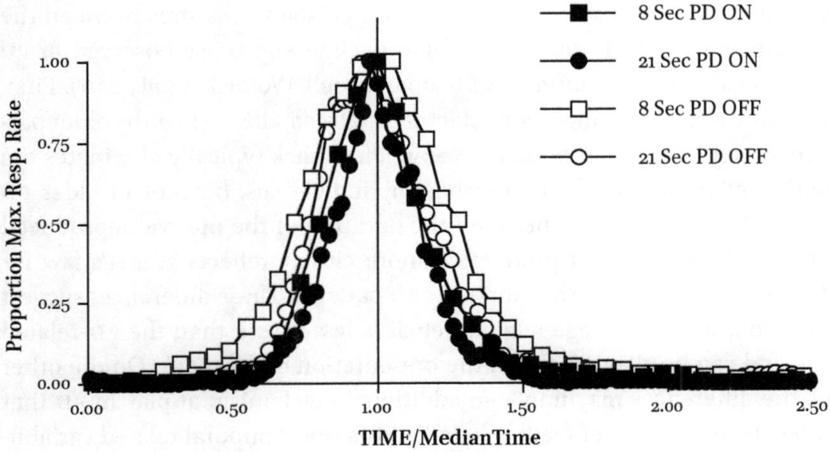

FIGURE 11.3 *Proportion maximum response rate for 8- and 21-s target durations for Parkinson's disease (*PD*) patients trained ON and OFF their levodopa medication. Data are plotted in time relative to the median time and normalized as proportions of the maximum response rate/frequency. Peak functions superimpose in the ON medication state, but not in the OFF medication state – indicating a violation of the scalar property of interval timing with lowered dopaminergic function.*

target durations result in overestimation of all intervals during subsequent reproduction. In contrast, retrieving the trace of two or more different target durations results in migration, a pattern of bidirectional errors such that reproductions of one of a pair of target durations are unusually close to the reproductions of the second. In addition to migration, DA-deficient retrieval is accompanied by a violation of the scalar property of interval timing, whereas with unidirectional shifts, the scalar property holds. It is important to note here that estimation of continuous attributes other than time (e.g., line length) is spared in PD in an analogous "encode/decode" paradigm (Malapani and Fairhurst, 2002). Moreover, patients with other kinds of brain damage, such as focal cerebellar lesions, do not show the temporal distortions seen in PD (Ivry, 1996; Malapani et al., 1998b, 2002). Taken together, this evidence strongly suggests that the basal ganglia and their cortical targets are an important aspect of the neural basis of interval timing in humans, and seem to involve both a slowing of "clock speed" and a "memory migration" component when PD patients are withdrawn from their dopaminergic medication (Buhusi and Meck, 2005; Fournet et al., 2000; Meck and Benson, 2002).

Migration in temporal memory has also been reported in healthy aged participants, when no further feedback or training is available. Although migration in normal aging appears similar to the migration observed in PD patients when tested OFF medication, there are distinguishable features between the effects attributable to normal age-related decline and those observed in PD patients (Lustig, 2003; Rakitin and Malapani, 2008; Wojtecki et al., 2011). First, the absolute size of the migration effect is usually smaller in healthy older participants than in older PD patients. Second, feedback typically eliminates the migration effect observed in older healthy individuals, but not in older PD patients. Third, the relation between the duration of the interval reproduced and the variability of that production more closely reflects Weber's law for healthy older participants than for older PD patients. These differences suggest that at a minimum, the age-related deficit is less severe than the PD-related deficit, and can be alleviated with the presentation of feedback. On the other hand, the differences may imply an additional mechanism at play in PD that obviates the usefulness of feedback and adds a non-temporal related variability component (Rakitin and Malapani, 2008; Rakitin et al., 2005). Yet another difference between the two groups was that all PD patients tested OFF medication show migration, whereas the effect is considerably more variable among healthy aged participants. Aged participants, for example, can be divided into sub-groups that show: (a) rightward shifts for both durations, (b) migration towards the mean for both durations, or (c) non-systematic shifts for both durations consistent with random variation in veridical timekeeping (Rakitin et al., 2005). These data raise the possibility that the observation of the "migration effect" in apparently healthy older adults may be an early, sensitive measure of age-associated cognitive decline, further suggesting that interval-timing paradigms may be an important tool for early detection of PD (Jurkowski and Meck, 2003). This hypothesis can be further tested by administering aged participants levodopa or placebo treatments as a function of training and test sessions in the PI and other timing procedures (Rakitin et al., 2006).

The neural basis of the PD-migration effect and the other temporal-processing deficits exhibited by PD patients are still poorly understood. The "migration effect" has been reported to be more evident in left hemi-Parkinsonian patients (Koch et al., 2005) – a finding that supports the lateralization of interval timing to the right hemisphere (Ferrandez et al., 2003; Harrington et al., 1998a,b; Meck, 2005; Meck and Malapani, 2004; van Wassenhove et al., 2008). In addition to dopaminergic medication (e.g., L-dopa), timing deficits in PD are alleviated by repetitive transcranial magnetic stimulation (rTMS) of the right dorsolateral prefrontal cortex (DLPFC), suggesting that impaired evidence-accumulation and decision-making processes contribute to distortions in timing and time perception (Allman and Meck, 2012; Jones et al., 2011; Kim and Lee, 2011; Koch et al., 2003, 2004; Philiastides et al., 2011; Rahnev et al., 2011). Traditional accounts of timing and time perception have used a pacemaker-accumulator

TABLE 11.1 *Temporal migration effects in Parkinson's disease.*

PD migration studies	Reproduction durations	Train-Test DA replacement therapy conditions	Feedback conditions/notes
Dušek et al. (2012)	5, 5.95, 7.07, 8.41, 10, 11.89, 14.14, and 16.81 s	ON-OFF	No feedback
Gu et al. (2015)	7 s and 14 s	ON-OFF	25, 50, and 100% peak trials with ITI feedback
Gu et al., (2015)	7, 7.875, 9.625, 10.5, 11.375, 12.24, 13.125, and 14 s	ON-OFF	No feedback
Koch et al. (2004)	5 and 15 s	ON-OFF	No feedback – DBS and DA therapy
Koch et al. (2005)	5 and 15 s	ON-OFF	No feedback – left and right hemi-PD*
Koch et al. (2008)	500ms and 2 s	ON-OFF	No feedback**
Malapani et al. (1998b)	8 and 21 s, 21 s alone	ON-OFF	FT trials + 66% peak trials with ITI feedback during both ON and OFF
Malapani et al., (2002)	6 and 17 s	ON-ON; ON-OFF; OFF-ON; OFF-OFF	FT trials + 75% peak trials with ITI feedback during ON, no feedback during OFF
Wojtecki et al. (2012)	5 and 15 s	Controls vs PD OFF	No feedback – DBS frequency varied

DA = dopamine; DBS = deep brain stimulation; FT = fixed-time; ITI = intertrial interval; PD = Parkinson's disease

* Only left hemi-PD patients showed the full PD-migration effect. Right hemi-PD patients showed an overproduction for the "short" (5 s) duration (Koch et al., 2005).

** Failed to see the full PD-migration effect when the "short" duration was in the milliseconds range (500ms) and the "long" duration was in the seconds range (2 s), although the 2-s duration did shift leftward (Koch et al., 2008).

model to explain how the internal clock might function (Gibbon et al., 1984; Shea-Brown et al., 2006; Treisman, 1963). These models have been replaced by more physiologically-realistic theories that rely on the coincidence-detection of cortical oscillations by striatal medium spiny neurons – referred to as the SBF model of interval timing (Agostino et al., 2011; Allman and Meck, 2012; Buhusi and Oprisan, 2013; Matell and Meck, 2004; Matell et al., 2003; Oprisan and Buhusi, 2011, 2013). According to the SBF model, thalamo-cortico-striatal circuits learn to detect specific durations by associating feedback with particular patterns of cortical oscillations. Interestingly, pathological alterations of neuronal oscillations have recently been implicated in the pathophysiology of PD symptoms (Brown, 2003; Brown et al., 2001; Eusebio et al., 2012; Hutchison et al., 2004; Timmermann et al., 2003, 2004). Importantly, a number of recent studies have demonstrated that timing deficits in PD are alleviated by deep brain stimulation of the subthalamic nucleus (Koch et al., 2004) with differential effects of stimulation frequency (Wojtecki et al., 2011). These findings support the importance of specific oscillatory processes in that stimulation of the subthalamic nucleus differentially modulates motor and timing functions depending on stimulation frequency (Timmermann et al., 2004; Wojtecki et al., 2006). A summary of the PD-migration effects described above is provided in Table 11.1.

4. Effects of Corrective Feedback on Interval Timing

In PI timing experiments there are generally two types of feedback given, FT trials that demonstrate the target duration and ITI histograms that inform participants of the relative accuracy and precision of responses made during the preceding peak trial (Malapani et al., 1998b, 2002; Lake and Meck, 2013; Lustig and Meck, 2005; Rakitin and Malapani, 2008; Rakitin et al., 1998, 2005). It can, thus, be postulated that corrective feedback has two components: (1) a memory component in which participants use the FT trial color change to update their memory of the target duration and (2) a threshold-setting component in which participants use the ITI histogram for purposes of correcting errors in start and stop times (see Church et al., 1994; MacDonald et al., 2012). Although one could argue that FT trials might also affect threshold setting, the findings from a number of experiments suggest that these two types of feedback have distinct effects on timing behavior (Lustig and Meck, 2005; Rakitin and Malapani, 2008; Rakitin et al., 1998).

Lustig and Meck (2005) investigated the effects of ITI feedback in the PI procedure (7- and 14-s target durations) as a function of the probability of feedback (25, 50, or 100%) and the blockade of DA receptors by haloperidol. Within the 25% feedback condition the effect on peak time of the number of "trials since feedback" – which reflects the decay of feedback effects over time – was also investigated. The results showed an interaction between the feedback and drug conditions such that a proportional over-production of peak time was observed for the 7- and 14-s target durations under chronic haloperidol that increased as the probability of feedback was reduced as illustrated in Figure 11.4. Moreover, a trial-by-trial analysis demonstrated that reproduction of the target interval becomes later in time as feedback becomes more remote. This effect occurs for both control and drug treatment, but was exaggerated under haloperidol. This pattern of results raises some interesting questions about the mechanisms by which feedback has its effects, and the interactions of those effects with dopaminergic function. Feedback is often thought to have its impact by updating the representation of the target duration in memory (see discussions by Matell and Meck, 2004; Meck, 1996; Meck and Benson, 2001). That is, feedback serves to alert the participant that the amount of time they have associated with the target duration is incorrect, and on subsequent trials they should change the amount of time before responding in order to adjust for the discrepancy (Meck, 2002a,b). For example, in the current experiment, where DA receptor blockage and distance from feedback led to an overproduction of time, the ITI feedback would alert participants that they were waiting too long and that on the next trial they should respond earlier. Consequently, one possible explanation for the pattern of results found here is that the representation of the target duration and/or the corrective effects of feedback in working memory decay over time, and that this decay is exaggerated by the blockade of DA receptors. This interpretation would be consistent with previous findings that chronic haloperidol administration downregulates the D1 sub-type of DA receptors and impairs working memory (Castner et al., 2000), and that DA release is important for the correction effects of feedback observed in many tasks (e.g., Schultz, 1998; Schultz and Dickinson, 2000 – but see Kim et al., 2014).

Another possibility is that both feedback and haloperidol had direct effects on the clock stage given that DA antagonists such as haloperidol have been shown to decrease "global" clock speed during the timing of all trials in the PI procedure (e.g., Buhusi and Meck, 2002; MacDonald and Meck, 2005, 2006; Meck, 1986; Lake and Meck, 2013; Williamson et al., 2008). As a consequence, the dynamic and proportional rightward shifts observed as a function of the number of "trials since feedback" can be viewed as "local" effects on clock speed if one assumes that the clock slows down as the DA-releasing effects of feedback wear off and is temporarily speeded up again when feedback is delivered (e.g., Allman and Meck, 2012; Kawashima et al., 2012; Koepp et al., 1998; Matell and Meck, 2004). This proposal doesn't distinguish between positive

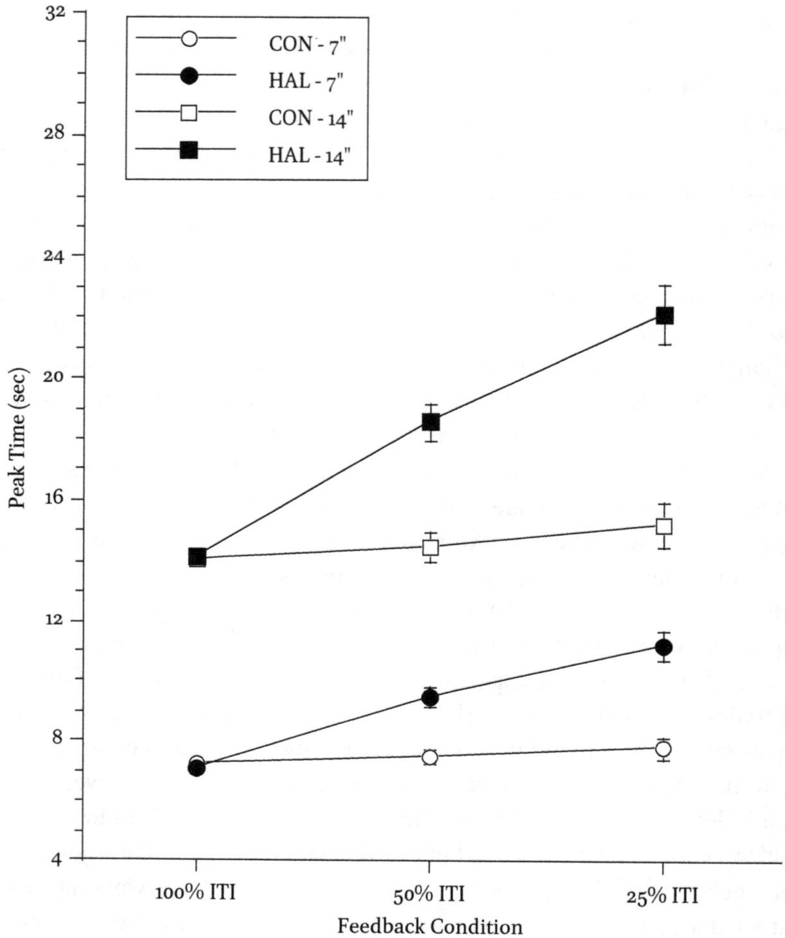

FIGURE 11.4 *Peak times (means ± SE) for participants in the control (CON) and haloperidol (HAL) conditions tested with 7- and 14-s target durations in the peak-interval procedure. Data are plotted as a function of the probability of intertrial interval (ITI) feedback (25, 50, and 100%).*

and negative feedback and assumes that the information provided by the ITI feedback is equally "stimulating" (at least in healthy participants) regardless of whether it indicates that they are doing well or have made an error. It is, of course, possible to separate these effects under conditions where participants make numerous errors (i.e., do not place their responses within the bracketed window displayed in the ITI histogram), but in practice young control participants do not make many errors of this sort even under the 25% feedback condition. It's interesting to note that when aged participants are tested without feedback following an initial training session with feedback, there is a migration effect regardless of the type and amount of feedback given during training. Moreover, it was observed that false and inconsistent feedback induces migration in the elderly of an order of magnitude similar to the pd-migration effect (Rakitin and Malapani, 2008; Rakitin et al., 2005).

Importantly, PD patients evaluated ON and OFF their dopaminergic medication using the same ITI feedback probabilities (25, 50, and 100%) and target durations (7 and 14 s) as Lustig and Meck (2005) showed effects different from young participants evaluated under haloperidol. When age-matched controls were trained with 7- and 14-s target durations, they showed a tendency to shift rightward as ITI feedback became more infrequent. In contrast, PD patients in both the ON and OFF medication states showed a "migration effect" that increased as ITI feedback became more infrequent such that the 7-s target duration was over-produced and the 14-s target duration was under-produced. This "migration effect" however, was substantially larger in the PD-OFF medication state than in the PD-ON medication state (in which significant migration was only observed for the 25 and 50% ITI conditions) as illustrated in Figure 11.5. Relating these effects to clock speed, memory encoding/decoding, and error processing presents an interesting challenge to understanding the cognitive deficits associated with PD and DA depletion (see Nagano-Saito Atsuko et al. 2008).

5 Mismatch Theory and the Error-Related Negativity

Error processing reflects the brain's ability to detect and adjust to incorrect behaviors (e.g., Falkenstein et al., 2000). Initially, investigators hypothesized that error processing involved a comparison between a participant's actual response and the representation of the correct response stored in memory (e.g., Nieuwenhuis et al., 2002). Studies investigating this "mismatch theory" found an electrophysiological component of error processing, the error-related negativity (ERN; Falkenstein et al., 2000; Herrmann et al., 2004; Holroyd and Coles, 2002; Nieuwenhuis et al., 2002). The ERN is a negative shift in the recorded electroencephalogram (EEG) believed to be the unconscious aspect of error processing and appears approximately 80ms following an error (Holroyd and Coles, 2002). The ERN is typically measured at the peak height of the negative deflection.

According to the initial mismatch theory of error processing, the ERN follows an error in order to represent the error as a difference between the actual response and the correct response. For correct responses, little if any

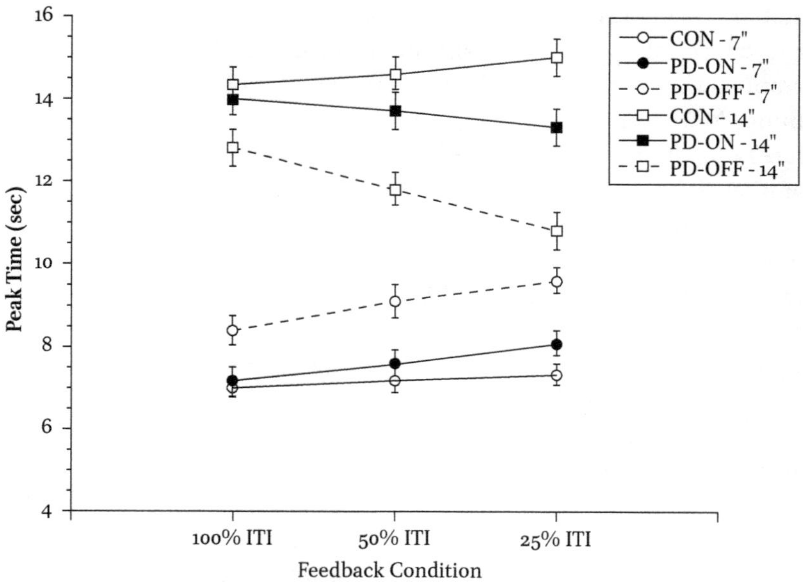

FIGURE 11.5 *Peak times (means ± SE) for Parkinson's disease (PD) patients trained ON and OFF their dopaminergic medication and aged-matched controls using 7- and 14-s target durations in the peak-interval procedure. Data are plotted as a function of the probability of intertrial interval (ITI) feedback (25, 50, and 100%).*

difference should exist between the actual response and the correct response. Thus, the ERN should appear following *errors only*, but not correct responses. However, experimental evidence has shown that a comparatively less negative ERN followed correct responses in addition to the ERN that followed errors (Herrmann et al., 2004). The ERN following correct responses refuted the initial mismatch theory, and a refined version of the theory was developed to explain the presence of the ERN following both correct and incorrect responses. The refined mismatch theory combines the ERN with a reinforcement-learning algorithm referred to as the "method of temporal differences," which relies upon the mesencephalic DA system for updating (Holroyd and Coles, 2002). According to the refined mismatch theory, the ERN should occur when the mesencephalic DA system produces a negative reinforcement signal (less DA release following an error than during baseline) to the anterior cingulate cortex (ACC), the region of the brain believed to generate the ERN (Nieuwenhuis et al., 2002; San Martín, 2012). This negative reinforcement signal represents the DA system's response to negative feedback when an error is made. Negative feedback occurs when the participant is notified that he/she made an error;

however, even in the absence of explicit information, it is possible that the DA system can still transmit negative feedback following the error. Specifically, in a task that involves judging different signal durations, for example, the range and other statistics defining the distribution of signal durations might become stored in the participant's reference memory and, thus, provide internal feedback. This type of feedback is also thought to depend on the levels of DA in cortico-striatal circuits (Buhusi and Meck, 2005; Cools et al., 2002; de Vries et al., 2010; Nieuwenhuis et al., 2004; Wise, 2004).

5.1 *Dopamine Effects on the Error-Related Negativity*

After an error occurs, the ACC detects the extent to which DA levels decrease, and accordingly, the amplitude of the ERN depends on *how much* the actual response differs from the correct response (Holroyd and Coles, 2002). The ERN amplitude for correct responses reflects the smallest possible difference between the actual response and the correct response, and the ERN of an error is expected to have a more-negative amplitude than the ERN of a correct response. The refined mismatch theory thereby accounts for the less negative ERN found after correct responses by proposing that the ERN amplitude is regulated by the amount of DA released in specific pathways. The amount of DA released by the mesocortical and nigrostriatal pathways also depends on biological factors. Natural degeneration of the mesocortical and nigrostriatal pathways occurs as healthy adults age (Fearnley and Lees, 1991; Rubin, 1999) and severe degeneration of the mesocortical and nigrostriatal pathways occurs as a result of PD. Thus, the presumed reduction in DA in these groups should result in a lowered response to negative feedback. To test these predictions concerning the effects of DA level on error processing, past studies of the ERN compared healthy young adults to healthy aged adults (Kolev et al., 2005) or healthy aged adults to PD patients (Holroyd et al., 2002; Ito and Kitagawa, 2006) by comparing correct responses with errors and by comparing the ERN amplitude for errors across groups.

5.2 *Age Effects on the Error-Related Negativity*

Kolev et al. (2005) used a choice reaction time task to examine differences in error processing due to age and found that the ERN amplitude was less negative in correct responses compared to incorrect responses for both young and aged adults, as consistent with studies of the initial mismatch theory (Herrmann et al., 2004). In a comparison between groups, however, the ERN amplitude was more negative in young adults compared to aged adults. These results were interpreted as support for the refined mismatch theory because age is expected to decrease DA levels. The more negative ERN amplitude in young adults was also interpreted to indicate that error detection is impaired by age (Kolev et al., 2005).

5.3 *Parkinson's Disease Effects on the Error-Related Negativity*

Holroyd et al. (2002) tested aged adults and PD patients OFF dopaminergic medication using a choice reaction time task and found that the ERN amplitude was similar across aged adults and PD patients. The ERN amplitude of correct responses was not directly compared to incorrect responses within each group. Instead, the ERN of correct responses was subtracted from the ERN of incorrect responses for comparisons between groups. Because the ERN was still present after the subtraction, the ERN amplitude of correct responses was assumed to be less negative than the ERN amplitude of incorrect responses. The similar ERN amplitude across aged adults and PD patients tested OFF medication was interpreted to indicate that PD did not reliably impair error detection (Holroyd et al., 2002). In contrast, Ito and Kitagawa (2006) tested aged adults and PD patients ON dopaminergic medication using a lexical decision task and found that the ERN amplitude was less negative in correct responses compared to incorrect responses for both aged adults and PD patients, as consistent with studies of the initial mismatch theory. Between groups, the ERN amplitude was more negative in aged adults compared to PD patients. These results were interpreted to support the refined mismatch theory even though dopaminergic medication is thought to increase DA levels in PD patients and make them more similar to age-matched controls. The more negative ERN amplitude in aged adults was also interpreted to show that PD impaired error detection (Ito and Kitagawa, 2006). As a consequence, it is still an open question as to the relationship between error processing, DA levels, and the ON and OFF medication states of PD.

6. Affective-processing Hypothesis of the Error-Related Positivity

The error positivity (Pe) is a positive deflection in the EEG believed to be the conscious component of error processing (Overbeek et al., 2005). The Pe appears approximately 200–500ms following an error and is measured at the peak height of the positive deflection (Falkenstein et al., 2000, 2005). According to the affective processing hypothesis of error processing, the Pe reflects post-error processing in the form of subjective/emotional assessment (Falkenstein et al., 2000) and is independent of the ERN (Overbeek et al., 2005). The amplitude of the Pe is believed to indicate the emotional significance of an error (Falkenstein et al., 2000). A study of the affective processing hypothesis compared participants who made numerous errors with participants who made relatively few errors and found that the amplitude of the Pe was more positive in participants who made few errors compared to participants who made many errors. The results were interpreted to support the affective processing hypothesis because errors were assumed to have less emotional significance for participants who made many errors compared to participants who made few errors – although the Pe has not been investigated following correct responses (Falkenstein et al., 2000).

The Pe follows the ERN in time, but is considered independent of the ERN because the Pe and ERN are generated by different areas of the ACC. The ERN appears in the medial prefrontal region of the ACC while the Pe appears in a more rostral region of the ACC (Herrmann et al., 2004). Also, unlike the ERN, the Pe is not present in all errors. For example, Ehlis et al. (2005) tested the affective processing hypothesis using a negative response feedback task, in which participants were sometimes told that they responded incorrectly when they actually responded correctly (false feedback). This study found that the Pe was present following incorrect responses but was *not* present following negative response feedback, possibly because participants expressed certainty that they had responded correctly. The assumed independence of the Pe from the ERN is also supported by evidence that the DA system does not appear to affect the Pe. Falkenstein et al. (2005) found that the Pe was similar in healthy aged adults compared to PD patients evaluated OFF dopaminergic medication even though PD reduces DA levels beyond the effects of normal aging. Moreover, Ito and Kitagawa (2006) found that the Pe amplitude was more positive in healthy aged adults compared to PD patients tested ON dopaminergic medication, even though medication is thought to increase DA levels in relevant brain circuits in PD patients. Although the ERN and Pe have been studied in lexical decision (Kolev et al., 2005), choice reaction time (Ito and Kitagawa, 2006), and negative feedback tasks (Ehlis et al., 2005), no previous research has investigated the ERN and the Pe using temporal ordinal comparison tasks (e.g., Coull et al., 2004; Harrington et al., 2010; Rao et al., 2001; Van Rijn et al., 2011 – see also Larson and Clayson, 2011).

7 Effects of Dopamine-related Error Processing on Interval Timing

Like all errors, errors associated with interval timing should be followed by an error-correction signal from the DA system. This error correction signal represents the DA system's response to negative feedback. Even in the case where no feedback follows correct or incorrect responses, the ERN may still occur in situations where the experimental design allows for storage of the relevant stimuli in reference memory (see Cicchini et al., 2012). Consequently, when participants are trained to compare two sequentially presented signal durations in order to determine which signal is longer (Coull et al., 2004, 2012, 2013) the base durations are likely encoded during the early training trials and stored in the participant's reference memory for the rest of the session (Lustig and Meck, 2005; Matell and Meck, 2004). Thus, the cortico-striatal circuits may provide a form of "endogenous" feedback by comparing the current clock reading with previous durations stored in temporal memory and subsequently generate an ERN when judgments/responses do not match with those values previously stored in memory – this despite the fact that the judgment could have been made based upon the information presented within the current trial (Galea et al., 2012; Gu and Meck, 2011). The use of this type of statistical analysis of previous trials in the control of timing behavior on the current trial provides strong support for models based Bayesian decision theory (Körding and Wolpert, 2006).

In an unpublished study, we examined the effects of PD and dopaminergic medication on interval timing and error processing using an ordinal comparison procedure with hue and duration stimulus dimensions (Chen, 2007; see also Coull et al., 2004, 2012, 2013; Gu and Meck, 2011; Harrington et al., 2010). Psychophysical and EEG/ERP data were collected in PD patients ON and OFF dopaminergic medication. Based on previous studies of the ERN, the ERN was hypothesized to be present following both correct and incorrect responses, with a less negative ERN amplitude predicted for correct responses compared to incorrect responses (Beste et al., 2006, 2007; Falkenstein et al., 2005). Based on the refined mismatch theory, participants with increased DA levels were hypothesized to show higher ERN amplitudes. Thus, PD patients tested ON medication were hypothesized to show more negative ERN amplitudes compared to PD patients tested OFF medication. As predicted, we observed a reliable diminution in the ERN for PD patients tested OFF their dopaminergic medication compared to when these same patients were tested ON medication. In contrast, there were no consistent changes in the Pe as a function of dopaminergic medication in PD patients as illustrated in Figure 11.6. These results confirm the previous findings of Falkenstein et al. (2001, 2005) and support the view that deficient error processing may underlie the distortions in temporal memory observed in the PD-migration effects reviewed above.

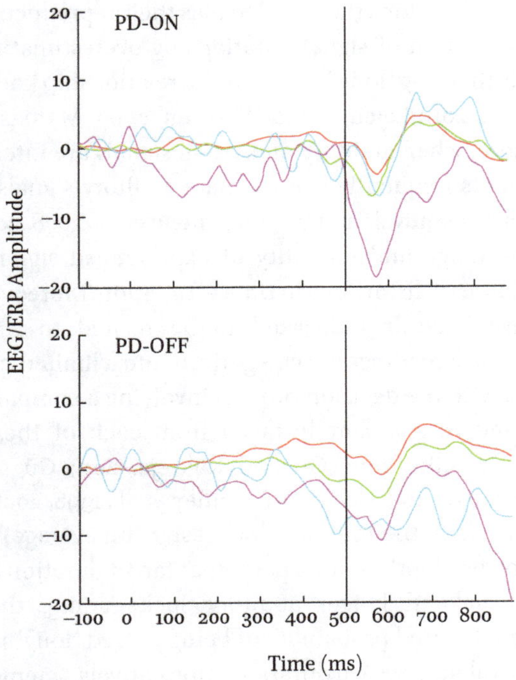

FIGURE 11.6 *Ordinal comparison task using hue and duration stimulus combinations in the manner of Coull et al. (2004). The duration data presented here represent the grand average response-locked event-related potentials (ERPs) at electrode Fz after correct left-handed (green) and right-handed (red) responses and incorrect left-handed (magenta) and right-handed (cyan) responses in patients with Parkinson's disease (PD) in the ON and OFF medication states. The 0-ms time point represents the time of the "shorter" or "longer" classification (key press) determined by a left-handed or right-handed response. The vertical line represents the peak latency of the error related negativity (ERN) in the PD patients. The ERN is smaller in the OFF medication state than in the ON medication state for the left-handed (magenta) incorrect response with no difference in the Pe between conditions (see Falkenstein et al., 2001, 2005). Interestingly, no handedness/lateralization effects were observed for the hue condition in the PD patients as a function of the ON/OFF states.*

8 Vierordt's Law and the Migration of Temporal Memories

Subjective time can be affected by various factors, including memory load and temporal context (Buhusi and Meck, 2005, 2009a,b; Buhusi et al., 2002; Grondin, 2005; Jazayeri and Shadlen, 2010; Merchant et al., 2008b). For example, when participants are presented with a range of signal durations and are then asked to reproduce those durations they tend to bias their reproductions towards the mean of the distribution of signal durations by overestimating "short" durations and underestimating "long" durations – a relationship known as Vierordt's law (Gu and Meck, 2011; Lejeune and Wearden, 2009; Mamassian and Landy, 2010). Furthermore, when auditory and visual signals are intermixed within a session, participants frequently overestimate auditory signals and underestimate visual signals of equivalent duration (Cheng et al., 2008, 2011; Penney et al., 2000, 2005). The range and modality of experienced signal durations are known to contribute to distortions in time estimation. Moreover, both of these phenomena have been hypothesized to be related to "memory-mixing" whereby similar clock readings are categorized into a limited number of memory distributions with the decision process involving a comparison of the current clock reading with a sample taken from each of these distributions (Gamache and Grondin, 2010; Gu and Kukreja, 2011; Gu and Meck, 2011; Klapproth, 2009; Lustig and Meck, 2011; Penney et al., 1998, 2008). For example, if auditory signals drive the internal clock faster (on average) than visual signals and memory distributions for a particular target duration are a mixture of these "shorter" visual and "longer" auditory clock readings, then auditory signals will have an increased probability of being judged "long" relative to visual signals of an equivalent physical duration. Alternatively, "memory-mixing" can occur when different standard durations used within the same temporal context (e.g., test session) are combined into a single memory distribution, rather than separate memory distributions being maintained for each standard. As a consequence of the temporal context, "memory-mixing" can produce distortions in scaling of duration that are consistent with Vierordt's law and may be modifiable by training conditions (e.g., blocking of different standards or signal modalities and/or by providing feedback in order to encourage the formation of separate rather than mixed-memory distributions). This sort of "memory-mixing" is distinct from the time-order error, which refers to the influence of the order of presentation on the comparison of successively presented stimuli (see Allan, 1979; Brown et al., 2005; Eisler et al., 2008; Grondin, 2010). As described above, the mixing of unimodal stimuli in memory as a function of temporal context is consistent with Vierordt's law (Gu and Meck, 2011; Hairston and Nagarajan, 2007; Lejeune and Wearden, 2009; Mamassian and Landy, 2010), and the tendency for overestimating shorter durations and underestimating longer durations is one of the most robust temporal phenomena exhibited under a variety of conditions (Gu and Meck, 2011; Jazayeri and Shadlen, 2010; Lejeune and Wearden, 2009). However, the exact mechanism underlying this "memory-mixing" effect and whether it can account for all of the PD-migration effects observed in the "encode/decode" paradigm is currently unknown.

A recent series of investigations of temporal context has shown that a Bayesian model can explain the biased performance of participants when they are asked to reproduce or compare a recently presented signal duration (Cicchini et al., 2012; Gu and Meck, 2011; Jazayeri and Shadlen, 2010). This model incorporates the knowledge of the distribution of previous signal durations into the perception of the current signal duration, thus biasing the reproduction of the current interval towards the mean of the distribution. In this model, it is hypothesized that a tradeoff exists between accuracy and precision such that the distribution of signal durations (i.e., temporal context) is used to optimize performance by reducing temporal uncertainty at the cost of temporal accuracy. In this case, the implicit knowledge of the underlying distribution from which a sample is drawn would be useful when the current clock reading is uncertain due to the effects of noise and/or inattention, such as that which may result from dopaminergic drug administration (e.g., Lake and Meck, 2013) and/or neurological conditions that may affect dopaminergic function such as attention deficit hyperactivity disorder, Huntington's disease, PD, or obsessive compulsive disorder (e.g., Allman and Meck, 2012; Beste et al., 2007; Conners et al., 1996; Gu and Kukreja, 2011; Gu et al., 2008, 2011; Levin et al., 1996, 1998; Meck, 2005). This explains how the intermixing of a previous trial's signal durations with the perception of the current trial's signal duration could bias performance. Under certain conditions, however, this statistical analysis can provide an efficient strategy for reducing variability in the presence of high levels of uncertainty or noise (Cain et al., 2012; De Ridder et al., 2012). At present, this hypothesis is mainly supported by computer simulation rather than experimental results from behavioral or neurobiological studies.

Although the "memory-mixing" effect that we have described is highly reliable, large individual differences have been reported (Gu and Meck, 2011). Interestingly, healthy adult participants with a relatively high degree of "memory-mixing" display relatively low variation in the reaction times for their response classifications, whereas the opposite is true for participants with a low degree of "memory-mixing." Indeed, the "memory-mixing" index was negatively correlated with the observed variability in reaction time across participants. Although it is uncertain why variability in reaction time should be related to the level of "memory-mixing," one possibility is that participants who pay less attention to feedback utilize a more automatic and less variable process of sampling from a "mixed memory" distribution rather than engaging the more variable process of maintaining separate distributions under conditions of high temporal uncertainty (Gu and Meck, 2011). The observation of reliable individual differences in "memory-mixing" begs the question as to whether it is possible to increase or decrease the degree of "memory-mixing" by experimental manipulation. One can assume that if the degree of attention given to the encoding of the standard duration contributes to variation in the degree of "memory-mixing," then it would be possible to systematically alter the degree of "memory-mixing." Reward cues have been shown to increase the encoding of information through the activation of mesolimbic DA systems (Adcock et al., 2006; Koepp et al., 1998) and feedback tied to this dopaminergic system has been shown to improve the accuracy and precision of timing functions (Coull et al., 2011, 2012, 2013; Lustig and Meck, 2005; Malapani et al., 1998b; Meck, 2005). The assumption is that this feedback aids in the formation of modality-specific or standard-duration specific memory representations within the context of the psychophysical procedures described above. In order to determine whether feedback can affect the degree of "memory-mixing," participants were presented the word "WRONG" on the computer screen along with an aversive buzzer sound when they made an incorrect response and presented the word "CORRECT" along with a pleasant bell sound when they made a correct response. Following the implementation of this feedback procedure, some participants showed a dramatically reduced "memory-mixing" effect in ordinal comparison procedures (Gu and Meck, 2011). It will be necessary to conduct additional studies in order to determine the most reliable and effective form of feedback (e.g., valence and intensity) for the modulation of "memory-mixing" as a function of development (Allman et al., 2012; Droit-Volet and Meck, 2007) and different patient populations (Allman and Meck, 2012; Gu et al., 2008, 2011).

9 Bayesian Models of Interval Timing

Previous work has shown that participants readily adjust to the underlying statistics involving the distributions of event durations in order to optimize the accuracy of their timed responses. This behavior is consistent with the development of a representation of the experimentally presented distribution of signal durations (the prior) and the calculation of the error (loss function) as predicted by Bayesian integration (Acerbi et al., 2012; Miyazaki et al., 2005). Within the Bayesian framework, the responses of an ideal observer depend on the manner in which these internal representations encode the complexity of the underlying distributions of signal durations (e.g., uniform to highly skewed or bimodal distributions). It has recently been shown that duration reproduction times are affected by both the underlying distribution of signal durations as well as the type and probability of corrective feedback (Acerbi et al., 2012; Sohn and Lee, 2013) – although the tracking of the distribution becomes more difficult with higher-order statistical features (e.g., kurtosis and multimodality). Overall, these findings are in good agreement with a performance-optimizing Bayesian observer and actor model of decision-making suggesting that humans and other animals are sensitive to the temporal statistics of virtually all of the information presented within a specific temporal context. This statistical information influences the determination of temporal judgments even in cases where the current trial may simply require the reproduction of a signal duration that was presented to the participant just moments before the "go" signal to respond on the current trial (Acerbi et al., 2012; Cicchini et al., 2012; Jazayeri and Shadlen, 2010). Moreover, a Bayesian observer model, in which the subjective duration of an event is determined jointly by a prior and a likelihood function for timing, captures the temporal dynamics of accuracy and precision in a manner suggesting that the width of the prior, not the likelihoods, gradually decreases over training sessions, again indicating the relevance of prior knowledge for interval timing (Sohn and Lee, 2013).

Based on these prior arguments, a Bayesian-inspired model was used to simulate duration reproductions in the PI procedure using 7- and 14-s target durations (Lustig and Meck, 2005) as well as 8- and 21-s target durations (Malapani et al., 1998b) as illustrated in Figure 11.7 and Table 11.2. The model is designed to reduce reproduction errors under conditions where the time base is noisy as a result of a slowing in the mean clock speed and increased variability in its regulation – possibly as a result of DA deficiency and/or reduction in feedback. This is done by incorporating an adaptable central tendency prior as a probability density function centered at the mean of the distribution of target durations presented within all session blocks using the mean and standard deviation in the manner of Cicchini et al. (2012). The model did not associate different target durations with different discriminative stimuli because this feature was not present in the experimental procedures used for any of the PI data sets under study (e.g., Lustig and Meck, 2005; Malapani et al., 1998b, 2002; Rakitin et al., 1998, 2005). In this sense, the current model is a simplified version of the Bayesian least squares model used by Jazayeri and Shadlen (2010) and doesn't assume that all information is stored from previous trials and used to determine the prior even though this level of detail has been used in previous descriptions of interval timing (see Gibbon and Church, 1984, 1990, 1992).

Following Cicchini et al. (2012), we simulated the Bayesian model of estimated durations by incorporating information of previous target durations used within the same temporal context and a measurement of current trial's target duration. In this Bayesian model, the estimation of the current duration corresponds with the posterior distribution and this was calculated by combining the likelihood and prior functions, which represent the sensory

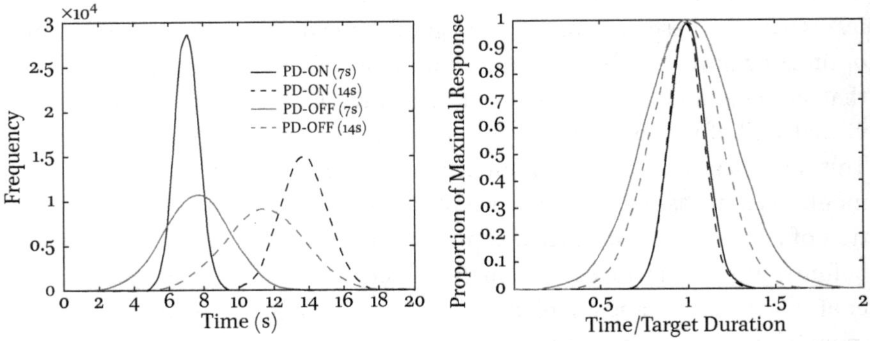

FIGURE 11.7 *Bayesian simulations of 7- and 14-s target durations in the peak-interval procedure for Parkinson's disease (PD) patients tested ON and OFF dopaminergic medication. Weber fractions for sensory measurement (Wm) were set as 0.1 in PD-ON medication state and 0.3 in PD-OFF medication state. Simulation of 200,000 trials demonstates veridical timing (i.e., response functions centered on the target durations) in the PD-ON medication state and a high degree of migration (i.e., overestimation of the shorter duration and underestimation of the longer duration – Vierordt's law) for the 7- and 14-s target durations in the PD-OFF medication state (left panel). Scaled distribution functions, i.e., proportion of maximal responding plotted on a relative time scale (right panel), show that the estimations of 7- and 14-s target durations superimpose on top of each other in the PD-ON medication state. In contrast, the response function for the 14-s target duration shows a narrower dispersion compared to the response function for the 7-s target duration in the PD-OFF medication state, indicating a violation of scalar property of interval timing (see Hinton and Rao, 2004; Malapani et al., 1998; Rakitin et al., 1998).*

measurement of the current target duration and the memory distribution of the previous target durations, respectively. The likelihood function was modeled as a Gaussian distribution with mean Ts (current target duration) and standard deviation (s.d.) WmTs – the measurement Weber fraction (Wm) multiplied by the target duration (Ts). According to these assumptions, increased variation will be correlated with the target duration(s), which corresponds to the scalar property of interval timing (Gibbon et al., 1984; Jazayeri and Shadlen, 2010). In addition, the Wm values for the PD-ON and PD-OFF medication states were set 0.1 and 0.3, respectively, in order to represent the increased noise observed in sensory measurement and/or uncertainty as a function of DA depletion.

The prior function was defined as a Gaussian distribution centered at the mean of the previous target duration(s), and the s.d. (width) for each target duration was calculated to optimize performance considering precision and accuracy. In other words, the optimal s.d. of the prior was set to minimize the

TABLE 11.2 *Bayesian model predictions for peak-interval measurements.*

	Measurement (Wm_ON = 0.1; Wm_OFF = 0.3)		Reproduction (Fixed Wp = 0.1)		Reproduction (Wp_ON = 0.1; Wp_OFF = 0.3)	
7 and 14 s	Peak	CV	Peak	CV	Peak	CV
ON (7 s)	7.07	0.10	7.07	0.14	7.07	0.14
ON (14 s)	13.89	0.10	13.38	0.14	13.38	0.14
OFF (7 s)	7.58	0.25	7.58	0.27	6.57	0.40
OFF (14 s)	11.62	0.19	11.36	0.22	10.86	0.36
8 and 21 s	Peak	CV	Peak	CV	Peak	CV
ON (8 s)	8.08	0.10	8.08	0.14	8.08	0.14
ON (21 s)	20.61	0.10	20.61	0.14	20.61	0.14
OFF (8 s)	8.48	0.27	8.08	0.29	6.87	0.41
OFF (21 s)	17.78	0.26	17.78	0.28	16.57	0.40

Peak = peak time – mode of timing functions obtained from the peak-interval procedure with 7- and 14-s target durations or 8- and 21-s target durations for Parkinson's disease patients tested ON and OFF dopaminergic medication. CV = coefficient of variation – defined as the ratio of the standard deviation of the peak functions to the mean of the function as represented by the obtained peak time. Wm = Weber fraction for the measurement phase of timing. Wp = Weber fraction for the production phase of timing.

mean squared error (MSE), which is equal to the sum of the variance and the squared bias of the estimated target duration. In terms of these calculations, the width of the prior gets narrower as the value of WmTs get larger. For example, the 14-s target duration in the PD-OFF medication state (TsWm = 14*0.3 = 4.2; prior width = 2.62) has a much narrower prior width compared to the 7-s target duration in PD-ON medication state (TsWm = 7*0.1 = 0.7; prior width = 4.9). In addition, if the temporal difference between the target durations gets smaller, the prior width also gets narrower. Therefore, the function for 7- and 14-s target durations will be more affected by their prior compared to that of 8- and 21-s target durations. The narrower width (s.d.) of the prior in those cases may indicate that the duration estimation relies more on the memory distribution of previous target durations compared to the sensory measurement of the current target duration. By relying more on the memory of previous target durations, performance will result in lower accuracy (biased toward the mean), but higher precision in the reproduced duration than would otherwise be possible given the level of uncertainly in the DA-depleted, PD-OFF medication state.

These estimated likelihood and prior functions were combined to determine the posterior function which corresponds to the distribution of estimated durations according to the Bayes' rule (Cicchini et al., 2012; Sohn and Lee, 2013). Simulation of 200,000 trials using 7- and 14-s target durations from the calculated posterior in PD-ON/OFF medication states according to the methods described by Gu et al. (2015) shows that the degree of migration toward the mean was increased in the PD-OFF medication state as illustrated in Figure 11.7. The estimated peak times for the two target durations were 7.07 and 13.89 s in the PD-ON medication state, and 7.58 and 11.62 s in the PD-OFF medication state – showing nearly veridical timing in the ON state and the "PD migration" effect in the OFF medication state. In addition to the migration of peak times, the CVs for the two target durations were reliably different from each other in PD-OFF medication state (0.25 for 7 s and 0.19 for 14 s) – more similar to counting than timing processes and a clear violation of the scalar property of interval timing (Hinton and Rao, 2004; Malapani et al., 1998b; Rakitin et al., 1998). In contrast, the CV's were observed to be constant around 0.1 for both the 7- and 14-s target durations in the in PD-ON medication state. Given the high levels of uncertainty in sensory measurement due, in part, to a slowed clock in the PD-OFF medication state, participants rely more on the previous target durations when making temporal judgments. By doing so, the variation in duration estimation is reduced at the cost of estimation biases – especially at longer target durations. Consequently, the violation of the scalar property of interval timing shown in the current Bayesian-model simulations is best understood as the by-product of incorporating prior information into the sensory measurement (Acerbi et al., 2012; Cicchini et al., 2012; Jazayeri and Shadlen, 2010; Shi et al., 2013).

The results of our simulations indicate a pattern of responding compatible with the observed PD-migration effects (e.g., Malapani et al., 1998b, 2002; Rakitin et al., 2005, 2008) as well as Vierordt's law (Gu and Meck, 2011; Lejeune and Wearden, 2009). The extent of the observed migration effect varied as a function of a parameter estimate of temporal precision, which can be related to the severity of DA depletion in normal aging and PD-ON/OFF medication states and/or the amount and type of feedback provided for responding (Frank, 2005; Lustig and Meck, 2005; Pessiglione et al., 2006; Volkow et al., 1998). As a consequence, this estimate of temporal precision can be viewed as similar to the differences observed for modality (auditory vs. visual) and musical training (controls vs. percussionists vs. strings) in the Bayesian-model simulations reported by Cicchini et al. (2012). These results suggest that drummers may be particularly sensitive to distortions in timing and time perception associated with PD and other types of basal ganglia dysfunction (Bilger, 2011). This Bayesian account is consistent with recent proposals suggesting that DA levels modulate the reliability and precision of sensory information. As a consequence, PD patients trained or tested in the OFF medication state have low DA levels and are therefore less confident about incoming sensory information and more reliant on top-down predictions (Galea et al., 2012). The major conclusion is that the underlying Bayesian likelihood function combined with a prior representing the distribution of target durations can be used to account for the migration effects observed as a function of normal aging and PD. Moreover, assuming that variability in measurement and production (e.g., clock and motor variability) increases in the PD-OFF medication condition, produces not only migration, but also a violation of the scalar property consistent with the observations from PD patients tested ON and OFF their DA replacement therapy (Malapani et al., 1998b, 2002).

10 Summary

Since the PD-migration effect was first identified by Malapani and colleagues (Malapani et al., 1998b), we now know that this phenomenon is the result of a DA-dependent dysfunction of retrieving temporal memories as opposed to a problem with encoding (Malapani et al., 2002). Moreover, although migration is observed in healthy older adults, it is of a smaller magnitude and shows a different pattern of variability in that the scalar property of interval timing is not violated as it is in PD patients taken off of their DA replacement therapy (Rakitin et al., 2005; Rakitin and Malapani, 2008). We also know that the PD-migration effect may be lateralized to the right hemisphere (Koch et al., 2005) and that deep brain stimulation of the sub-thalamic nucleus can reduce or increase these effects depending upon the frequency (and perhaps area) of stimulation (Koch et al., 2004; Wojtecki et al., 2011). Moreover, the size of PD-migration effect is modulated both by feedback and exogenously administered drugs thought to increase the effective level of DA – suggesting a common mechanism for these effects in terms of error processing and temporal precision (Gu et al., 2015). Taken together, these findings are accounted for using a Bayesian model that assumes that increases in temporal uncertainty associated with PD are compensated for by combining the memories from multiple durations in order to be more accurate in the mean of the interval(s) being timed rather than trying to identify the specific target durations that form the distribution (Gu and Meck, 2011, 2013; Shi et al., 2013; also see Cicchini et al., 2012; Griffiths et al., 2008; Knill and Pouget, 2004). In this manner DA-deficient individuals are able to optimize performance by reducing temporal uncertainty at the cost of temporal accuracy – thus, demonstrating the types of temporal distortions presented in clinical situations involving the psychopathological mind (e.g., Allman and Meck, 2012; Allman et al., 2012, 2014a,b).

References

Acerbi, Luigi, Daniel M. Wolpert, and Sethu Vijayakumar. 2012. "Internal Representations of Temporal Statistics and Feedback Calibrate Motor-Sensory Interval Timing" *PLoS Computational Biology* 8(11): e1002771.

Adcock, R. Alison et al. 2006. "Reward-Motivated Learning: Mesolimbic Activation Precedes Memory Formation" *Neuron* 50: 507–17.

Agostino, Patricia V., Diego A. Golombek, and Warren H. Meck. 2011. "Unwinding the Molecular Basis of Interval and Circadian Timing" *Frontiers in Integrative Neuroscience*, 5: 64.

Agostino, P.V. et al. 2013. "Acquisition of Response Thresholds for Timed Performance is Regulated by a Calcium-Responsive Transcription Factor, CaRF" *Genes, Brain and Behavior* 12: 633–44.

Allan, Lorraine G. 1979. "The Perception of Time" *Perception and Psychophysics* 26: 340–54.

Allman, Melissa J., and Warren H. Meck. 2012. "Pathophysiological Distortions in Time Perception and Timed Performance" *Brain* 135: 656–77.

Allman, Melissa J., Kevin A. Pelphrey, and Warren H. Meck. 2012. "Developmental Neuroscience of Time and Number: Implications for Autism and Other Neurodevelopmental Disabilities" *Frontiers in Integrative Neuroscience*, 6: 7.

Allman, Melissa J. et al. 2014a. "Properties of the Internal Clock: First- and Second-Order Principles of Subjective Time" *Annual Review of Psychology* 65: 743–71.

Allman, Melissa J., Bin Yin, and Warren H. Meck. 2014b. "Time in the Psychopathological Mind" In *Subjective Time: The Philosophy, Psychology, and Neuroscience of Temporality*, edited by Valtteri Arstila, and Dan Lloyd, 637–54. Cambridge, MA: MIT Press.

Artieda J. et al. 1992. "Temporal Discrimination is Abnormal in Parkinson's Disease" *Brain* 115: 199–210.

Balci, Fuat et al. 2009. "Timing Deficits in Aging and Neuropathology" In *Animal Models of Human Cognitive Aging*, edited by Jennifer L. Bizon, and Alisa Woods, 161–201. Totowa, NJ: Humana Press.

Benecke, R. et al. 1986. "Performance of Simultaneous Movements in Patients with Parkinson's Disease" *Brain* 109: 739–57.

Beste, Christian et al. 2006. "Error Processing in Huntington's Disease" *PLoS ONE* 1(1): e86.

Beste, Christian et al. 2007. "Time Processing in Huntington's Disease: A Group-Control Study" *PLoS ONE* 2(12): e1263.

Bilger, Burkhard. 2011. "The Possibilian: What a Brush with Death Taught David Eagleman About the Mysteries of Time and the Brain" *The New Yorker* 25 April, 54–60, 62–65.

Bloxham, C.A., D.J. Dick, and M. Moore. 1987. "Reaction Times and Attention in Parkinson's Disease" *Journal of Neurology, Neurosurgery, and Psychiatry* 50: 1178–83.

Brown, P. 2003. "Oscillatory Nature of Human Basal Ganglia Activity: Relationship to the Pathophysiology of Parkinson's Disease" *Movement Disorders* 18: 357–63.

Brown, R.G., and C.D. Marsden. 1988. "Internal Versus External Cues and the Control of Attention in Parkinson's Disease" *Brain* 111: 323–45.

Brown, Gordon D.A. et al. 2005. "Identification and Bisection of Temporal Durations and Tone Frequencies: Common Models for Temporal and Nontemporal Stimuli" *Journal of Experimental Psychology: Human Perception and Performance* 31: 919–38.

Brown, Peter et al. 2001. "Dopamine Dependency of Oscillations Between Subthalamic Nucleus and Pallidum in Parkinson's Disease" *Journal of Neuroscience* 21: 1033–8.

Bueti, Domenica, and Emiliano Macaluso. 2011. "Physiological Correlates of Subjective Time: Evidence for the Temporal Accumulator Hypothesis" *NeuroImage* 57: 1251–63.

Buhusi, Catalin V. 2003. "Dopaminergic Mechanisms of Interval Timing and Attention" In *Functional and Neural Mechanisms of Interval Timing*, edited by Warren H. Meck, 317–38. Boca Raton, FL: CRC Press.

Buhusi, Catalin V., and Warren H. Meck. 2002. "Differential Effects of Methamphetamine and Haloperidol on the Control of an Internal Clock" *Behavioral Neuroscience* 116: 291–7.

———. 2005. "What Makes us Tick? Functional and Neural Mechanisms of Interval Timing" *Nature Reviews Neuroscience* 6: 755–65.

———. 2009a. "Relative Time Sharing: New Findings and an Extension of the Resource Allocation Model of Temporal Processing" *Philosophical Transactions of the Royal Society – London B* 364: 1875–85.

———. 2009b. "Relativity Theory and Time Perception: Single or Multiple Clocks?" *PLoS ONE* 4(7): e6268.

Buhusi, Catalin V., and Sorien A. Oprisan. 2013. "Time-Scale Invariance as an Emergent Property in a Perceptron with Realistic, Noisy Neurons" *Behavioural Processes* 95: 60–70.

Buhusi, Catalin V., Aya Sasaki, and Warren H. Meck. 2002. "Temporal Integration as a Function of Signal/Gap Intensity in Rats (*Rattus norvegicus*) and Pigeons (*Columba livia*)" *Journal of Comparative Psychology* 116: 381–90.

Cain, Matthew S. et al. 2012. "A Bayesian Optimal Foraging Model of Human Visual Search" *Psychological Science* 23: 1047–54.

Castner, Stacy A., Graham V. Williams, and Patricia S. Goldman-Rakic. 2000. "Reversal of Antipsychotic-Induced Working Memory Deficits by Short-Term Dopamine D1 Receptor Stimulation" *Science* 287: 2020–2.

Chen, Kelsey. 2007. "Electrophysiological Components of Error Processing in Age and Parkinson's Disease" Honors thesis submitted to Duke University, unpublished.

Cheng, Ruey-Kuang, and Warren H. Meck. 2007. "Prenatal Choline Supplementation Increases Sensitivity to Time by Reducing Non-Scalar Sources of Variance in Adult Temporal Processing" *Brain Research* 1186: 242–54.

Cheng, Ruey-Kuang, Christopher J. MacDonald, and Warren H. Meck. 2006. "Differential Effects of Cocaine and Ketamine on Time Estimation: Implications for Neurobiological Models of Interval Timing" *Pharmacology Biochemistry and Behavior* 85: 114–22.

Cheng, Ruey-Kuang et al. 2008. "Prenatal Choline Availability Differentially Modulates Timing of Auditory and Visual Stimuli in Aged Rats" *Brain Research* 1237: 167–75.

Cheng, Ruey-Kuang et al. 2011. "Categorical Scaling of Duration as a Function of Temporal Context in Aged Rats" *Brain Research* 1381: 175–86.

Church, Russell M., Warren H. Meck, and John Gibbon. 1994. "Application of Scalar Timing Theory to Individual Trials" *Journal of Experimental Psychology: Animal Behavior Processes* 20: 135–55.

Cicchini, Guido Marco et al. 2012. "Optimal Encoding of Interval Timing in Expert Percussionists" *Journal of Neuroscience* 32: 1056–60.

Conners, C.K. et al. 1996. "Nicotine and Attention in Adult Attention Deficit Hyperactivity Disorder (ADHD)" *Psychopharmacology Bulletin*, 32: 67–73.

Cools, Roshan et al. 2002. "Dopaminergic Modulation of High-Level Cognition in Parkinson's Disease: The Role of the Prefrontal Cortex Revealed by PET" *Brain* 125: 584–94.

Coull, Jennifer T. et al. 2004. "Functional Anatomy of the Attentional Modulation of Time Estimation" *Science* 30: 1506–8.

Coull, Jennifer T., Ruey-Kuang Cheng, and Warren H. Meck. 2011. "Neuroanatomical and Neurochemical Substrates of Timing" *Neuropsychopharmacology Reviews* 36: 3–25.

Coull, Jennifer T. et al. 2012. "Dopamine Precursor Depletion Impairs Timing in Healthy Volunteers by Attenuating Activity in Putamen and Supplementary Motor Area" *Journal of Neuroscience* 32: 16704–15.

Coull, Jennifer T. et al. 2013. "Dopaminergic Modulation of Motor Timing in Healthy Volunteers Differs as a Function of Baseline DA Precursor Availability" *Timing and Time Perception* 1: 77–98.

Cui, Xu. 2011. "Hyperbolic Discounting Emerges From the Scalar Property of Interval Timing" *Frontiers in Integrative Neuroscience* 5: 24.

De Ridder, Dirk, Sven Vannests, and Walter Freeman. 2012. "The Bayesian Brain:" Phantom Percepts Resolve Sensory Uncertainty" *Neuroscience and Biobehavioral Reviews* 44:4–15., in press. <http://dx.doi.org/10.1016/j.neubiorev.2012.04.001>.

De Vries, Meinou H. et al. 2010. "Increasing Dopamine Levels in the Brain Improves Feedback-Based Procedural Learning in Healthy Participants: An Artificial-Grammar-Learning Experiment" *Neuropsychologia* 48: 3193–7.

Droit-Volet, Sylvie, and Warren H. Meck. 2007. "How Emotions Colour Our Perception of Time" *Trends in Cognitive Sciences* 11: 504–13.

Dušek, Petr et al. 2012. "Abnormal Activity in the Precuneus During Time Perception in Parkinson's Disease: An fMRI Study" *PLoS ONE* 7(1): e29635.

Eagleman, David M. 2008. "Human Time Perception and Its Illusions" *Current Opinion in Neurobiology* 18: 131–136.

Ehlis, Ann-Christine et al. 2005. "Monitoring of Internal and External Error Signals" *Journal of Psychophysiology* 19: 263–9.

Eisler, Hannes, Anna D. Eisler, A., and Åke Hellström. 2008. "Psychophysical Issues in the Study of Time Perception" In *Psychology of Time,* edited by Simon Grondin, 75–109. Bingley: Emerald Group Publishing.

Eusebio, Alexandre, Hayriye Cagnan, and Peter Brown. 2012. "Does Suppression of Oscillatory Synchronisation Mediate Some of the Therapeutic Effects of DBS in Patients with Parkinson's Disease?" *Frontiers in Integrative Neuroscience* 6: 47.

Evarts, E.V., H. Teravainen, and D.B. Calne. 1981. "Reaction Time in Parkinson's Disease" *Brain* 104: 167–86.

Falkenstein, Michael et al. 2000. "ERP Components on Reaction Errors and Their Functional Significance: A Tutorial" *Biological Psychology* 51: 87–107.

Falkenstein, Michael et al. 2001. "Action Monitoring, Error Detection, and the Basal Ganglia: An ERP Study" *NeuroReport* 12: 157–61.

Falkenstein, Michael et al. 2005. "Error Processing in Parkinson's Disease: The Error Positivity (Pe)" *Journal of Psychophysiology* 19: 305–10.

Fearnley, Julian M., and Andrew J. Lees. 1991. "Ageing and Parkinson's Disease: Substantia Nigra Regional Selectivity" *Brain* 114: 2283–301.

Ferrandez, A.M. et al. 2003. "Basal Ganglia and Supplementary Motor Area Subtend Duration Perception: An fMRI Study" *NeuroImage* 19: 1532–44.

Fournet, Nathalie et al. 2000. "Working Memory Functioning in Medicated Parkinson's Disease Patients and the Effect of Withdrawal of Dopaminergic Medication" *Neuropsychology* 14: 247–53.

Frank, Michael J. 2005. "Dynamic Dopamine Modulation in the Basal Ganglia: A Neurocomputational Account of Cognitive Deficits in Medicated and Nonmedicated Parkinsonism" *Journal of Cognitive Neuroscience* 17: 51–72.

Galea, Joseph M. et al. 2012. "Action Reprogramming in Parkinson's Disease: Response to Prediction Error is Modulated by Levels of Dopamine" *Journal of Neuroscience* 32: 542–50.

Gamache, Pierre-Luc, and Simon Grondin. 2010. "The Lifespan of Time Intervals in Reference Memory" *Perception* 39: 1431–51.

Gibbon, John. 1992. "Ubiquity of Scalar Timing with a Poisson Clock" *Journal of Mathematical Psychology* 35: 283–93.

Gibbon, John, and Russell M. Church. 1984. "Sources of Variance in an Information Processing Theory of Timing" In *Animal Cognition,* edited by Herbert L. Roitblat, Thomas G. Bever, and Herbert S. Terrace, 465–88. Hillsdale, NJ: Erlbaum.

———. 1990. "Representation of Time" *Cognition* 37: 23–54.

———. 1992. "Comparison of Variance and Covariance Patterns in Parallel and Serial Theories of Timing" *Journal of The Experimental Analysis of Behavior* 57: 393–406.

Gibbon, John, and Chara Malapani. 2001. "Neural Basis of Timing and Time Perception" *Encyclopedia of Cognitive Science,* 401–406.

Gibbon, John, Russell M. Church, and Warren H. Meck. 1984. "Scalar Timing in Memory" *Annals of The New York Academy of Sciences* 423: 52–77.

Gibbon, John et al. 1997. "Toward a Neurobiology of Temporal Cognition: Advances and Challenges" *Current Opinion in Neurobiology* 7: 170–84.

Griffiths, Thomas L., Charles Kemp, and Joshua B. Tenebaum. 2008. "Bayesian Models of Cognition" In *CsmbridgeThe Handbook of Computational Psychology*, edited by Ron Sun, 1–49. Cambridge, UK: Cambridge University Press.

Grondin, Simon. 2005. "Overloading Temporal Memory" *Journal of Experimental Psychology: Human Perception and Performance* 31: 869–79.

Grondin, Simon. 2010. "Timing and Time Perception: A Review of Recent Behavioral and Neuroscience Findings and Theoretical Directions" *Attention Perception and Psychophysics* 72: 561–82.

Gu, Bon-Mi and Keshav Kukreja. 2011. "Obsessive-Compulsive Disorder and Memory-Mixing in Temporal Comparison: Is Implicit Learning the Missing Link?" *Frontiers in Integrative Neuroscience* 5: 38.

Gu, Bon-Mi, and Warren H. Meck. 2011. "New Perspectives on Vierordt's Law: Memory-Mixing in Ordinal Temporal Comparison Tasks" *Lecture Notes in Computer Science* 6789 LNAI: 67–78.

Gu, Bon-Mi, Hedderik van Rijn, and Warren H. Meck. 2015. "Oscillatory Multiplexing of Neural Population Codes for Interval Timing and Working Memory" *Neuroscience & Biobehavioral Reviews* 48: 160–85.

Gu, Bon-Mi et al 2008. "Neural Correlates of Cognitive Inflexibility During Task-Switching in Obsessive-Compulsive Disorder" *Brain* 131: 155–64.

Gu, Bon-Mi et al. 2011. "Quinpirole-Induced Sensitization to Noisy/Sparse Periodic Input: Temporal Synchronization as a Component of Obsessive-Compulsive Disorder" *Neuroscience* 179: 143–50.

———. 2015. "Bayesian Models of Interval Timing and Distortions in Temporal Memory as a Function of Dopamine Levels in Young, Aged, and Parkinson's Disease Patients" Submitted.

Hallett, Mark, and Shahram Khoshbin. 1980. "A Physiological Mechanism of Bradykinesia" *Brain* 103: 301–14.

Hairston, Ilana S., and Srikantan S. Nagarajan. 2007. "Neural Mechanisms of the Time-Order Error: An MEG study" *Journal of Cognitive Neuroscience* 19: 1163–74.

Harrington, Deborah L., Kathleen Y. Haaland, and Neal Hermanowicz. 1998a. "Temporal Processing in the Basal Ganglia" *Neuropsychology* 12: 3–12.

Harrington, Deborah L., Kathleen Y. Haaland, and Robert T. Knight. 1998b. "Cortical Networks Underlying Mechanisms of Time Perception" *Journal of Neuroscience* 18: 1085–95.

Harrington, Deborah L. et al. 2010. "Neural Modulation of Temporal Encoding, Maintenance, and Decision Processes" *Cerebral Cortex* 20: 1274–85.

Harrington, Deborah L. et al. 2011a. "Neural Underpinnings of Distortions in the Experience of Time Across Senses" *Frontiers in Integrative Neuroscience* 5: 32.

Harrington, Deborah L. et al. 2011b. "Neurobehavioral Mechanisms of Temporal Processing Deficits in Parkinson's Disease" *PLoS ONE* 6(2): e17461.

Harrington, Deborah L. et al. 2014. "Dissociation of Neural Mechanisms for Intersensory Timing Deficits in Parkinson's Disease" *Timing and Time Perception* 2: 17–43.

Herrmann, Christoph S., MAtthias H.J. Munk, and Andreas K. Engel. 2004. "Cognitive Functions of Gamma-Band Activity: Memory Match and Utilization" *Trends in Cognitive Sciences* 8: 347–55.

Hinton, Sean C. 2003. "Neuroimaging Approaches to the Study of Interval Timing" In *Functional and Neural Mechanisms of Interval Timing* edited by Warren H. Meck, 419–38. Boca Raton, FL: CRC Press.

Hinton, Sean C., and Stephen M. Rao. 2004. "'One-Thousand *One*...One-Thousand *Two*...': Chronometric Counting Violates the Scalar Property in Interval Timing" *Psychonomic Bulletin and Review* 11: 24–30.

Hinton, Sean C., and Warren H. Meck. 1997. "How Time Flies: Functional and Neural Mechanisms of Interval Timing" In *Time and Behaviour: Psychological and Neurobiological Analyses*. edited by Christopher M. Bradshaw and Elmer Szabadi, 409–57. New York: Elsevier.

Hinton, Sean C., and Warren H. Meck. 2004. "Frontal-Striatal Circuitry Activated by Human Peak-Interval Timing in the *Supra*-Seconds Range" *Cognitive Brain Research*, 21: 171–82.

Höhn, Sophie et al. 2011. "Behavioral and *In Vivo* Electrophysiological Evidence for Presymptomatic Alteration of Prefronto-Striatal Processing in the Transgenic Rat Model for Huntington Disease" *Journal of Neuroscience* 31: 8986–97.

Holroyd, Clay B., and Michael G.H Coles 2002. "The Neural Basis of Human Error Processing: Reinforcement Learning, Dopamine, and the Error-Related Negativity" *Psychological Review* 109: 679–709.

Holroyd, Clay B. et al. 2002. "Spared Error-Related Potentials in Mild to Moderate Parkinson's Disease" *Neuropsychologia* 40: 2116–24.

Hutchison, William D. et al. 2004. "Neuronal Oscillations in the Basal Ganglia and Movement Disorders: Evidence From Whole Animal and Human Recordings" *Journal of Neuroscience* 24: 9240–3.

Ito, Junko, and Junko Kitagawa. 2006. "Performance Monitoring and Error Processing During a Lexical Decision Task in Patients with Parkinson's Disease" *Journal of Geriatric Psychiatry and Neurology* 19: 46–9.

Ivry, Richard B. 1996. "The Representation of Temporal Information in Perception and Motor Control" *Current Opinion in Neurobiology* 6: 851–7.

Ivry, Richard B., and R. Eliot Hazeltine. 1995. "Perception and Production of Temporal Intervals Across a Range of Durations: Evidence for a Common Timing Mechanism" *Journal of Experimental Psychology: Human Perception and Performance* 21: 3–18.

Ivry, Richard B., and Steven W. Keele. 1989. "Timing Functions of the Cerebellum" *Journal of Cognitive Neuroscience* 1: 137–52.

Ivry, R.B., S.W. Keele, and Diener, H.C. 1988. "Dissociation of the Lateral and Medial Cerebellum in Movement Timing and Movement Execution" *Experimental Brain Research* 73: 167–80.

Jazayeri, Mehrdad, and Michael N. Shadlen. 2010. "Temporal Context Calibrates Interval Timing" *Nature Neuroscience* 13: 1020–6.

Jahanshahi, Marjan et al. 2006. "The Substantia Nigra Pars Compacta and Temporal Processing" *Journal of Neuroscience*, 26: 12266–73.

Jahanshahi, Marjan et al. 2010a. "Dopaminergic Modulation of Striato-Frontal Connectivity During Motor Timing in Parkinson's Disease" *Brain* 133: 727–45.

Jahanshahi, Marjan et al. 2010b. "Medication Impairs Probabilistic Classification Learning in Parkinson's Disease" *Neuropsychologia* 48: 1096–103.

Jones, Catherine R.G. et al. 2011. "Modeling Accuracy and Variability of Motor Timing in Treated and Untreated Parkinson's Disease and Healthy Controls" *Frontiers in Integrative Neuroscience*, 5: 81.

Jones, Catherine R.G., and Marjan Jahanshahi. 2011. "Dopamine Modulates Striato-Frontal Functioning During Temporal Processing" *Frontiers in Integrative Neuroscience*, 5: 70.

Jones, Catherine R.G., and Marjan Jahanshahi. 2014. "Contributions of the Basal Ganglia to Temporal Processing: Evidence from Parkinson's Disease" *Timing and Time Perception*, 2: 87–127.

Jones, Catherine R.G. et al. 2008. "Basal Ganglia, Dopamine and Temporal Processing: Performance on Three Timing Tasks On and Off Medication in Parkinson's Disease" *Brain and Cogn*ition 68: 30–41.

Jones, Catherine L. et al. 2011. "Under Pressure: Response Urgency Modulates Striatal and Insula Activity during Decision- Making under Risk" PLoS ONE 6(6): e20942.

Jurkowski, Anita J., and Warren H. Meck. 2003. "Could Interval Timing Be Used as an Early Marker for PD?" Presentation given at the Udall Parkinson's Disease Research Center of Excellence Symposium, Duke University, Durham, NC, September 10–11, 2003.

Jurkowski, Anita J., and Mark Stacy. 2005. "Classification and Clinical Features of Movement Disorders" In *Movement Disorders: Genetics and Models*, edited by Mark S. LeDoux, 1–12. San Diego, CA: Elsevier Academic Press.

Jurkowski, Anita J., Elizabeth Stepp, and Steven A. Hackley. 2005a. "Variable Foreperiod Deficits in Parkinson's Disease: Dissociation Across Reflexive and Voluntary Behaviors" *Brain and Cognition* 58: 49–61.

Jurkowski, Anita J., Elizabeth Stepp, and Steven A. Hackley. 2002. "Dissociation in Reflexive and Voluntary Behaviors for Interval Timing in Aged and Parkinsonian Adults" *Journal of Cognitive Neuroscience*, S, 129–E37.

Jurkowski, Anita J. et al. 2005b. "Dopamine Levels in Young, Aged, and Parkinson's Adults Predict Behavioral and Electrophysiological Measures of Color and Time

Perception" Program No. 329.47 *Neuroscience 2005 Abstracts*. Washington, D.C., Society for Neuroscience. Online.

Kawashima, Shoji et al. 2012. "Changes in Striatal Dopamine Release Associated with Human Motor-Skill Acquisition" *PLoS ONE* 7(2): e31728.

Keele, Steven et al. 1985. "Do Perception and Motor Production Share Common Timing Mechanisms? A Correlational Analysis" *Acta Psychologica* 60: 173–91.

Kim, Namsoo et al. 2014. "Striatal Firing Rate Reflects Head Movement Velocity" *European Journal of Neuroscience*, 40: 3481–90.

Kim, Soyoun, and Daeyeol Lee. 2011. "Prefrontal Cortex and Impulsive Decision Making" *Biological Psychiatry* 69: 1140–6.

Klapproth, Florian. 2009. "Single-Modality Memory Mixing in Temporal Generalization: An Effect Due to Instructional Ambiguity" *NeuroQuantology* 7: 8–94.

Knill, David C., and Alexandre Pouget. 2004. "The Bayesian Brain: The Role of Uncertainty in Neural Coding and Computation for Perception and Action" *Trends in Neuroscience* 27: 712–9.

Koch, Giacomo et al. 2003. "Underestimation of Time Perception After Repetitive Transcranial Magnetic Stimulation" *Neurology* 60: 1844–6.

Koch, Giacomo et al. 2004. "Subthalamic Deep Brain Stimulation Improves Time Perception in Parkinson's Disease" *NeuroReport*, 15: 1071–3.

Koch, Giacomo et al. 2005. "Memory for Time Intervals is Impaired in Left Hemi-Parkinson Patients" *Neuropsychologia* 43: 1163–7.

Koch, Giacomo et al. 2008. "Impaired Reproduction of Second But Not Millesecond Time Intervals in Parkinson's Disease" *Neuropsychologia* 46: 1305–13.

Koch, Giacomo Massimiliano Oliveri, and Carlo Caltagirone. 2009. "Neural Networks Engaged in Milliseconds and Seconds Time Processing: Evidence From Transcranial Magnetic Stimulation and Patients with Cortical or Subcortical Dysfunction" *Philosophical Transactions of the Royal Society B* 364: 1907–18.

Koepp, M.J. et al. 1998. "Evidence For Striatal Dopamine Release During a Video Game" *Nature* 393: 266–8.

Kolev, Vasil, Michael Falkenstein, and Juliana Yordanova. 2005. "Aging and Error Processing: Time-Frequency Analysis of Error-Related Potentials" *Journal of Psychophysiology* 19: 289–97.

Kotz, Sonia A., Michael Schwartze. 2011. "DIfferential Input of the Supplementary Motor Area to a Dedicated Temporal Processing Network: Functional and Clinical Implications" *Frontiers in Integrated Neuroscience*, 5, 86.

Körding, Konrad P., and Daniel M. Wolpert. 2006. "Bayesian Decision Theory in Sensorimotor Control" *Trends in Cognitive Sciences* 10: 319–26.

Lake, Jessica I., and Warren H. Meck. 2013. "Differential Effects of Amphetamine and Haloperidol on Temporal Reproduction: Dopaminergic Regulation of Attention and Clock Speed" *Neuropsychologia* 51: 284–92.

Larson, Michael J., and Peter E. Clayson. 2011. "The Relationship between Cognitive Performance and Electrophysiological Indices of Performance Monitoring" *Cognitive, Affective, & Behavioral Neuroscience* 11: 159–71.

Lee, Eun-Young et al. 2010. "Visual Working Memory Deficits in Patients With Parkinson's Disease are Due to Both Reduced Storage Capacity and Impaired Ability to Filter Out Irrelevant Information" *Brain* 133: 2677–89.

Lejeune, Helga, and John H. Wearden. 2009. "Vierordt's *The Experimental Study of the Time Sense* (1868) and its Legacy" *European Journal of Cognitive Psychology* 21: 941–60.

Levin, E.D. et al. 1996. "Nicotine Effects on Adults with Attention-Deficit/Hyperactivity Disorder" *Psychopharmacology* 123: 55–63.

Levin, Edward D. et al. 1998. "Transdermal Nicotine Effects on Attention" *Psychopharmacology* 140: 135–41.

Lieberman, Philip et al. 1992. "Speech Production, Syntax Comprehension, and Cognitive Deficits in Parkinson's Disease" *Brain and Language* 43: 169–89.

Lustig, Cindy. 2003. "Grandfather's Clock: Attention and Interval Timing in Older Adults" In *Functional and Neural Mechanisms of Interval Timing*, edited by Warren H. Meck, 261–93. Boca Raton, FL: CRC Press.

Lustig, Cindy, and Warren H. Meck. 2001. "Paying Attention to Time as One Gets Older" *Psychological Science* 12: 478–84.

———. 2005. "Chronic Treatment with Haloperidol Induces Working Memory Deficits in Feedback Effects of Interval Timing" *Brain and Cognition* 58: 9–16.

———. 2011. "Modality Differences in Timing and Temporal Memory Throughout the Lifespan" *Brain and Cognition* 77: 298–303.

Lustig, Cindy, Matthew S. Matell, and Warren H. Meck. 2005. "Not 'Just' a Coincidence: Frontal-Striatal Synchronization in Working Memory and Interval Timing" *Memory* 13: 441–8.

MacDonald, Christopher J., and Warren H. Meck. 2004. "Systems-Level Integration of Interval Timing and Reaction Time" *Neuroscience and Biobehavioral Reviews* 28: 747–69.

———. 2005. "Differential Effects of Clozapine and Haloperidol on Interval Timing in the Supraseconds Range" *Psychopharmacology* 182: 232–44.

———. 2006. "Interaction of Raclopride and Preparatory-Interval Effects on Simple Reaction-Time Performance" *Behavioural Brain Research* 175: 62–74.

MacDonald, Christopher J., Ruey-Kuang Cheng, and Warren H. Meck. 2012. "Acquisition of 'Start' and 'Stop' Response Thresholds in Peak-Interval Timing is Differentially Sensitive to Protein Synthesis Inhibition in the Dorsal and Ventral Striatum" *Frontiers in Integrative Neuroscience* 6: 10.

MacDonald, Christopher J. et al. 2014. "Retrospective and Prospective Views on the Role of the Hippocampus in Interval Timing and Memory for Elapsed Time" *Timing and Time Perception* 2: 51–61.

Malapani, Chara, and Stephen Fairhurst. 2002. "Scalar Timing in Animals and Humans" *Learning and Motivation* 33: 156–76.

Malapani, Chara, and Brian C. Rakitin. (2003). "Interval Timing in the Dopamine-Depleted Basal Ganglia: From Empirical Data to Timing Theory" In *Functional and Neural Mechanisms of Interval Timing*, edited by Warren H. Meck, 485–514. Boca Raton, FL: CRC Press.

Malapani, Chara et al. 1994. "Impaired Simultaneous Cognitive Task Performance in Parkinson's Disease: A Dopamine Related Dysfunction" *Neurology* 44: 319–26.

Malapani, Chara et al. 1998a. "Cerebellar Dysfunction of Temporal Processing in the Seconds Range in Humans" *Neuroreport*, 9: 3907–12.

Malapani, Chara et al. 1998b. "Coupled Temporal Memories in Parkinson's Disease: A Dopamine-Related Dysfunction" *Journal of Cognitive Neuroscience* 10: 316–31.

Malapani, Chara, Bernard Deweer, and John Gibbon. 2002. "Separating Storage From Retrieval Dysfunction of Temporal Memory in Parkinson's Disease" *Journal of Cognitive Neuroscience* 13: 311–22.

Mamassian, Pascal, and Michael S. Landy. 2010. "It's That Time Again" *Nature Neuroscience* 13: 914–6.

Maricq, Andres V., and Russell M. Church. 1983. "The Differential Effects of Haloperidol and Methamphetamine on Time Estimation in the Rat" *Psychopharmacology* 79: 10–15.

Matell, Matthew S., and Warren H. Meck. 2000. "Neuropsychological Mechanisms of Interval Timing Behaviour" *BioEssays* 22: 94–103.

———. 2004. "Cortico-Striatal Circuits and Interval Timing: Coincidence-Detection of Oscillatory Processes" *Cognitive Brain Research* 21: 139–70.

Matell, Matthew S., George R. King, and Warren H. Meck. 2004. "Differential Adjustment of Interval Timing by the Chronic Administation of Intermittent or Continuous Cocaine" *Behavioral Neuroscience* 118: 150–6.

Matell, Matthew S., Melissa Bateson, and Warren H. Meck. 2006. "Single-Trials Analyses Demonstrate That Increases in Clock Speed Contribute to the Methamphetamine-Induced Horizontal Shifts in Peak-Interval Timing Functions" *Psychopharmacology* 188: 201–12.

Matell, Matthew S., Warren H. Meck and Miguel A.L. Nicolelis. 2003. "Interval Timing and the Encoding of Signal Duration by Ensembles of Cortical and Striatal Neurons" *Behavioral Neuroscience* 117: 760–73.

Matthews, William J., Warren H. Meck. 2014. "Time Perception: The Bad News and the Good" *Wiley Interdisciplinary Reviews: Cognitive Science* 5: 429–46.

Meck, Warren H. 1983. "Selective Adjustment of the Speed of Internal Clock and Memory Processes" *Journal of Experimental Psychology: Animal Behavior Processes* 9: 171–201.

———. 1986. "Affinity for the Dopamine D2 Receptor Predicts Neuroleptic Potency in Decreasing the Speed of an Internal Clock" *Pharmacology Biochemistry and Behavior* 25: 1185–9.

———. 1988. "Hippocampal function is required for feedback control of an internal clock's criterion" *Behavioral Neuroscience* 102: 54–60.
——— 1996. "Neuropharmacology of Timing and Time Perception" *Cognitive Brain Research* 3: 227–42.
———. 2002a. "Choline Uptake in the Frontal Cortex is Proportional to the Absolute Error of a Temporal Memory Translation Constant in Mature and Aged Rats" *Learning and Motivation* 33: 88–104.
———. 2002b. "Distortions in the Content of Temporal Memory: Neurobiological Correlates" In *Animal Cognition and Sequential Dehavior: Behavioral, Biological, and Computational Perspectives*, edited by Stephen B. Fountain, Michael D. Bunsey, Joseph H. Danks, and Michael K. McBeath, 175–200. Boston, MA: Kluwer Academic Press.
———. 2003. *Functional and Neural Mechanisms of Interval Timing*. Boca Raton, FL: CRC Press LLC.
———. 2005. "Neuropsychology of Timing and Time Perception" *Brain and Cognition* 58: 1–8.
———. 2006a. "Frontal Cortex Lesions Eliminate the Clock Speed Effect of Dopaminergic Drugs on Interval Timing" *Brain Research* 1108: 157–67.
———. 2006b. "Neuroanatomical Localization of an Internal Clock: A Functional Link Between Mesolimbic, Nigrostriatal, and Mesocortical Dopaminergic Systems" *Brain Research* 1109: 93–107.
———. 2006c. "Temporal Memory in Mature and Aged Rats is Sensitive to Choline Acetyltransferase Inhibition" *Brain Research* 1108: 168–75.
Meck, Warren H., and Aimee M. Benson. 2002. "Dissecting the Brain's Internal Clock: How Frontal-Striatal Circuitry Keeps Time and Shifts Attention" *Brain and Cognition* 48: 195–211.
Meck, Warren H., and Chara Malapani. 2004. "Neuroimaging of Interval Timing" *Cognitive Brain Research* 21: 133–7.
Meck, Warren H., and Katherine E. Angell. 1992. "Repeated Administration of Pyrithiamine Leads to a Proportional Increase in the Remembered Durations of Events" *Psychobiology* 20: 39–46.
Meck, Warren H., and Russell M. Church. 1987. "Cholinergic Modulation of the Content of Temporal Memory" *Behavioral Neuroscience* 101: 457–64.
Meck, Warren H. et al. 2012a. "Gene-Dose Dependent Effects of Methamphetamine on Interval Timing in Dopamine-Transporter Knockout Mice" *Neuropharmacology* 62: 1221–9.
Meck, Warren H, Valérie Doyère, and Agnès Gruart. 2012b. "Interval Timing and Time-Based Decision Making" *Frontiers in Integrative Neuroscience* 6: 13.
Meck, Warren H., Russell M. Church, and Gary L. Wenk. 1986. "Arginine Vasopressin Inoculates Against Age-Related Increases in Sodium-Dependent High Affinity Choline Uptake and Discrepancies in the Content of Temporal Memory" *European Journal of Pharmacology* 130: 327–31.

Meck, Warren H., Russell M. Church, and Matthew S. Matell. 2013. "*Hippocampus, time, and memory* – A Retrospective Analysis" *Behavioral Neuroscience* 127: 642–54.

Meck, Warren H., Fariba N. Komeily-Zadeh, and Russell M. Church. 1984. "Two-Step Acquisition: Modification of an Internal Clock's Criterion" *Journal of Experimental Psychology: Animal Behavior Processes* 10: 297–306.

Meck, Warren H., Trevor B. Penney, and Viviane Pouthas. 2008. "Cortico-Striatal Representation of Time in Animals and Humans" *Current Opinion in Neurobiology* 18: 145–52.

Merchant, Hugo, Wilbert Zarco, and Luis Prado. 2007. "Do We Have a Common Mechanism for Measuring Time in the Hundreds of Millisecond Range? Evidence From Multiple-Interval Timing Tasks" *Journal of Neurophysiology* 99: 939–49.

Merchant, Hugo et al. 2008a. "Interval Timing and Parkinson's Disease: Heterogeneity in Temporal Performance" *Experimental Brain Research* 184: 233–48.

Merchant, Hugo et al. 2008b. "The Context of Temporal Processing Is Represented in the Multidimensional Relationships between Timing Tasks" *PLoS ONE* 3(9): e3169.

Merchant, Hugo, Deborah L. Harrington, and Warren H. Meck. 2013a. "Neural Basis of the Perception and Estimation of Time" *Annual Review of Neuroscience* 36: 313–36.

Merchant, Hugo et al. 2013b. "Interval Tuning in the Primate Medial Premotor Cortex as a General Timing Mechanism" *Journal of Neuroscience* 33: 9082–96.

Miyazaki, Makoto, Daichi Nozaki, and Yasoichi Nakajima. 2005. "Testing Bayesian Models of Human Coincidence Timing" *Journal of Neurophysiology* 94: 395–9.

Nagano-Saito Atsuko et al. 2008. "Dopamine Depletion Impairs Frontostrital Functional Connectivity During a Set-Shifting Task" *Journal of Neuroscience* 28: 3697–706.

Nakamura, Ryuichi, Hiroshi Nagasaki, and Hirotaro Narabayashi. 1978. "Disturbances of Rhythm Formation in Patients with Parkinson's Disease: Part 1. Characteristics of Tapping Response to the Period Signals" *Perceptual and Motor Skills* 46: 63–75.

Nichelli, Paolo, David Alway, and Jordan Grafman. 1996. "Perceptual Timing in Cerebellar Degeneration" *Neuropsychologia* 34: 863–72.

Nieuwenhuis, Sander et al. 2002. "A Computational Account of Altered Error Processing in Older Age: Dopamine and the Error-Related Negativity" *Cognitive, Affective, and Behavioral Neuroscience* 2: 19–36.

Nieuwenhuis, Sander et al. 2004. "Reinforcement-Related Brain Potentials from Medial Frontal Cortex: Origins and Functional Significance" *Neuroscience and Biobehavioral Reviews* 28: 441–8.

O'Boyle, Donald J., John S. Freeman, and Frederick W.J. Cody. 1996. "The Accuracy and Precision of Timing of Self-Paced, Repetition Movements in Subjects with Parkinson's Disease" *Brain* 119: 51–70.

Oprisan, Sorien A., and Catalin V. Buhusi. 2011. "Modeling Pharmacological Clock and Memory Patterns of Interval Timing in a Striatal Beat-Frequency Model with Realistic, Noisy Neurons" *Frontiers in Integrative Neuroscience* 5: 52.

———. 2013. "How Noise Contributes to Time-Scale Invariance of Interval Timing" *Physical Review E* 87: 052717.

Overbeek, Therese J.M., Sander Nieuwenhuis, and K. Richard Ridderinkhof. 2005. "Dissociable Components of Error Processing: On the Functional Significance of the Pe Vis-a-vis the ERN/Ne" *Journal of Psychophysiology* 19: 319–29.

Parker, K.L. et al. 2013. "Executive Dysfunction in Parkinson's Disease and Timing Deficits" *Frontiers in Integrative Neuroscience* 7: 75.

Pastor, M.A. et al. 1992. "Time Estimation and Reproduction is Abnormal in Parkinson's Disease" *Brain* 115: 211–25.

Penney, Trevor B. et al. 1998. "Memory Mixing in Duration Bisection" In *Timing of Behavior: Neural, Psychological and Computational Perspectives,* edited by David A. Rosenbaum, and Charles E. Collyer, 165–93. Cambridge, MA: MIT Press.

Penney, Trevor B., John Gibbon, and Warren H. Meck. 2000. "Differential Effects of Auditory and Visual Signals on Clock Speed and Temporal Memory" *Journal of Experimental Psychology: Human Perception and Performance* 26: 1770–87.

Penney, Trevor B. et al. 2005. "Interval-Timing Deficits in Individuals at High Risk for Schizophrenia" *Brain and Cognition* 58: 109–18.

Penney, Trevor B., John Gibbon, and Warren H. Meck. 2008. "Categorical Scaling of Duration Bisection in Pigeons (*Columba livia*), Mice (*Mus musculus*), and Humans (*Homo sapiens*)" *Psychological Science* 19: 1103–9.

Perbal, Séverine et al. 2005. "Effects of Internal Clock and Memory Disorders on Duration Productions and Reproductions in Patients with Parkinson's Disease" *Brain and Cognition* 58: 35–48.

Pessiglione, Mathias et al. 2006. "Dopamine-Dependent Prediction Errors Underpin Reward-Seeking Behavior in Humans" *Nature* 442: 1042–5.

Philiastides, Marios G. et al. 2011. "Causal Role of Dorsolateral Prefrontal Cortex in Human Perceptual Decision Making" *Current Biology* 21: 980–3.

Pouthas, Viviane, and Séverine Perbal. 2004. "Time Perception Depends on Accurate Clock Mechanisms as Well as Unimpaired Attention and Memory Processes" *Acta Neurobiologiae Experimentalis* 64: 367–85.

Rahnev, Dobromir, Hakwan Lau, and Floris P. de Lange. 2011. "Prior Expectation Modulates the Interaction Between Sensory and Prefrontal Regions in the Human Brain" *Journal of Neuroscience* 31: 10741–8.

Rakitin, Brian C., and Chara Malapani. 2008. "Effects of Feedback on Time Production Errors in Aging Participants" *Brain Research Bulletin* 75: 23–33.

Rakitin, Brian C., and Yaakov Stern. 2006. "Parkinson Disease" *Encyclopedia of Cognitive Science*, 458–462.

Rakitin, Brian C. et al. 1998. "Scalar Expectancy Theory and Peak-Interval Timing in Humans" *Journal of Experimental Psychology: Animal Behavior Processes* 24: 15–33.

Rakitin, Brian C., Yaakov Stern, and Chara Malapani. 2005. "The Effects of Aging on Time Reproduction in Delayed Free-Recall" *Brain and Cognition* 58: 17–34.

Rakitin, Brian C. et al. 2006. "Single-Dose Levodopa Administration and Aging Independently Disrupt Time Production" *Journal of Cognitive Neuroscience* 18: 376–87.

Rao, Stephen M., Andrew R. Meyer, and Deborah L. Harrington. 2001. "The Evolution of Brain Activation During Temporal Processing" *Nature Neuroscience* 4: 317–23.

Ray, Debajyoti, and Peter Bossaerts. 2011. "Positive Temporal Dependence of the Biological Clock Implies Hyperbolic Discounting" *Frontiers in Integrative Neuroscience* 5: 2.

Roberts, Seth. 1981. "Isolation of an Internal Clock" *Journal of Experimental Psychology: Animal Behavior Processes* 7: 242–68.

———. 1993. "Use of Independent and Correlated Measures to Divide a Time-Discrimination Mechanism Into Parts" *Attention and performance*, XIV, 589–610, MIT Press, Cambridge, MA.

Roy, Eric A. et al. 1993. "Movement Sequencing Disorders in Parkinson's Disease" *International Journal of Neuroscience* 73: 183–94.

Rubin, David C. 1999. "Frontal-Striatal Circuits in Cognitive Aging: Evidence for Caudate Involvement" *Aging, Neuropsychology, and Cognition* 6: 241–59.

San Martín, Réne. 2012. "Event-Related Potential Studies of Outcome Processing and Feedback-Guided Learning" *Frontiers in Human Neuroscience* 6: 304.

Schirmer Annett. 2004. "Timing Speech: A Review of Lesion and Neuroimaging Findings" *Cognitive Brain Research* 21: 269–87.

Schultz, Wolfram. 1998. "Predictive Reward Signal of Dopamine Neurons" *Journal of Neurophysiology* 80: 1–27.

Schultz, Wolfram, and Anthony Dickinson. 2000. "Neuronal Coding of Prediction Errors" *Annual Review of Neuroscience* 23: 473–500.

Shea-Brown, Eric et al. 2006. "A Firing Rate Model of Parkinson's Deficits in Interval Timing" *Brain Research* 1070: 189–201.

Shi, Zhuanghua, Russell M. Church, and Warren H. Meck. 2013. "Bayesian Optimization of Time Perception" *Trends in Cognitive Sciences* 17: 556–64.

Smith, Jared G. et al. 2007. "The Effect of Parkinson's Disease on Time Estimation as a Function of Stimulus Duration Range and Modality" *Brain and Cognition* 64: 130–43.

Sohn, Hansem, and Sang-Hun Lee. 2013. "Dichotomy in Perceptual Learning of Interval Timing: Calibration of Mean Accuracy and Precision Differ in Specificity and Time Course" *Journal of Neurophysiology* 109: 344–62.

Stelmach, George E., and Charles J. Worringham. 1988. "The Control of Bimanual Aiming Movements in Parkinson's Disease" *Journal of Neurology, Neurosurgery, and Psychiatry* 51: 223–31.

Tarantino, Vincenza et al. 2010. "The Time Course of Temporal Discrimination: An ERP Study" *Clinical Neurophysiology* 121: 43–52.

Teki, Sundeep, Manon Grube, and Timothy D. Griffiths. 2012. "A Unified Model of Time Perception Accounts for Duration-Based and Beat-Based Timing Mechanisms" *Frontiers in Integrative Neuroscience* 5: 90.

Timmermann, Lars et al. 2003. "The Cerebral Oscillatory Network of Parkinsonian Resting Tremor" *Brain* 126: 199–212.

Timmermann, Lars et al. 2004. "Ten-Hertz Stimulation of Subthalamic Nucleus Deteriorates Motor Symptoms in Parkinson's Disease" *Movement Disorders* 19: 1328–33.

Treisman, Michel. 1963. "Temporal Discrimination and the Indifference Interval. Implications for a Model of the 'Internal Clock'" *Psychological Monographs: General and Applied* 77(13): 1–31.

Van Rijn, Hedderik et al. 2011. "Contingent Negative Variation and its Relation to Time Estimation: A Theoretical Evaluation" *Frontiers in Integrative Neuroscience* 5: 91.

Van Rijn, Hedderik, Bon-Mi Gu, and Warren H. Meck. 2014. "Dedicated Clock/Timing-Circuit Theories of Time Perception and Timed Performance" *Advances in Experimental Medicine and Biology* 829: 75–99.

Van Wassenhove, Virginie et al. 2008. "Distortions of Subjective Time Perception Within and Across Senses" *PLoS ONE* 3(1): e1437.

Volkmann, J. et al. 1992. "Impairment of Temporal Organization of Speech in Basal Ganglia Diseases" *Brain and Language* 43: 386–99.

Volkow, Nora D. et al. 1998. "Association Between Decline in Brain Dopamine Activity With Age and Cognitive and Motor Impairment in Healthy Individuals" *American Journal of Psychiatry* 155: 344–9.

Wearden, J.H., and Helga Lejeune. 2008. "Scalar Properties in Human Timing: Conformity and Violations" *The Quarterly Journal of Experimental Psychology* 61: 569–587.

Wearden, J.H., and Rachel Grindrod. 2003. "Manipulating Decision Processes in the Human Scalar Timing System" *Behavioural Processes* 61: 47–56.

Williamson, Lauren L. et al. 2008. "'Speed' Warps Time: Methamphetamine's Interactive Roles in Drug Abuse, Habit Formation, and the Biological Clocks of Circadian and Interval Timing" *Current Drug Abuse Reviews* 1: 203–12.

Wing, Alan M., Steven Keele, and David I. Margolin. 1984. "Motor Disorder and the Timing of Repetitive Movements" *Annals of The New York Academy of Sciences* 423: 183–92.

Wise, Roy A. 2004. "Dopamine, Learning and Motivation" *Nature Reviews Neuroscience* 5: 483–94.

Wojtecki, Lars et al. 2006. "Frequency-Dependent Reciprocal Modulation of Verbal Fluency and Motor Functions in Subthalamic Deep Brain Stimulation" *Archives of Neurology* 63: 1273–6.

Wojtecki, Lars et al. 2011. "Modulation of Human Time Processing by Subthalamic Deep Brain Stimulation" *PLoS ONE* 6(9): e24589.

Yin, Bin, and Andrew B. Troger. 2011. "Exploring the 4th Dimension: Hippocampus, Time, and Memory Revisited" *Frontiers in Integrative Neuroscience* 5: 36.

Yin, Bin, and Warren H. Meck. 2014. "Comparison of Interval Timing Behaviour in Mice Following Dorsal or Ventral Hippocampal Lesions with Mice Having δ Opioid Receptor Gene Deletion" *Philosophical Transactions of the Royal Society – London B* 369: 20120466.

Yin, Bin, Nicholas A. Lusk, and Warren H. Meck. 2015. "Interval-Timing Protocols and Their Relevancy to the Study of Temporal Cognition and Neurobehavioral Genetics" In Valter Tucci (ed.) Neuro-Phenome: Cutting-Edge Approaches and Technologies in Neurobehavioral Genetics. New York, NY: WIley-Blackwell, in press.

CHAPTER 12

Aphasia as a Temporal Information Processing Disorder

Elzbieta Szelag[*,†], *Aneta Szymaszek*[*,†], *and Anna Oron*[*]

1 Introduction

In the Millennium issue of the *Brain and Language* journal, several articles emphasised the importance of studying temporal aspects of information processing and their close associations with language processing. One article even stated that "...on space, time and language: for the next century, timing is (almost) everything" (Osterhout, 2000).

Temporal information processing (TIP) mechanisms have been identified as the neural basis for many higher cognitive functions, such as language, memory, new learning, attention, emotional evaluation, motor control, executive system, planning of activities or decision making. All these functions can be characterized by specific temporal dynamics at different ranges of TIP. Such temporal constraints can be observed both within millisecond and multisecond time ranges, which create pre-semantically defined (i.e., prior to any semantic evaluation) temporal 'windows' for cognitive processing (Pöppel, 1997, 2004, 2009). Such temporal dynamics appear to be a general property of human cognition. It provides a structure not only for language, but also for any perceptual or motor act, as well as any functional state characterized apparently by spatiotemporal patterns of behaviour, which are evidenced in neuroimaging studies.

Strong experimental evidence supporting this conceptual notion comes from the language domain in which temporal structure has been well documented by every day observations of fluent speech, as well as linguistic and psychophysical data. Using different techniques and populations, experimental studies have consistently indicated that the left hemispheric superiority for processing verbal information may reflect more specific control of temporal cues for which human speech is one example (Szelag et al., 2008, 2011; Wittmann et al., 2004). This evidence can be supported by clinical data, indicating that certain left hemispheric lesions selectively affect temporal mechanisms (see Table 12.2).

* Laboratory of Neuropsychology, Nencki Institute of Experimental Biology, Warsaw, Poland.
† University of Social Sciences and Humanities, Warsaw, Poland.

In the present chapter, we concentrate on language functions, especially on TIP and acquired language disorders following injuries to the brain, resulting in aphasia. As many different methodologies have been employed in studies on TIP, we will also discuss theoretical perspectives in relation to these findings. The purpose of this chapter is to summarize the results of studies described in the existing literature focusing on TIP in aphasia and to illustrate these data with the results of our own studies.

2 Language Function

Language is a highly specialized mental function that is a key factor in social interaction and the evolution of human culture. The broader definition of language in Cognitive Neuroscience is *"...a symbolic system used to communicate concrete or abstract meanings, irrespective of the sensory modality employed or the particular means of expression..."* (Purves et al., 2008, p. 511). On the other hand, human speech can be defined as vocalized concrete acts of language used in order to communicate (e.g., Price, 2004).

It has been commonly accepted that human language is a complex system, thus, it is not monolithic. Several language functions or sub-systems are incorporated within the framework of network models of language. Accordingly, language includes spoken (expression) or heard (reception) modalities, communication in writing or reading (visual forms), as well as sign language. Although verbal output differs significantly among the ca. 6000 languages existing worldwide, every given natural language contains specific phonology, vocabulary, grammar, and syntactic rules.

For neurolinguists, the classical dichotomy assumes differentiation between expression (production) and reception (understanding). *Language expression* and our spoken utterances incorporate several processes and sub-functions involved in the production of the verbal output, such as: conceptual preparation including planning a sequence of the speech acts, articulation, verbal fluency, lexicon (lexical encoding), syntax (grammatical encoding), phonology (phonological encoding), prosody, suprasegmental encoding, pragmatics, self-monitoring and repair (Blanken et al., 1993). On the other hand, *language reception* comprises several processes and mechanisms that allow the listener to understand the speaker's communicative intentions, as well as allow the speaker to understand his or her own speech. The linguistic processing comprises acoustic-phonetic analysis of the speech signal, phonology (phonological decoding), vocabulary (lexical selection), sentences, syntax (grammatical decoding), and, finally, interpretation of discourse.

It should be stressed that these processes do not work sequentially one after another. On the contrary, parallel processing occurs in both speaking and listening. Direct feedback interactions between particular components of speech production or speech understanding processes have been heavily debated. There is only some evidence on the existence of highly limited feedback from phonological to grammatical encoding in speech production (Blanken et al., 1993).

Despite diversity of speech sounds, grammar, and syntax among different mother tongues, some important generalization regarding the neuroanatomical representation and neural mechanisms underlying language communication are possible and, thus, reviewed below.

3 Epidemiology, Definitions, and Clinical Syndromes of Aphasia

The term *aphasia* denotes acquired disturbances of language processing, resulting from brain damage. Aphasia is a complex language disorder and refers to deficits apparent in speech, writing or reading, caused by an injury to brain regions specialized for these functions. This definition contains four crucial aspects, namely: (1) aphasia is acquired; (2) aphasia is a consequence of brain damage; (3) aphasia affects language processes; (4) other cognitive functions are relatively spared.

Several types of aphasia classification and classes of aphasic syndromes have been proposed, among other by Luria, Geschwind, Goodglass, Kertesz, and other researchers. The following are the main aphasic syndromes: *global aphasia, Broca's aphasia, Wernicke's aphasia, amnestic aphasia, conduction aphasia, and transcortical aphasias*. The classic location of brain damage in these aphasia types as well as the main language deficits are summarized in Table 12.1.

Nowadays, aphasia results predominantly from brain stroke and approximately 60% of stroke patients suffer from aphasia. The most frequent cause of aphasia is ischemic infarction in the territory of supply of the left middle cerebral artery. Other causes are haemorrhages, tumour, trauma, cerebral infection or atrophic pathology.

It is estimated that up to 270.000 patients with stroke-related aphasia are diagnosed worldwide each year (Szaflarski et al., 2011). Only a minority of these patients recover completely from their language deficits. Examinations of typical recovery curves reported in the literature have revealed that only 25% of patients have a chance for full restoration of disturbed language functions. These data indicate that aphasia still constitutes a major medical and social

TABLE 12.1 *The summary of main language deficits observed in particular aphasic syndromes.*

Type of aphasia	Classic location of brain damage	Particular language functions			
		spontaneous speech	comprehension	repetition	naming
Global	Extensive damage to the LH, including both anterior and posterior speech areas	non – fluent	impaired	impaired	impaired
Broca's	Inferior part of the premotor cortex (BA 44 and 45)	non- fluent (agrammatism)	preserved	impaired	impaired
Wernicke's	Posterior part of the STG (BA 22) and surrounding cortices	fluent	impaired	impaired	impaired
Amnestic	Different parts of temporal, parietal lobes, also other areas of the LH	fluent	preserved	preserved	impaired
Conduction	Arcuate fasciculus	fluent	preserved	impaired	impaired
Transcortical motor	Frontal cortex anterior and/ or superior to the Broca's area, SMA	non – fluent	preserved	preserved	impaired
Transcortical sensory	Posterior speech area	fluent	impaired	preserved	impaired

Abbreviations: LH: left hemisphere; BA: Brodmann area; SMA: supplementary motor area; STG: superior temporal gyrus.

problem in modern society. This should inspire researchers worldwide to elaborate new techniques of language rehabilitation.

4 The Neuroanatomical Basis of Language

Current knowledge on understanding cerebral representation of language comes from 4 main groups of evidence: (1) clinical observations of language deficits following acquired lesions to the brain; (2) electrophysiological mapping of brain areas as an adjunct to various neurosurgical procedures; (3) 'split-brain' studies of patients whose cerebral hemispheres were surgically disconnected as a treatment of epilepsy; (4) experimental studies employing many different laboratory techniques (e.g., psychophysical, behavioural, molecular, biochemical, and neuroimaging).

4.1 *Clinical Observations of Language Deficits*

An important key generalization is that in the vast majority of individuals the left brain hemisphere is a primary locus of language communication. Accumulating evidence has indicated such cerebral representation in 96% of right-handed and in ca. 70% of left-handed individuals (e.g., Springer and Deutsch, 1998). Another generalization is that brain areas involved in language are closely related to the sensory (auditory and visual) and motor regions that process non-speech signals, as well as generate the non-speech motor acts (Springer and Deutsch, 1998).

4.2 *Electrophysiological Mapping of Brain Areas*

Further insight into more detailed mapping of language functions within the left hemispheric cortical areas comes from electrical stimulation of cerebral cortex during surgery by Penfield and Roberts (1959). The reason for such surgery is usually brain tumour or searching for epileptic focus before removing diseased brain tissue. Such stimulations were performed in awake and conscious patients and allowed the surgeon to localize cortical areas that interfere with speech during stimulation. These observations allowed exact location of language and indicated that several left hemispheric regions control different language functions. These large regions are usually outside the *'classic'* language area described by Paul Broca and Carl Wernicke in the late 1800s (Blanken et al., 1993). Despite individual differences in the borders of the region supporting language, it comprises usually parts of the left hemispheric frontal, temporal, and parietal lobes, specifically, large regions of perisylvian cortices for both language production and comprehension.

4.3 *'Split-brain' Studies*

Commissurotomy refers to complete surgical sections of the corpus callosum, anterior commissure, dorsal, and ventral hippocampal commissures and, in some cases, the *massa intermedia*. It resulted in total isolation of cortical areas of two hemispheres. Following complete cerebral commissurotomy, the clear hemispheric specialization was observed as the 'split-brain' phenomenon provided clear evidence on independent contribution of each hemisphere to language processing as well as special behaviour and capacity of isolated hemispheres.

In summary, the left hemisphere represented language competence observed in speaking, hearing, reading or writing, but some limited right hemispheric capacities in language comprehension and production were also evidenced (Zaidel, 2001; Zaidel et al., 2003).

4.4 *Experimental Studies*

The results of detailed mapping of language functions derived from brain stimulation during surgery were further supported by non-invasive brain imaging techniques, using electrophysiological methods, positron emission tomography (PET), functional magnetic resonance imaging (fMRI), magnetoencephalography (MEG) or suppression of language functions during transcranial magnetic stimulation (TMS). All these modern techniques showed activity of various areas in different language tasks, indicating a broader picture of the way language is processed.

4.5 *Summary of Cortical Regions Supporting Language*

A detailed characteristic of cerebral representation of language is beyond the scope of the present chapter but has been described for example by Scott (2012) and Turken and Donkers (2011). We summarized briefly the main role of cortical regions supporting language communication, focussing only on clinical-pathological data of patients with lesions of the specific brain areas. Next, we referred these clinical data to the neural basis for language production and comprehension. It is evident that the clinical data were further confirmed by results of modern neuroimaging studies. Such approach is justified by the topic of the present chapter in which we concentrate on the dissociation between different ranges of TIP or 'temporal windows' (i.e., processing of information within millisecond *vs.* multisecond time domains) in case of expressive and receptive deficits observed in particular aphasia types. In this approach, therefore, we concentrate on the left frontal lobe where the motor functions are carried out controlling *language expression*, as well as on the left posterior regions of the temporal lobe in which the auditory functions are represented, controlling *language reception*. Moreover, in language communication subcortical structures are involved, e.g., basal ganglia, insula, cingulate gyrus or cerebellum, as well as other structures supporting writing, reading, verbal memory, word finding, repetition, verbal fluency, programming of longer verbal utterances, affective control, and articulation. The right hemispheric contribution seems also important.

The neural basis for producing language comprises Broca's area, i.e., inferior part of the premotor cortex (inferior frontal gyrus) of the left frontal lobe. This region includes Brodmann areas 44 and 45. Initial evidence on the involvement of this region in producing language was provided by the French anatomist and neurologist – Paul Broca in 1861. He described a patient who suffered from acquired brain damage to this region, resulting in disability to produce fluent speech. The lesion's location was confirmed by Broca in post-mortem brain examination.

The importance of this region is vividly evidenced in patients suffering from motor aphasia (other names: efferent motor aphasia, Broca's aphasia, production aphasia, or nonfluent aphasia), a common neurologic disorder, resulting in inability to organize and control the linguistic content of the utterance with relatively preserved auditory comprehension. The typical patient suffering from a lesion in the Broca's area cannot express thoughts appropriately, because of the loss of the meaningful fluent verbal output, reduced phrase length (so called 'telegraphic style' of the verbal output) and agrammatism, i.e., disrupted rules of grammar and syntax, even though the verbal apparatus and articulatory movements remained intact.

The core symptom in Broca's aphasia is a disruption of brain mechanisms responsible for producing phonemes, as well as combining phonemes into morphemes and words into sentences. Thus, in Broca's aphasia the efferent pathway from the Broca's area to the peripheral articulatory apparatus is disrupted. Generally, such patients produce structurally incorrect utterances, nonsense syllables, transpose words with some sensible meaning. Thus, a listener may have an impression that the patients knew what they intended to say.

Broca's aphasia encompasses interactions of several impairments, including defects in linguistic formulation, motor programming, initiation and maintenance of verbal output, and affective control. Thus, selective damage to the posterior inferior frontal cortex does not produce the full motor syndrome, but rather Broca's aphasia is caused by extensive damage to several neighbouring frontal regions.

On the other hand, an important contribution indicating the neuroanatomical basis of language comprehension was made first by the German psychiatrist – Carl Wernicke in 1874. He described for the first time a patient who produced fluent utterances with appropriate grammar and syntax, but could not understand heard (or written) communications addressed to him by other speakers. These deficits were caused by damage to the left hemispheric temporal lobe, more specifically, to the posterior part of the superior temporal gyrus, which is referred to, in honour of its discoverer, as Wernicke's area. According to the Brodmann map it comprises area 22, but often also surrounding temporal and parietal cortices. The lesions in this part of the brain caused Wernicke's aphasia (other names: receptive aphasia, sensory aphasia or fluent aphasia).

In contrast to Broca's aphasia, patients suffering from lesions affecting Wernicke's area cannot understand the speech of other people, as well as their own speech. As a consequence, utterances generated by such patients have little (or even not at all) sense because phonemes, syllables, and words are not correctly linked. In more severe forms of this disorder, the verbal output is named *'jargon aphasia'*, as the listener cannot understand it at all. Wernicke's aphasia can be characterized by fluent speech, adequate syntax and grammar, but inappropriate choice of words (paraphasia) and little spontaneous repetitions. A core symptom of Wernicke's aphasia is disruption of phonemic hearing, i.e., the ability of analysis and synthesis of speech sounds (phonological decoding). For example, a patient with such disruption cannot properly decode phonemes (consonants and vowels) and, as a consequence, is not able to properly decode words and sentences. The impaired phonological hearing results, therefore, in disordered comprehension. The schematic description of classic aphasic syndromes is given in Figure 12.1 (compare also Table 12.2 and Section 7, below).

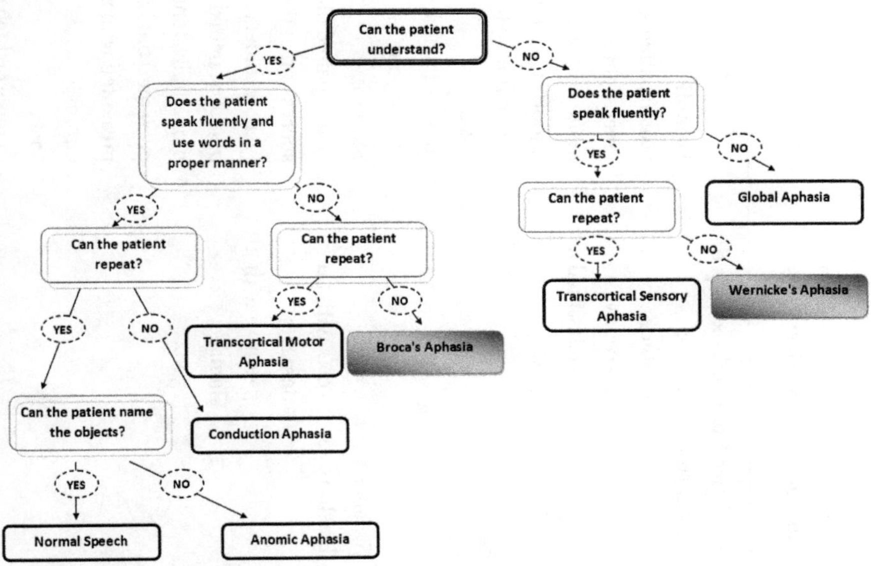

FIGURE 12.1 *The differential diagnosis of main aphasic syndromes.*

TABLE 12.2 *Summary of results reported in the existing literature regarding deficient temporal processing in aphasic patients.*

No	Author	Subjects	Method	Results	Conclusions
	Evidence from milliseconds timing				
1	Efron, 1963.	n = 16: 6 motor AP; 4 receptive AP; 1 unclassified AP; 5 NC.	**Stimuli:** *visual*: paired 5ms pulses of either red or green light presented in rapid succession; *auditory*: paired 10ms tones (250Hz and 2500Hz) presented with 13 different ISIs varied from 0 to 600ms; **Task:** to report which stimulus in the pair was the first one.	100 % correctness was achieved in NC for ISIs from 62 to 162ms and in AP for ISI from 162 to 600ms. Worsened performance in motor AP on the auditory task, in receptive AP on the visual task.	
2	Swisher & Hirsh, 1972.	n = 30: 10 AP with LH damage; 5 non AP with RH damage; 15 NC.	**Stimuli:** *visual*: paired diodes presented in rapid succession at the same or different places (yellow- blue; yellow- yellow); *auditory*: paired clicks presented monaurally, tones (300 Hz, 3000 Hz) presented binaurally. The stimuli were presented with SOAs varied from: 20 to 640ms; **Task:** to report which stimulus in the pair was the first one.	NC: TOT = 20ms for stimuli presented at different places; TOT = 30 – 40ms for stimuli presented at the same place; fluent AP: required the longest gap to order stimuli, especially the auditory ones; non AP: no deficits on the visual tasks, but required longer intervals to order tonal stimuli.	Affected timing, depending on the subject group.

3	Tallal & Newcombe, 1978.	n = 26: 10 AP with LH damage; 10 non AP with RH damage; 6 NC.	**Stimuli:** paired 75-ms complex tones composed of frequencies within the speech range and presented in rapid succession with different ISIs; **Task:** to report the order of stimuli presented in a pair.	ISI of 428ms: no group differences; ISIs from 8 to 305ms: poorer performance in AP than in other groups.	
4	v. Steinbüchel et al., 1999a.	n = 65: 7 non – fluent AP with LH damage; 16 fluent AP with LH damage; 9 non AP with LH damage; 8 non AP with RH pre – central damage; 8 non AP with RH post – central damage; 17 NC.	**Stimuli:** paired 1-ms clicks presented monaurally; **Task:** to report the order of stimuli presented in a pair.	Fluent AP: increased TOT in comparison to all other groups.	Affected timing, depending on the subject group.
5	v. Steinbüchel et al., 1999b.	n = 62: 7 non – fluent AP with LH damage;	**Stimuli:** paired 1-ms clicks presented monaurally;	Fluent AP: increased TOT in comparison to all other groups.	

TABLE 12.2 *Summary of results reported in the existing literature regarding deficient temporal processing in aphasic patients* (cont.)

No	Author	Subjects	Method	Results	Conclusions
		14 fluent AP with LH damage; 9 non AP with LH damage; 9 non AP with RH pre – central damage; 6 non AP with RH post – central damage; 17 NC.	**Task:** to report the order of stimuli presented in a pair.		
6	Wittmann et al., 2004.	n = 30: 6 non – fluent AP with LH damage; 12 fluent AP with LH damage; 6 non AP with RH pre – central damage; 6 non AP with RH post – central damage.	**Stimuli:** paired 1-ms clicks presented monaurally; **Task:** to report the order of stimuli presented in a pair.	Fluent AP: increased TOT in comparison to all other groups.	

7	Fink et al., 2006.	n = 38: 8 anomic AP; 2 Wernicke's AP; 4 Broca's AP; 5 non- classified AP; 19 NC.	**Stimuli:** Exp. 1 – paired 1-ms clicks presented monaurally; Exp. 2 – paired tones of 800Hz and 1200Hz presented binaurally; **Task:** Exp. 1, 2 – to report the order of presented stimuli.	AP: increased TOT in both tasks.	Affected timing, depending on the subject group.
8	Stefanatos et al., 2007.	n = 13: 5 Broca's AP; 7 anomic AP; 1 Wernicke's AP.	**Stimuli:** noise segments modelled after formant in speech of different length, separated by a silent gap of 10, 20, 40, or 80ms duration; **Task:** to detect the gap between stimuli and ascertain if two stimuli were the same or different.	AP: impaired gap detection task.	Affected timing in all groups of AP.
9	Sidiropoulos et al., 2010.	n = 6: 1 conduction AP with both LH ischemic infarction and an older, contralateral parieto – occipital cerebrovascular lesion; 5 NC.	**Stimuli:** Exp. 1 paired two noise segments of identical spectral energy distribution ('intra- channel condition'); Exp. 2 pairs of different noise segments: broadband and filtered noise of different length ('inter- channel condition');	AP: Exp. 1 – intact temporal resolution in 'intra- channel condition'; Exp. 2 – reduced performance in 'inter- channel condition'.	Affected or preserved timing in AP, depending on the experiment.

TABLE 12.2 *Summary of results reported in the existing literature regarding deficient temporal processing in aphasic patients* (cont.)

No	Author	Subjects	Method	Results	Conclusions
			Task: Exp. 1 – to detect the gap between two identical noise segments; Exp. 2 – to detect the gap between pairs of different noise segments.		
Evidence from multiseconds timing					
10	Szeląg et al., 1997.	n = 56: 6 Broca's AP with LH pre – central damage; 14 Wernicke's or amnestic AP with LH post- central damage; 9 non AP with LH damage; 6 non AP with RH pre – central damage; 6 non AP with RH post – central damage; 15 NC.	**Stimuli:** metronome beats presented at 9 different frequencies; **Task:** to accentuate mentally every 2nd, 3rd or other beat to create subjective rhythmic patterns.	Broca's AP: disrupted temporal integration evidenced by a different strategy in mental accentuation based on counting; Other groups: a combination of mental counting and constant time strategies.	Affected timing only in Broca's AP.

11	v. Steinbüchel et al., 1999a.	n = 65: 7 non – fluent AP with LH damage; 16 fluent AP with LH damage; 9 non AP with LH damage; 8 non AP with RH pre – central damage; 8 non AP with RH post – central damage; 17 NC.	**Stimuli:** Exp. 1 *visual*: ambiguous figures: Necker Cube or Rubin Vase; *auditory*: tones of different frequencies, reversible syllables; Exp. 2 metronome beats presented at 9 different frequencies; **Task:** Exp. 1 – to press a button at each perspective switching for a presented ambiguous figure; Exp. 2 – to accentuate mentally every 2nd, 3rd or other beat to create a subjective rhythmic pattern; Exp. 3 – to tap in a self- paced tempo; Exp. 4 – to tap in a maximum tempo.	Exp. 1: longer duration of spontaneous reversal rate in patients than in NC; Exp. 2: AP apply integration by counting differently than all other groups; Exp. 3: impaired self- paced tapping tempo in AP; Exp. 4: no differences between patients and NC in a maximum tapping tempo.	Affected timing in AP in both hundreds of milliseconds and a few seconds time domains.
12	Wittmann et al., 2001.	n = 61: 22 AP with LH damage; 14 non AP with RH damage; 9 non AP with LH subcortical damage; 16 NC.	**Task:** Exp. 1 – to tap in a self- paced tempo; Exp. 2 – to tap in a maximum tempo.	Exp. 1: impaired in AP; Exp. 2: no differences between patients and NC.	Affected timing in AP in hundreds of milliseconds time range.

TABLE 12.2 *Summary of results reported in the existing literature regarding deficient temporal processing in aphasic patients* (cont.)

No	Author	Subjects	Method	Results	Conclusions
13	Kagerer et al., 2002.	n = 43: 5 Broca's AP; 10 Wernicke's, amnestic AP; 9 non AP with LH damage; 5 non AP with RH pre- central damage; 5 non AP with RH post – central damage; 9 NC.	**Stimuli:** *visual:* 1.green oblong of 80 x 50 cm^2, with equal luminance 2.19th century painting depicting a pastoral scene *auditory:* 1. 500 Hz tone 2. noise of running water Stimuli presented in blocks with different standard intervals varied from 1000ms to 5500ms. **Task:** to reproduce standard interval.	No differences between patients and NC in reproduction of short intervals (1000 – 2500ms); Impaired reproduction of long intervals (3000 – 5500ms) in non AP with RH damage both pre- – and post – central, in comparison to other groups.	Affected timing in non AP with RH damage in a few seconds time range.

Abbreviations: LH – left hemisphere, RH – right hemisphere, AP -aphasic patients, non AP – non aphasic patients, NC – normal controls, TOT – temporal – order – threshold, ISI – inter – stimulus – interval, SOA – stimulus – onset – asynchrony

Despite the general features of neuroanatomical basis of language described above, a vast number of subsequent studies using various methods have confirmed the contribution of many additional areas of both hemispheres, as well as subcortical structures to verbal communication.

Taking into account the other language-sensitive regions, the traditional model of language representation assuming two well-defined parts supporting expression and reception needs to be modified. Although it seems very useful in the clinical practice and there is a number of evidence supporting such dissociation of language function, the current thinking on the traditional model of language is undergoing some fine-tuning, considering three groups of evidence: (1) neuroimaging data indicating the involvement of many other brain regions, beside the Broca's and Wernicke's areas, in language communication; (2) individual differences in representation of language production and reception far beyond the Broca's and Wernicke's area; (3) neuroanatomical representation of various language functions (see above) and phenomena, like bilingualism or sign language.

For instance, using neuroimaging it is possible to define the exact areas damaged by stroke or trauma and to relate these structural lesions to deficient language functions, which often do not cover the deficits expected by the traditional model. Moreover, data from individual patients indicated that lesions to the Broca's area not always led to Broca's aphasia. The deficits characteristic for this syndrome were also observed in case of brain damage to many other brain regions far beyond the region traditionally associated with the production of speech. Similar observations concern Wernicke's aphasia and disrupted auditory comprehension. It is interesting to note that brain imaging techniques (MRI, fMRI or PET) have revealed that anatomical damage does not necessarily cover the functional damage, which usually extends beyond the border of structural damage. At this point it should be mentioned that these neuroimaging techniques can map brain activity, but not necessarily linguistic function directly.

In summary, many brain regions control language communication. It would be a question for further studies what is their contribution to particular language functions for verbal communication, moreover, to what extent these areas are specific to language. To answer these questions, many psycholinguists seem to concentrate not only on the structures (localization) of particular language functions, but also on the neural mechanisms underlying language processing in real time, thus, on the temporal structure of linguistic processing. Focusing on temporal mechanisms underlying different language functions would allow us to understand temporal coordination, segmentation and integration involved in expression and comprehension. In this approach, studies on TIP in different time domains can indicate the activity of the dynamic network in which human language is rooted in norm and pathology. If the activity of the timing mechanisms operating at different ranges differs reliably in patients and controls, it may interfere with the associated mental activity.

5 Temporal Constraints of Speech

Both everyday observations and experimental data indicate temporal constraints of human speech. Its temporal organization is multileveled. There is strong evidence supporting the thesis that our language communication is rooted in TIP, i.e., in temporal 'integration' and temporal 'chunking' or 'segmentation' on several time scales. One can distinguish the milliseconds (shorter time scale) and multiseconds (longer time scale) temporal constraints, which are controlled by a high- or low-frequency processing system, respectively (Pöppel, 2004; Szelag et al., 2004; Szelag, 2011). The former system concerns predominantly phonological encoding/decoding and syllabification, whereas the latter one concerns rather lexical selection and sentence production/reception.

It is important to note that language is not an isolated system, therefore, similar time scales can be distinguished in several key aspects of our behaviour, e.g., in motor control, as well as in many other mental functions which are strongly related to speech expression and/or reception (Pöppel 1997; Szelag et al., 2009). Therefore, we are of the opinion that timing is a critical factor for our verbal communication and provides an important insight into how the brain parses language in norm and pathology.

It is commonly known that normal production and understanding of language involves highly skilled coordination of several processes. Similar temporal levels or 'processing platforms' are incorporated in both speech expression (encoding) and reception (decoding). In phonological *encoding*, for instance, we produce the phonological forms, in particular, their segmental structure. Thus, central to phonological encoding are phonemes or syllables as the basic units for articulation produced by the speaker successively. On the other hand, in phonological *decoding* these phonological forms have to be perceived by the listener. Thus, these strings of successive phonemes and syllables perceived by the listener are also central to phonological decoding and speech reception.

The processing of temporal information in the speech signal has been investigated explicitly over the last few decades. The *high-frequency processing* system, i.e., a time domain of some tens of milliseconds, seems to be relevant to phonemic awareness (labelled also as phonemic hearing). It is defined as a unique human ability to analyse and synthesise speech sounds. These processes are fundamental with respect to auditory comprehension. In the time platform of about 20–40ms, information about the place of articulation in stop-consonants (p, b, t, d etc.) is contained. Formant transitions, characterized as short sound waveforms that change frequency across a time interval of ca. 40ms vary, according to the place of articulation. Spectrographic analyses of our verbal utterances clearly indicated that rapid formant transitions in stop-consonants (p, b, t, d etc.) are limited in time up to 40ms. Because of the specific structure of our articulators these stop-consonants cannot be prolonged in the fluent speech because a forthcoming vowel sounds immediately.

On the other hand, fricatives (w, f, s, z etc.) or vowels (a, o, u etc.) are characterized by a structure less limited in time and can sound in the fluent speech much longer, even up to 200ms. Despite important individual differences in the speech expression rate, particular speech sounds come on average as fast as 10 to 15 phonemes per second. The temporal structure of the verbal output is usually automatic (controlled by procedural memory) and often happens without any attentional or motor control of 'how' we say things. In contrast, 'what' we say is fully consciously controlled.

Another example of temporal constraints with respect to phonemic awareness on the millisecond time scale comes from the Voice-Onset-Time (VOT) paradigm. In the perception of voicing contrasts of stop-consonants in syllables like DO *vs.* TO, a difference in duration of about 40ms between the burst of the air and the onset of laryngeal pulsing in articulation of the initial consonant, is defined as the VOT. It distinguishes voiced (/b/, /d/, /g/ etc.) from voiceless stop-consonants (e.g., /p/, /t/, /k/ etc.).

Such evidence supports the thesis that the high-frequency processing system is related to phoneme processing and phonemic hearing, thus, the crucial processes with respect to auditory comprehension.

Furthermore, there exists the *low-frequency processing* system, corresponding to the time range of a few seconds, which is related to the duration of sentences in fluent speech. For example, our average syllabic rate is ca. 3–4 per seconds, whereas words are usually articulated at a speed of approx. 2 per second. At this rate we are able to retrieve lexical items from a mental lexicon that on average contains probably tens of thousands items. Although there are some so-called 'fast speakers' who are able to produce even up to 7 words per second, it is usually difficult to follow such fast speech tempo, which does not fit the typical tempo.

Another source of evidence comes from analysis of the temporal structure of fluent speech in many languages, like Polish, Russian, German, English or Chinese (Pöppel, 1997). Several data have indicated that in many languages the duration of phrases in fluent speech is limited in time up to a few seconds.

Thus, spontaneous flow of speech is temporally segmented or 'chunked' in a few seconds time domain. This phenomenon is called 'semantic parsing'. Usually, particular phrases are followed by hesitation pauses that are necessary both for the listener to process the information provided by the speaker, as well as for the speaker to prepare the next phrase.

At this point it seems interesting to note that similar temporal segmentation as observed in oral language has been confirmed in sign language. In such way of communication, consecutive signs (i.e., motor acts) are grouped within segments of a few seconds duration, followed by pauses when signs are not provided. The duration of particular signed segments corresponds to those observed in oral language, described above, with a little shorter durations of ca. 2s in sign language. The similar temporal chunking of speech expression, independently of the kind of articulation (using either the oral output or hand movements) suggests the existence of a common neural mechanism controlling human language communication, independently of the output from the system.

To summarize, both the millisecond and multisecond time ranges are crucial for our language communication. Hence, authors are of the opinion that the left hemispheric specialization for language derives from specialization for timing. Furthermore, language deficits of various etiologies in children and adults have been associated with timing disorders on both mentioned time scales. Language processing depends, thus, on temporal constraints of the brain on several levels, which may be affected after the brain injury.

In the traditional view different language disorders (e.g., aphasia, specific language impairment) or reading and writing disorders (dyslexia) have been viewed as distinct clinical syndromes. Recently, this viewpoint has been reformulated because of a close link between these syndromes. Neuropsychological profiles taking into account the coexistence of specific temporal processing disorders and phonological impairments show some similarities between particular syndromes. In general, they are characterized by deficient TIP in both, the millisecond and multisecond ranges. Because of the central relevance of TIP to our understanding of the neural basis of cognitive systems, one may define a typical TIP and abnormalities in such processing in human populations suffering from impaired cognitive function (Szelag and Skolimowska, 2012).

6 Deficits in Processing Temporal Information in Aphasic Patients

Studies of patients suffering from aphasia following left hemispheric brain damage provide an important contribution to our understanding of TIP.

The first evidence of this type was provided by Efron (1963) and later by Swisher and Hirsh (1972). They observed that aphasic patients displayed parallel deficits in both speech comprehension and rapidly changing temporal information.

These observations were confirmed in several further reports, including our own studies. Table 12.2 summarized the results of studies conducted using different patient populations and experimental techniques. Looking at Table 12.2, we may conclude that TIP within both the millisecond (points 1–9) and multi-second (points 10–13) time ranges is usually disturbed in patients with aphasia following left hemispheric brain damage.

There exist also some experimental evidence on the high frequency processing system in normal healthy volunteers. These studies are often focussed on temporal order of incoming events and measurement of temporal-order threshold. It reflects an ability to perceive the relation *before-after* for stimuli presented in rapid succession in order to be brought into a sequence by approximately the same value of some tens of milliseconds for visual, auditory, and tactile modality (Hirsh and Sherrick 1961; Kanabus et al., 2002; Zampini et al., 2003, 2005). The correspondence of this value across different senses suggests a common central mechanism for vision, hearing or touch and it may be concluded that the neural process underlying sequential ability is activated to identify primordial events, independently of the sensory modality (Pöppel, 2009). Such ability allows us to reduce complexity of incoming events, which stimulate incessantly our senses, and to create our conscious percepts.

Strong experimental evidence provided in Table 12.2 (points 1–9) may suggest a disruption of this high-frequency process following left hemispheric brain damage, resulting in disordered language communication, as a consequence of primordial event disruption. Moreover, the data on temporal constraints of human speech on the milliseconds level discussed in the previous section have indicated the importance of this time platform with respect to speech reception. Hence, disordered processing of information on this level following brain injury may underlie language problems, especially, problems in phonemic hearing and auditory comprehension. These problems often constitute a core symptom after posterior lesions to the left hemisphere, resulting in receptive aphasia (see above for aphasia classification).

On the other hand, Table 12.2 presents also data indicating deficient timing on low frequency processing, i.e., in the domain of hundreds of milliseconds or a few seconds (points 9–13). Considering temporal constraints of speech (see Section 5), the former time window is related to the level of syllables, whereas, the latter one may be important with respect to the fluent verbal output and expression of phrases and sentences. It requires the ordering of successive syllables and words into logic strings, which build fluent utterances and provide a structure for language communication.

According to the hierarchical model of time perception proposed by Ernst Pöppel (1997, 2004; Szelag et al., 2004), the existence of temporal integration on a few seconds scale creates a high-frequency level of complexity reduction. In fact, in fluent verbal communication we usually do not concentrate so much on particular phonemes, syllables or words, but on logic utterances of what we say or what we hear. Thus, the primordial events identified on a level of higher temporal resolution (some tens of milliseconds) are next sequentially linked together into a perceptual *'gestalt'*, representing an operating range of a few seconds. This process is also labelled in the literature as 'temporal binding' or 'temporal integration' and it is expected to be important for our conscious experience, of which verbal communication may be one example.

Although reports on TIP on the supra-second level are fewer, Table 12.2 shows that using different experimental paradigms, like subjective accentuation of metronome beats (points 10–11), spontaneous reversal rate of ambiguous figures (point 11), finger tapping in a self-pace and maximum tempi (point 12) or reproduction of temporal intervals (point 13), a disruption of temporal binding can be observed, which has been also evidenced following injury to the brain, predominantly to the anterior parts of left hemisphere, resulting in motor aphasia. The patients suffering from this syndrome had deficient binding operations that probably underlie problems with the fluent verbal output (i.e., effortful, nonfluent speech, compare above), but also some comprehension deficits, especially for longer utterances, which need to integrate and hold the incoming information for a few seconds.

To sum up, acquired language disorders following brain damage are strongly related to deficient timing on several levels, which create a neural timeframe for verbal communication. Thus, the superiority of the left hemisphere for the processing of verbal information may reflect more primary specialization for TIP.

7 Relationships between Disrupted Timing and Specific Aphasia Syndromes

It seems also interesting to note, that in our studies on disrupted timing in aphasia, an important dissociation between specific language deficits (in phonemic hearing *vs.* effortful nonfluent verbal output) and temporal deficits (at the sub- or supra-second level) were observed (v. Steinbüchel et al., 1999b; Szelag et al., 1997; Wittmann et al., 2001). Specific left hemispheric lesions selectively damaged temporal mechanisms operating either within the millisecond or multi-second time scales (Table 12.2, points 4–6, 10–12). The former range was assessed with auditory temporal-order-threshold (Figure 12.2 whereas, the latter one with subjective accentuation paradigm, which corresponded to the upper limit of the temporal integration mechanism, operating in a time window of a few seconds (Figure 12.3). In patients with left hemispheric post-central lesions, suffering from Wernicke's aphasia (characterized by deficits in phonemic hearing and poor comprehension), deficient temporal-order-thresholds were reported (Figure 12.2). Those patients needed a significantly longer gap between successive events to correctly report their before-after relation. These timing deficits corresponded to their language

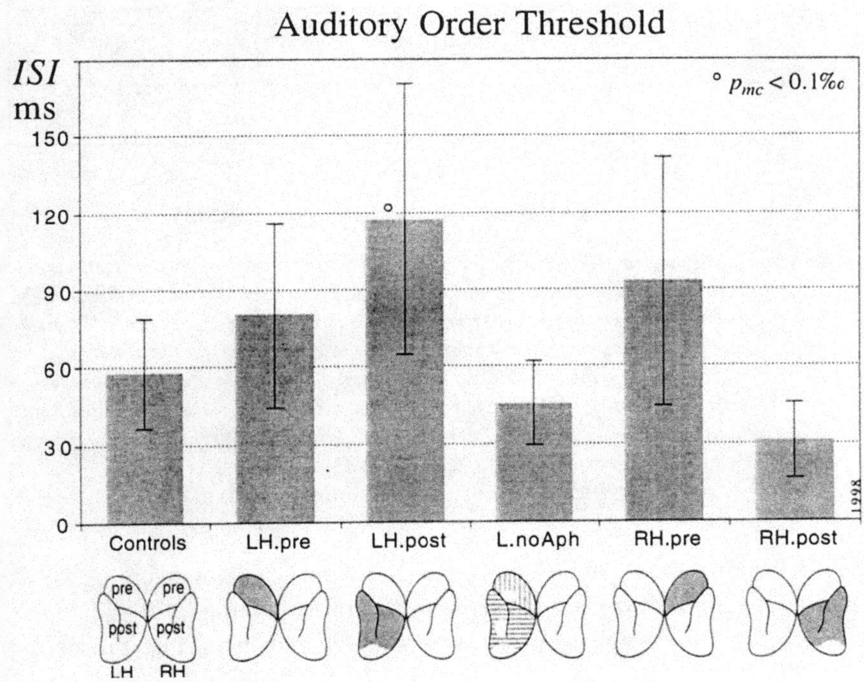

FIGURE 12.2 *Auditory order thresholds (mean and standard deviations) for the five patient groups with focal brain injuries and for an orthopaedic control group are shown:* LH. *pre-anterior left hemisphere (pre-central) with non-fluent aphasia;* LH. *post-posterior left hemisphere (post-central) with fluent aphasia (*p<1% for group differences as compared with controls and L. noAph; statistical calculation with Scheffe post-hoc test); L. noAph-left-sided subcortical lesions without aphasia;* RH. *pre-anterior right hemisphere (pre-central);* RH. *post-posterior right hemisphere (post-central). Reprinted from Neuroscience Letters, 264, v. Steinbüchel N., Wittmann M., Strasburger H., Szelag E. "Auditory temporal-order judgment is impaired in patients with posterior regions of the left hemisphere" 168–71.*

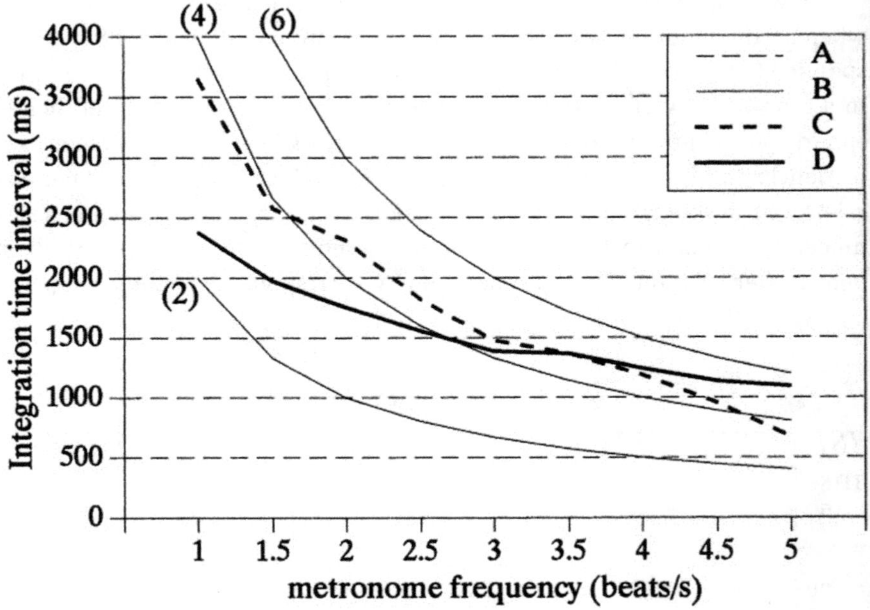

FIGURE 12.3 *The measured integration interval length (MIIL) plotted against the metronome frequency using three different strategies in Broca's aphasia and other patients. Integration (A) by time; (B) by number; (C) in Broca's aphasic patients; (D) in all the remaining patients. Standard deviation values (in ms) for the consecutive frequencies in the Broca's aphasic patients: 1964, 911, 887, 591, 439, 363, 408, 491, 406; in the other subjects: 1191, 698, 535, 370, 338, 370, 363, 394, 401. Reprinted from Neuroscience Letters, 235, Szelag E.v. Steinbüchel N., Pöppel E. "Temporal processing disorders in patients with Broca's aphasia" 33–36.*

deficits, because limited duration of consonants and vowels in incoming words could not be properly decoded because of disrupted timeframe.

In contrast, Broca's aphasic patients were unaffected on this high processing level (Figure 12.2). They were characterized by relatively preserved comprehension but nonfluent speech and displayed deficits within the supra-second processing range (Figure 12.3).

Using subjective accentuation paradigm of monotonous metronome beats, normal healthy volunteers can integrate the information up to a maximum of ca. 3 seconds. In this task they applied a combination of integration strategies based on perceptual grouping of separate beats in a constant time (related to the upper limit of temporal integration, Figure 12.3 dashed parallel lines A reflecting integration within e.g., 1000, 2000 or 4000 s time windows) and automatic counting up to 2, 4 or 6 s (Figure 12.3 solid lines B labelled by numbers 2, 4 or 6). In contrast, in Broca's aphasic patients the constant time strategy was disrupted, but the accentuation was based predominantly on automatic mental counting (Figure **12.3** bold dashed line C). Moreover, the upper limit of integration in Broca's aphasic patients exceeded the typical frames characteristic for normal healthy volunteers. This suggests deficient temporal integration in motor aphasic patients in a time window of a few seconds, but relatively preserved automatic mental counting. The data obtained implied that some highly structured syntactic abilities related to temporal integration are located in the anterior language area (Broca's area).

The evidence provided by Szelag et al. (1997) and v. Steinbüchel et al. (1999); Table 12.2 point 10–11; Figure **12.2** and **12.3**) showed that a disruption in temporal binding is a core symptom in case of nonfluent speech and agrammatism, dominating in motor aphasia. In fact verbal expression requires information to be integrated and held for a few seconds. The disruption of temporal integration at the level of seconds may cause symptoms typical for motor aphasia.

Furthermore, aphasic patients demonstrated impaired self-paced (personal) tempo of finger tapping, indicating the dominant control of left hemisphere in voluntarily timed actions in the special time scale of hundreds of milliseconds involved in motor behaviour (Wittmann et al., 2001; Figure **12.4**). These timing deficits may be related to disrupted TIP linked to the syllable level, independently of the type of aphasia.

To conclude, we argue that TIP in aphasics may be selectively affected at either sub- or supra-second time range, depending on the localisation of lesion and observed disfluency pattern. We also postulate that a disruption of timing mechanisms leads to phonological and/or syntactical disorders, commonly observed in aphasic patients.

These results may have some practical applications with respect to new neurorehabilitation methods. As TIP processing is impaired in aphasic patients, in our recent studies we verified the effectiveness of a new method of aphasia therapy which is concentrated not on the linguistic level, but on the pre-semantic one and concerns the training of neuronal mechanisms underlying both TIP and language (Szelag et al., 2014). These methods address both the receptive and expressive language functions, in particular, auditory speech comprehension, phonemic hearing and nonfluent verbal output.

Our recent studies are addressed greater understanding of the relationship between timing and cognition (Szelag et al., 2011). Using Fast For Word Language® (FFW) program, we showed for the first time significant benefits of such temporal training on broad aspects of cognitive function in the elderly beyond 65 years of life (Szelag and Skolimowska, 2012). Following such training, we indicated stable positive effects not only in TIP, but also in divided and

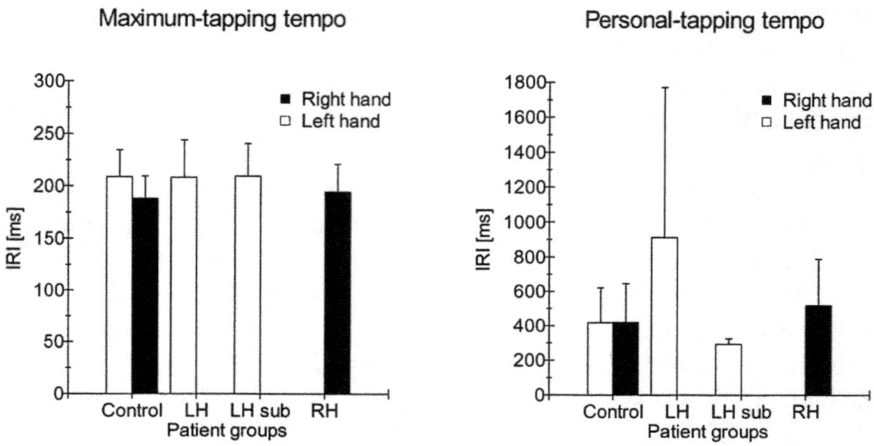

FIGURE 12.4 *Maximum-tapping tempo and personal-tapping tempo as mean inter-response interval (IRI) over the orthopaedic control group and the brain-injured patient groups with left-hemispheric cortical lesions (LH), with left-hemispheric lesions in predominantly subcortical regions (LH sub) and with right-hemispheric cortical lesions (RH). Data for the brain-injured patients are only presented for the hand ipsilateral to the lesion site. Reprinted from Cognitive Brain Research, 10, Wittmann M., v. Steinbüchel N., Szelag E. "Hemispheric specialisation for self-paced motor sequences" 341–44.*

vigilance of attention, matching complex patterns, and working memory span. These results show a new impact of temporal training not only on deficient language function, but also on age-related cognitive decline in the senior population.

On the basis of these results we offer a promising direction for neurorehabilitation and have recently initiated an innovative rehabilitation computer program that addresses improvement of a broad range of cognitive functions in children and adults.

8 Concluding Remarks

Aphasia is a complex language disorder and refers to deficits apparent in speech, writing, and reading caused by an injury to brain areas, specialized for these functions. It is commonly known that the expression and reception of language have defined temporal dynamics and involve highly skilled coordination of several processes. Language processing depends, thus, on the temporal constraints of information processing which may be affected after injury to the brain.

In this chapter we summarize the existing literature studies on TIP as a key to understand language disturbances in aphasic syndromes. The literature data are illustrated with examples from our studies, indicating parallel language and timing deficits following damage to specific brain regions. We showed dissociation between receptive and expressive language deficits associated with a disruption of specific timing mechanisms controlling information processing on milliseconds or multiseconds time domains.

To conclude, in aphasic patients TIP may be selectively affected either within the sub- or supra-second range, depending on the lesion location and observed disfluency pattern. A disruption of timing mechanisms leads to phonological disorders and/or effortful nonfluent speech. Finally, we have indicated that the specific nonverbal training in temporal processing can ameliorate auditory comprehension in aphasic patients. These studies provide an important contribution to our understanding of TIP as a neural basis of language, as well as other cognitive function.

Acknowledgments

Supported by grants 507/1/N-DFG/2009/0 and INNOTECH/IN1/30/159041/NCBR/12. We wish to thank Argiro Vatakis and Melissa Allman for the helpful reading of this chapter.

References

Blanken, Gerhard et al. 1993. *Linguistic Disorders and Pathologies*, Berlin – New York (Walter de Gruyter).

Efron, Robert. 1963. "Temporal Perception, Aphasia and Déjà vu" *Brain* 86: 403–24.

Fink, Martina, Jan Churan, and Marc Wittmann. 2006. "Temporal Processing and Context Dependency of Phoneme Discrimination in Patients with Aphasia" *Brain and Language* 98: 1–11.

Hirsh, Ira J., and Carl E. Sherrick. 1961. "Perceived Order in Different Sense Modalities" *Journal of Experimental Psychology* 62: 423–32.

Kagerer, Florian A. et al. 2002. "Cortical Involvement in Temporal Reproduction: Evidence for Differential Roles of the Hemispheres" *Neuropsychologia* 40(3): 357–66.

Kanabus, Magdalena et al. 2002. "Temporal Order Judgement for Auditory and Visual Stimuli" *Acta Neurobiologiae Experimentalis* 62: 263–70.

Osterhout, Lee. 2000. "On space, time, and language: For the next century, timing is (almost) everything" *Brain and Language* 71: 175–7.

Penfield, Wilder, and Lamar Roberts. 1959. *Speech and Brain Mechanisms*, New York: Princeton University Press.

Pöppel, Ernst. 1997. "A Hierarchical Model of Temporal Perception" *Trends in Cognitive Sciences* 1: 56–61.

———. 2004. "Lost in time: A historical frame, elementary processing units and the 3-second window" *Acta Neurobiologiae Experimentalis* 64: 295–302.

———. 2009. "Pre-Semantically Defined Temporal Windows for Cognitive Processing" *Philosophical transactions of the Royal Society, Series B, Biological Sciences* 364: 1887–96.

Price, Cathy. 2004. "An Overview of Speech Comprehension and Production" In *Human brain function* edited by Richard Frackowiak, Karl J Friston, Christopher D Frith. Raymond J Dolan, Cathy Price, Semir Zeki, John T. Ashburner and William D Penny, 517–32. Oxford: Elsevier.

Purves, Dale et al. 2008. *Principles of Cognitive Neuroscience* Sunderland – Massachusetts USA: Sinauer Assoc., Inc. Publishers.

Scott, Sophie K. 2012. "The neurobiology of speech perception and production – can functional imaging tell us anything we did not already know?" *Journal of Communication Disorders* 45: 419–25.

Sidiropoulos, Kyriakos et al. 2010. "Temporal Processing Capabilities in Repetition Conduction Aphasia" *Brain and Cognition* 73(3): 194–202.

Springer, Sally P., and Georg Deutsch. 1989. *Left Brain, Right Brain* New York: H. Freeman and Co.

Stefanatos, Gerry A. et al. 2007. "Fine Grain Temporal Analysis in Aphasia: Evidence from Auditory Gap Detection" *Neuropsychologia* 45(5): 1127–33.

Swisher, Linda, and Ira J. Hirsh. 1972. "Brain Damage and the Ordering of Two Temporally Successive Stimuli" *Neuropsychologia* 10: 137–52.

Szaflarski, Jerzy P. et al. 2011. "Poststroke Aphasia Recovery Assessed with Functional Magnetic Resonance Imaging and a Picture Identification Task" *Journal of Stroke and Cerebrovascular Disease* 20(4): 336–45.

Szelag, Elzbieta et al. 2004. "Individual Differences in Temporal Information Processing in Humans" *Acta Neurobiologiae Experimentalis* 64: 349–66.

Szelag, Elzbieta et al. 2008. "Cortical Representation of Time and Timing Processes" In *Neuronal correlates of thinking*, edited by Eduardo Kraft, Balazs Guylas and Ernst Pöppel, 185–99. Berlin, Springer-Verlag.

Szelag, Elzbieta et al. 2011. "Hearing Loss and Auditory Processing Disorders: Clinical and Experimental Perspectives" In *Culture and Neural Frames of Cognition and Communication* edited by Shihui Han and Ernst Pöppel, 153–68. Berlin: Springer.

Szelag, Elzbieta et al. 2009. "Neuropsychological Rehabilitation from Brain-Behavior Perspective" In *New Ideas in Studying and Supporting Development of Exceptional People* edited by Ewa Pisula and Piotr Tomaszewski, 49–63.Warszawa: Wydawnictwo Uniwersytetu Warszawskiego.

Szelag, Elzbieta, Nicole v Steinbüchel, and Ernst Pöppel. 1997. "Temporal Processing Disorders in Patients with Broca's Aphasia" *Neuroscience Letters* 235: 33–6.

Szelag, Elzbieta, and Justyna Skolimowska. 2012. "Cognitive Function in Elderly Can be Ameliorated by Training in Temporal Information Processing" *Restorative Neurology and Neuroscience* 30: 419–34.

Szelag, Elzbieta et al. 2014 "Training in Temporal Information Processing Ameliorates Auditory Comprehension in Aphasic Patients: A Randomized Controlled Pilot Study" *Journal of the Neurological Sciences* 338: 77–86.

Tallal, Paula, and Freda Newcombe. 1978. "Impairment of Auditory Perception and Language Comprehension in Dysphasia" *Brain and Language* 5: 13–24.

Turken, And U., and Nina F. Dronkers. 2011. "The neural architecture of the language comprehension network: converging evidence from lesion and connectivity analysis" *Frontiers in Systems Neuroscience* 5(1): 1–20.

von Steinbüchel Nicole, Marc Wittmann, and Elzbieta Szelag. 1999a. "Temporal Constraints of Perceiving, Generating and Integrating Information: Clinical Evidence" *Restorative Neurology and Neuroscience* 14: 167–82.

von Steinbüchel Nicole, . 1999b. "Auditory Temporal-Order Judgement is Impaired in Patients with Cortical Lesions in Posterior Regions of the Left Hemisphere" *Neuroscience Letters* 264: 168–71.

Wittmann Marc et al. 2004. "Effects of Brain-Lesion Size and Location on Temporal Order Judgement" *Neuroreport* 15(15): 2401–5.

Wittmann Marc Nicole v Steinbüchel, and Elzbieta Szelag. 2001. "Hemispheric Specialisation for Self-Paced Motor Sequences" *Cognitive Brain Research* 10: 341–4.

Zaidel, Eran. 2001. "Hemispheric Specialization for Language in the Split Brain" In *Handbook of neuropsychology: Language and aphasia*, edited by François Boller and Jordan Grafman, 393–418. Amsterdam: Elsevier.

Zaidel, Eran et al. 2003. "The Callosal Syndromes" In *Clinical Neuropsychology* edited by Kenneth M Heilman and Edward Valenstein, 347–403 (4th edition). New York: Oxford University Press, Inc..

Zampini, Massimiliano, David I. Shore, and Charles Spence. 2003. "Audiovisual Temporal Order Judgments" *Experimental Brain Research* 152(2): 198–210.

Zampini, Massimiliano et al. 2005. "Audiotactile Temporal Order Judgments" *Acta Psychologica* 118(3): 277–91.

CHAPTER 13

How Could Circadian Clock Genes Influence Short Duration Timing?

*Brad Nicholas**

1 Introduction

Neuropsychiatric disorders such as autism, schizophrenia, and bipolar disorder are often associated with sleep disturbance and other circadian rhythm anomalies (Barnard and Nolan, 2008). Parsimoniously, alterations to sleep and circadian rhythms were initially considered to be accompanying problems, rather than linked with causal processes. Latterly, better understanding of the biology of circadian rhythms and the pathologies of neuropsychiatric and neurodegenerative conditions, has indicated that disregulation of circadian rhythms and sleep disturbance may critically exacerbate the development of more serious symptoms (Benca et al., 2009; Pallier et al., 2007; Thome et al., 2011).

In autism and schizophrenia, behavioural timing deficits, short duration timing anomalies and circadian rhythm anomalies, have led some researchers to consider timing deficit as an intrinsic component of these disorders (Allman, DeLeon, and Wearden, 2011; Boucher, 2001; Ward, Kellendonk, Kandel, and Balsam, 2012; Wimpory, Nicholas, and Nash, 2002). In laboratory animals (fruit fly and hamster), mutations in specific genes that regulate the circadian cycle also affect certain short period oscillations measured in seconds and minutes. This chapter explores possible mechanisms for such short period phenomena and posits the hypothesis that clock genes exert effects on neural oscillator networks supporting short period timing phenomena. This hypothesis may be relevant to research that implicates clock genes in autism, schizophrenia, and Parkinson's disease (PD).

2 Circadian Rhythms

The rotational planetary motion of the Earth underpins the experience of day and night, and the daily oscillation in light-intensity and temperature presents

* School of Psychology, Bangor University, Wales, UK.

DOI 10.1163/9789004230699_014

a cycle of adversity and opportunity, a rhythm of repeating cycles of environmental challenge to life on Earth. It is perhaps hardly surprising, therefore, that cycles of about a day's length (circadian rhythms) of behaviour and physiology are observable in all organisms so far surveyed.

It has been suggested that the damaging effect of oxidative stress on cells, and particularly on the cell's genetic material (DNA), has driven selection leading to the evolution of organisms with the capacity to generate circadian rhythms in physiology and behaviour (Chen and McKnight, 2007; Edgar et al., 2012). In simple organisms such as yeast, growing in naturalistic conditions, oxidative metabolism is temporally segregated from cell division, when the DNA is particularly vulnerable to damage. Circadian rhythms of oxidative metabolism are primarily dependent on evolutionarily conserved peroxiredoxin molecules that function in a reduction-oxidation (redox) clock that generates cycles of oxidative and reductive biochemical states within the cell (Edgar et al., 2012; O'Neill and Reddy, 2011). In mammals, this redox clock is thought to mesh with a light entrainable circadian clock based on transcription factor feedback loops formed from the so-called clock genes and their protein products. Circadian rhythms can be maintained by either system alone, although their simultaneous operation appears to be required for robust cycles (Edgar et al., 2012), that are perhaps optimally advantageous.

Uchida, Hirayama, and Nishina (2010) note the presence of clock genes in signalling pathways common to the processes of cell division, the DNA damage response, and the photo-oxidative stress response triggered by ultraviolet light. They hypothesise that a cell-intrinsic yet light entrainable circadian clock would protect an organism (and particularly its genetic material) from the damaging effects of reactive oxygen species (ROS) produced by ultraviolet light. This clock would place UV vulnerable processes like cell division, into the hours of darkness. They emphasize that this ability would have given a selective advantage to primitive organisms with such a clock that arose far back in evolutionary time.

In higher animals behaviours such as falling asleep, awakening, feeding etc., tend to occur at particular times of the day and begs the question of whether these circadian rhythms occur: i) in response to cyclic environmental stimuli or ii) due to an innate clock-like capacity that is maintained in the absence of environmental time-setting cues (zietgebers) such as light, temperature cycles, and feeding regimes. The latter case was established through investigations with model organisms such as the fruit fly *Drosophila* (Pittendrigh, 1966) and through human research showing that people display innate clock like properties with periods of approximately 24 hours, even when isolated from all external time cues including social cues (Aschoff, 1965; Aschoff et al., 1971). More recently Mills, Minors, and Waterhouse (1974) conducted a similar study to the original "Bunker" experiments of Aschoff. In this investigation, participants spent between 5 to 13 days in an isolation unit without any knowledge of time of day. Although there was some variation between participants, the majority showed cycles of physical activity and alterations in body temperature with a period of 23–27 hours, a typical range of circadian rhythm. Given that there were no environmental cues to affect these subjects, this supports innate clock-like capacity.

Neurological investigations demonstrate the primary role of small, discrete brain regions in the maintenance of circadian rhythms in animals that are the likely neurobiological loci of an innate clock like capacity. For example, the *Drosophila* 'clock' comprises a group of about 150 brain neurons out of a total of approximately 100,000 in the animal's central nervous system (Nitbach and Taghert, 2008). The neural network of this brain clock shows a spontaneous circadian rhythm of neural activity that is dependent on the expression of clock genes. This oscillation in these pacemaker cells is stable for weeks of constant darkness, unlike non-pacemaker cells where the oscillations damp-out in a matter of days or less. The clock is directly sensitive to light due to the presence of CRYPTOCHROME, a blue-light-sensitive protein that is present in many of the clock neurons. The ventral lateral neurons of the *Drosophila* brain clock may also receive neural input from the eyes of the fly, via the visual fibres (Helfrich-Forster, 2002). These two mechanisms synchronise the 24 h oscillation of clock gene expression in the clock neurons with the day-night cycle, and in turn, through hormone and neural output signals, the brain clock imposes a circadian rhythm on the activity patterns of the fly.

The central organismal pacemaker in mammals is the suprachiasmatic nuclei (SCN) of the anterior hypothalamus. The phase of the self-sustaining 24 h cycles of neural activity in the SCN is set by neural input from the retina, to which the SCN is linked via the retinohypothalamic tract (RHT). The action of light on primarily melanopsin photoreceptors of the retina leads to a neural stimulus being transmitted along the RHT, supported by the neurotransmitters pituitary adenylate cyclase activating peptide (PACAP) and glutamate (Berson, Dunn, and Takao, 2002; Hattar et al., 2002; Panda et al., 2002). The phase shifting properties of this stimulus on the circadian clock of the SCN depends on the subjective time that the stimulus is delivered to the SCN. Light early in the subjective night phase delays the SCN clock, while light late in the subjective night phase advances the SCN clock. The SCN exerts its influence as a master-clock by emitting neuronal and hormonal signals, in a phase-typical manner, that entrain the circadian rhythms of the peripheral organs (Balsalobre et al., 2000; Hastings, Reddy, and Maywood, 2003).

The phase of circadian rhythms in mammals can be reset by other factors in addition to light. Feeding times (Stokkan et al., 2001), temperature changes (Brown et al., 2002), social cues (Aschoff et al., 1971; Levine, Funes, Dowse, and Hall, 2002; Mistlberger and Skene, 2004), and hormonal signals (Balsalobre et al., 2000), can each entrain or modulate circadian rhythms. However, even though the SCN shows a 24 h cycle of neuronal activity and output, not all the peripheral oscillators are in phase with this central clock. This is likely to be due to differences in the specific transduction pathways between SCN and each peripheral oscillator. Within the mammalian brain, for example, certain regions show circadian rhythms of activity, and the phase of one region with respect to another can vary (Feillet et al., 2008).

From the 1970s onwards, the revolution in the development of molecular biology techniques for gene manipulation, spurred investigation into the molecular and genetic aspects of circadian clocks. The period gene (*per*) in *Drosophila melanogaster* was the first example of a behavioural gene to be defined and cloned. *Per* is shown to regulate circadian rhythms in the fly and has become a focus of research on the mechanism and evolution of circadian clocks. The products of other so-called clock genes that interact with the *per* product (PERIOD protein), form a set of proteins that make up the core of the fly's circadian molecular clock. The molecular phylogeny of *per* and other circadian clock genes demonstrate a high degree of evolutionary conservation. Further work in mammals established that the circadian molecular clock was mechanistically similar and led to the notion of a conserved mechanism for

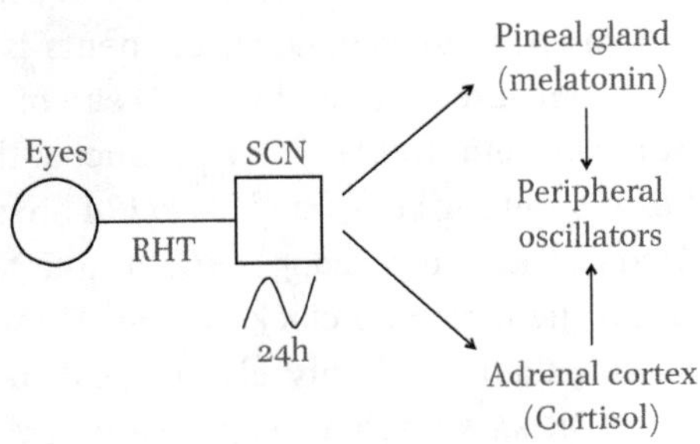

FIGURE 13.1 *A schematic and simplified representation of the role of the suprachiasmatic muclei (SCN) of the mammalian brain as the primary circadian pacemaker. The action of light on melanopsin containing cells of the retina can reset the phase of the 24 h oscillations in neural activity within the SCN. The SCN regulates (black solid arrows) the production of certain hormones such as melatonin and cortisol that show characteristic variation in level over the day-night cycle. Cortisol and melatonin convey information of circadian phase to peripheral clocks and, thus, couple the timing of metabolic processes and behaviours to SCN time.*

circadian rhythm generation amongst animals that was reflected in conserved molecular structures of clock genes and their protein products (Looby and Loudon, 2005).

3 The Circadian Molecular Clock is a System of Gene-protein Feedback Loops

The core circadian molecular clock comprises a suite of clock genes and their protein products operating as a system of transcription-translation auto-regulatory feedback loops. A heterodimer consisting of interacting proteins CLOCK and ARNTL (also known as BMAL1) is the main positive element of the circadian molecular clock and this heterodimer induces the transcription of clock controlled genes as well as the negative elements of the feedback loop (Bunger et al., 2000; Gekakis et al., 1998). The negative elements: PER1, PER2, PER3, CRY1, CRY2, and NR1D1 (also known as *REV-ERBα*), repress the action of the CLOCK/ARNTL activator and derive ~24 h oscillations in the levels of the CLOCK/ARNTL activator via transient repression by the PER/CRY complex that is released via protein turnover and translocatory effects on the negative complex (Shearman et al., 2000).

Oscillating activity of the CLOCK/ARNTL heterodimer is coupled to rhythmic expression of clock controlled genes, induced by the binding of CLOCK/ARNTL to circadian regulatory elements (E-boxes) in the promoter regions of the clock controlled genes (Hogenesch, Gu, Jain, and Bradfield, 1998; Honma et al., 2002; Kozlov et al., 2007; Matsuo et al., 2003; Mercer et al., 2009). Second of three putative circadian regulatory elements is the D-box. Genes that contain D-Box elements are targeted by the D site of albumin promoter (albumin D-box) binding protein (DPB) that is a product of the clock controlled *DBP* gene. Thus, D-Box containing genes such as *RORA* are second order clock controlled genes (Ueda et al., 2002, 2005). *RORA* and NR1D1 operate in a subsidiary circuit within the circadian clock system. These proteins regulate genes that contain a circadian regulatory element called the RevErbA/ROR binding element or RRE. RORA stimulates expression of ARNTL and *CLOCK* (and a homologue of *CLOCK, NPAS2*), via RREs while *NR1D1* is E-box regulated and NR1D1 represses ARNTL via RREs. NR1D1 and RORA compete for RRE elements in the ARNTL promoter although having opposite effects on transcription; RORA acting as an enhancer and NR1D1 as a repressor of the gene (Crumbley and Burris, 2011; Crumbley, Wang, Kojetin, and Burris, 2010; Preitner et al., 2002; Sato et al., 2004; Yin and Lazar, 2005; Yin, Wang, Klein, and Lazar, 2006).

Clock controlled genes are therefore represented primarily by genes containing E-Box, D-Box, and RRE circadian regulatory elements. Some of these clock-controlled genes encode transcription factors, so that directly or indirectly at least 15% of the whole mammalian transcriptome cycles in a circadian manner (though a different set of genes may be cycling in each of the different tissues at any one time; Akhtar et al., 2002; Yang et al., 2007).

Chromatin modification plays a central role in the mechanism of transcriptional activation/repression seen in the circadian molecular clock. CLOCK is a histone acetyltransferase (Doi, Hirayama, and Sassone-Corsi, 2006) and is responsible for a circadian rhythm in histone acetylation on the promoters of *PER* and *CRY* (Etchegaray, Lee, Wade, and Reppert, 2003). The histone deacetylase SIRT1 acts upon histone H3 of the ARNTL promoter and other circadian promoters such that the circadian rhythm can be envisaged as involving circadian cycles of acetylation and de-acetylation on the promoters of clock genes (Asher et al., 2008; Nakahata et al., 2008). Cyclic methylation of histone H3 on *PER* and possibly *CRY* promoters by the polychrome protein, EZH2 is also shown to be required for the maintenance of circadian rhythm, further showing the importance of chromatin modification as intrinsic to the circadian clock mechanism (Etchegaray et al., 2006).

The negative elements of the system (*PER*s and *CRY*s) operate as a PER/CRY heterodimer that inhibits the activating effects of the CLOCK/ARNTL heterodimer. The binding of CLOCK to ARNTL facilitates binding to and repression of the CLOCK/ARNTL heterodimer by the PER/CRY complex (Hirayama et al., 2007). This complex also contains splicing factor SFPQ/PSF that recruits SIN3A and HDACs to the *PER1* promoter, where deacetylation reverses the activating effect of histone acetylation by CLOCK (Duong, Robles, Knutti, and Weitz, 2011).

Eventually protein turnover, facilitated by proteins that phosphorylate, and ubiquitinylate components of the *PER/CRY* complex, degrades the negative elements, and subsequent restoration of the activator complex allows the cycle to restart. Retinoic acid orphan receptor alpha, beta and gamma proteins (RORS) also facilitate by inducing expression of ARNTL, *NPAS2*, and *CLOCK* in a feed-forward circuit of the circadian molecular clock (Akashi and Takumi, 2005; Crumbley and Burris, 2011; Crumbley et al., 2010; Sato et al., 2004). The duration of migration of proteins and mRNA to and from the nucleus, respectively, together with the turnover time, broadly defines the ~24 h period of this oscillatory system.

This core clock mechanism is coupled to the day-night cycle by genes associated with resetting the clock. Cryptochrome 1 and 2 (*CRY1* and *CRY2*) play a critical role in matching solar time with subjective biological time by resetting the clock in response to light–dark cycles (Berson et al., 2002). Output pathways from the clock allow the organism to adopt a state of appropriate physiological readiness that anticipates the environmental demands associated with a particular time of day or night.

The recent discovery that neurotypical circadian clock function in mouse requires microRNAs (miR219 and miR132) adds a previously unanticipated level of complexity to the circadian molecular clock mechanism, involving clock-related gene regulation by RNA interference (Cheng et al., 2007). *MIR219* is a clock-controlled microRNA gene, while *MIR132* is light regulated and intriguingly, these microRNAs are regulators of synaptic plasticity (Impey et al., 2010). This important discovery promises new insight on the possible mechanisms behind the established effects of sleep and circadian phase on memory formation and the association of atypical circadian rhythms with neuropsychiatric disorders.

Genetic variation in clock genes is associated with neuropsychiatric disorder. For example, variation in: *CLOCK*, Attention Deficit Hyperactivity Disorder (ADHD; Kissling et al., 2008); *PER1* Autism (Nicholas et al., 2007); *PER2*, Familial Advances Sleep Phase Syndrome (Toh et al., 2001); *TIMELESS*, Depression (Utge et al., 2010); Seasonal Affective Disorder and Autism *NPAS2* (Johansson et al., 2003; Nicholas et al., 2007) indicate involvement of clock genes in sleep, mood, and cognitive disorders (Barnard and Nolan, 2008).

4 The Circadian Molecular Clock may be More than a 24 hour Timer

Research on the biological, genetic, and molecular aspects of mammalian timing mechanisms has progressed furthest for the circadian clock with the understanding of the mechanisms sub-serving interval timing and other short period phenomena remaining underdeveloped. The close scrutiny of the circadian molecular clock has lead to findings that suggest this timer may additionally operate on scales longer and shorter than the 24 h period. An analogy might be made with the mechanical wristwatch or clock, where the hour hand, minute hand and second hand show different time scales but all driven from the same internal mechanism of cogs and springs. This notion of a common molecular genetic mechanisms sub-serving different biological time scales is attractive. However, to date, evidence suggests that the circadian clock only influences short period timing phenomena such as interval timing and ultradian rhythms rather than directly timing these phenomena (Beaver and Giebultowicz, 2004; Buhusi and Meck, 2005; Loudon et al., 1994). Nevertheless, it is intriguing that mutation of the clock gene *period* (*per*) in *Drosophila* and mutation of the circadian clock regulating kinase, *CSNK1E* in hamster, can change the period of ultradian behaviours (Kyriacou and Hall, 1980; Loudon et al., 1994).

The following discussion moves from direct evidence for the involvement of clock genes in short period timing phenomena to considering what clock gene regulated mechanisms could support these effects. The final section of the chapter considers the relevance of clock genes to altered timing in neuropsychiatric disorders. The association of specific short period timing problems and alterations in time-sense in schizophrenia and autism for example, suggest a possible role for the circadian molecular clock in these disorders that gains some support from human genetic studies (Mansour et al., 2006; Nicholas et al., 2007).

The period gene is a particularly interesting candidate for a gene that regulates biological timing on different scales. Not only does this gene play a central role in the circadian molecular clock, as described above, but it also appears to influence the timing of developmental processes in the fruit fly, *Drosophila*, and temporal aspects of the male fly's "love song," a primitive form of social auditory communication. *Drosophila* males court females with gesture, grooming, and a courtship song. The male's song, produced by vibrating a wing, is comprised of two elements: sine song (a hum) and pulse song (a series of pulses). The rests between the pulses (the inter pulse intervals) show a cyclic modulation in length over the duration of the song. The frequency of the modulating signal (K&H cycle) is ~55 s and is regulated by the period gene (*per*; Alt et al., 1998; Kyriacou and Hall, 1980). Intriguingly, mutations in *per* that derive lengthened or shortened circadian locomotor rhythms also determine that the flies have long or short K&H cycles, respectively. Should male flies interrupt their song, they restart singing with the K&H cycle of the restarted song in phase with the K&H cycle of the initial portion of the song. This suggests that the *per*-determined ~55 s K&H cycle is a neuronal oscillation running as a background temporal structural element of the courtship song. Even though the fly's wings and flight muscles are primarily organs of locomotion, they also serve a special function in social communication. Correct song is a prerequisite to successful mating in Drosophila and song structure affects the receptivity of the *Drosophila* female (Konopka and Benzer, 1971; Konopka, Kyriacou, and Hall, 1996; Miyatake and Kanmiya, 2004; Ritchie, Halsey, and Gleason, 1999). In *Bactrocera cucurbitae* (melon fly), circadian mutants with short circadian locomotor rhythms also produce short pulse train intervals in the male's courtship song (Miyatake and Kanmiya, 2004). Thus, this pleiotropic effect of the circadian clock gene *period* on courtship song may be a characteristic of this gene. To date, as far as this author is aware, no investigations into the possible effects of period gene mutations on social timing systems in higher animals have been published. Nevertheless, the evolutionary conservation of the period gene suggests such studies may have interesting outcomes.

The function of period gene orthologes in timing mechanisms beyond circadian scales may be relevant to the biology and behaviour of higher animals because of the evolutionary conservation of the *period* gene. Drosophila *per* is seen to regulate developmental timing of emergence of larvae from pupa (Qiu and Hardin, 1996) and lin-42, the caenorhabidites orthologue of *Drosophila period* regulates the timing of molt (Monsalve, Van Buskirk, and Frand, 2011). Such definitive results are difficult to explore in *Homo sapiens*, nevertheless a SNP close to human *PER1* that is associated with activity rhythms and time of death, tentatively indicates a conserved role for the period proteins in a multiscale biological timing mechanism.

Because of the characteristic interconnectedness of the individual clock elements in circadian clock function and that the circadian molecular clock mechanism is supported by there being homologues of virtually all the clock genes (*PER* 1, 2, and 3, *CRY* 1 and 2, ARNTL 1 and 2, *CLOCK* and *NPAS2* etc.), it is difficult to tease apart timing function that derives from a functioning clock and a specific timing phenotype that is perhaps more accurately ascribed to a particular element of the circadian clock mechanism (*PER* for example). Nevertheless, and for the purposes of this chapter, the question remains: How does the circadian clock or its components influence short period timing phenomena?

Studies based on pharmacological interventions implicate dopamine (Coull, Cheng, and Meck, 2011; Meck, 1996; Williamson, Cheng, Etchegaray, and Meck, 2008) and NMDA signaling in timing (Cheng, MacDonald, and Meck, 2006; Coull et al., 2011; Hata, 2011) but exactly how these signaling molecules are involved in interval timing is unclear. Timing anomalies are noted in autism (Allman, 2011), schizophrenia (Ward et al., 2012), and other neuropsychiatric disorders, such as PD, where dopamine and/or NMDA signaling is also implicated. There are points where these signaling pathways interact with the circadian clock pathway, suggesting possible avenues of influence between the circadian and interval timing mechanisms that to date remain unexplored.

5 Dopamine, Glutamate Receptors, and Period Proteins

The main dopamine pathways of the brain are: the nigrostriatal pathway (motor programming, attention shifting, and error prediction; Dreher and Grafman, 2002; Middleton and Strick, 2000), mesocortical pathway (working memory;

Aalto et al., 2005; Cools et al., 2008; Kimberg, D'Esposito, and Farah, 1997), and mesolimbic pathway (motivation and reward; Pierce and Kumaresan, 2006). The dopamine antagonists, haloperidol, and remoxipride act differentially on these dopamine neural pathways and block the dopamine D2 receptor (*DRD2*). Initial experiments showed halopiridol affects all the dopamine pathways (Lidow and Goldmanrakic, 1994), while remoxipride leaves the nigrostriatal circuit unblocked (Gerlach and Casey, 1990). Rammsayer and colleagues (Rammsayer, 1993; Rammsayer and Classen, 1997) took advantage of this differential blocking in experiments with humans that implicate the nigrostriatal pathway in timing intervals of less than 500ms and the mesocortical and mesolimbic pathways in timing longer intervals. The specificity of this is challenged however, by the understanding that temperature affects interval timing (Wearden and Pentonvoak, 1995) and that haloperidol affects body temperature. Nevertheless, human genetic analysis of single nucleotide polymorphisms in the dopamine D2 receptor gene *DRD2* and the catechol-O-methyltransferase gene (*comt*, concerned with dopamine metabolism) further supports the concept of nigrostriatal dopamine pathways dealing with short duration timing and prefrontal dopamine pathways with longer duration timings (Wiener, Lohoff, and Coslett, 2011).

NMDA receptors encoded by the glutamate receptor, ionotropic, N-methyl D-aspartate genes (*GRIN*s) are also implicated in interval timing mechanisms (Cheng et al., 2006; Coull et al., 2011; Hata, 2011) and disruption of this signaling system is implicated in schizophrenia and autism (Bangash et al., 2011; Goff and Coyle, 2001). Activation of ionotropic glutamate receptors (*GRIN* and *GRIA* genes), function in resetting the circadian clock in the SCN, a process involving phosphorylation of cAMP response element binding protein (CREB; Schurov et al., 1999) and *PER1* expression (Moriya, Horikawa, Akiyama, and Shibata, 2000).

PER1, a cardinal circadian clock protein that when mutated produces short duration timing effects, interacts with molecules that are implicated in NMDA and dopamine signaling. RACK1 (Receptor for Activated C Kinase 1, encoded by the gene *GNB2L1*) and the PER1 Interacting Protein of the SCN (PIPS) both physically interact with PER1 (Hu et al., 2006; Matsuki, Kiyama, Kawabuchi, Okada, and Nagai, 2001). These two physical interactivities may implicate PER1 in short period dopamine related timing mechanisms, as pharmacological blockades to NMDA and dopamine pathways affect interval timing.

The possibilities are complex because RACK1 has regulatory function in the circadian clock, dopamine signaling, and NMDAR signaling. RACK1 regulates the circadian clock by interacting with BMAL1 (Robles et al., 2010) and/or by interacting with glycogen synthase kinase 3 alpha (GSK-3a; Zeidner, Buescher, and Phiel, 2011). RACK1 also forms complexes with the human dopamine transporter DAT, encoded by *SCL6A3*, and regulates phosphorylation of DAT in a Syntaxin 1a dependent mechanism. This interaction appears to modulate membrane trafficking of the dopamine transporter (Lee, Kim, Kim, and Lee, 2004). RACK1, acting as an inhibitor of phosphorylation of NMDA receptor subunits, also regulates NMDA receptor function and thereby NMDA receptor-mediated currents. Thus, PER1, acting as an adaptor protein, for RACK1, has the potential to modulate both NMDA and dopamine signaling pathways, possibly implicating PER1 and RACK1 in short duration timing processes via the neuromodualtory effects of glutamate and dopamine.

PIPS (the PER1 interacting protein of the SCN) was discovered in the rat and this protein is homologous to the human protein GPRASP1 (G protein-coupled receptor associated sorting protein 1). GPRASP1 regulates the number of active D2 receptors on the cell surface by determining that the receptor, post agonist activation, is targeted to lysosomes for degradation (as opposed to endosomes for recycling; Moser et al., 2010). It is therefore plausible that PER1 may play a role in down regulating the D2 receptor. Consequently, misregulation of *PER1* might be associated either with hyper-D2R phenotypes or with hypo-D2R phenotypes, dependent on whether the PIPS-PER1 interaction facilitates or impedes PIPS-D2R binding.

In addition to the specific effects of per1 on short period timing and possible mechanisms described above, the circadian clock may have a more generalized effect on timing by generating circadian oscillations in the level of molecules that influence timing. For example, genes of the dopamine signaling pathway influence short duration timing and some of these genes are circadian clock controlled.

However, there is no clear-cut overlap between the few genes known to produce interval-timing anomalies and genes that contain conserved circadian regulatory elements. Table 13.1 gives the names of genes known to affect interval timing however, none of them contain conserved circadian regulatory elements in their promoters. Neither do these genes show circadian patterns of expression in the prefrontal cortex according to Yang et al. (2007). Circadian regulation of genes of the dopamine pathway may be imposed by circadian clock regulated microRNAs targeting dopamine pathway genes. For example, conserved microRNA target sites within the 3'UTR of the dopamine transporter gene *SLC6A3* (DAT) match the seed sequences of the circadian microRNAs MIR132 and MIR219. Furthermore, circadian variation in the overall level of dopamine in the dopamine neurons of the ventral tegmental area (mouse) is reported to be due to *CLOCK* and E-box mediated regulation of tyrosine hydroxylase (*TH*), the rate-limiting enzyme in the dopamine synthesis pathway (McClung et al., 2005).

6 Clock Genes Exert Diurnal Effects on Neuronal Architecture

Even though the period gene influences short duration timing phenomena such as K&H cycles and mating durations (at least in flies; Beaver and Giebultowicz, 2004; Kyriacou and Hall, 1980), it is unlikely that the gene-protein feedback loops of the circadian molecular clock could directly derive these second-scale oscillations. Neural oscillations as emergent properties of neural networks are more likely to be the substrate for such short period phenomena. However, an explanation is still required, of how clock genes might affect neural oscillators operating in milliseconds, seconds, and minutes.

The following hypothesis attempts to draw together threads of evidence from electrophysiology and molecular genetics to derive a model that suggests how clock genes might play a role in fine-tuning the frequency of certain neural oscillator circuits. From this point of view, mutations in clock genes that influence circadian rhythms by their effect on core circadian clock genes, also have effects on clock gene regulated processes that are additional to circadian timekeeping. In this context, attention is drawn to pleiotropy (multi functionality) of certain clock genes. For example, the protein encoded by the *NPAS2* gene, as well as acting as a circadian clock element, can also function as a carbon monoxide sensor protein (Dioum et al., 2002). Similarly, PER1 acts as an adaptor molecule in DNA damage response pathways additionally to its role in clock function (Sancar, Lindsey-Boltz, Unsal-Kacmaz, and Linn 2004).

TABLE 13.1 *Conserved circadian control elements are not present in genes that affect interval timing.*

Timing genes	**Gene symbol**	**(Conserved) E-Box/D-Box/RRE**
Dopamine receptor D2	DRD2	no
Dopamine transporter (DAT)	SLC6A3	no
Catechol-O-methyltransferase	COMT	no
Brain-specific L-proline transporter (PROT)	SLC6A7	no

Drawing again on the analogy of the mechanical wristwatch, such "moonlighting" clock genes may be thought of in terms of particular cogs in a fancy watch, where the cog, as well as being essential to the timing mechanism, also drives decorative animations on the watch-face that have no direct function in telling the time.

Factors that affect the frequency of neural oscillators include the number of neurotransmitter receptors on the dendritic arbors and the degree of inter-arboization between the cells of the oscillator circuit. For example, Gamma (γ, 20–80Hz) oscillations in a dendrodendritic inhibitory feedback circuit composed of mitral and granule cells of the mouse olfactory bulb occur in response to certain odors. Mutation of GABAAR-α1 subunit that alters the signal strength of the inhibitory limb of the circuit also alters the frequency of the oscillator (within limits). Computer modeling based on these and other experimental findings shows that the effect of GABAAR-α1 loss can be mimicked by a reduction in the number of inhibitory synapses (Lagier et al., 2007; Matsuoka, 2011). Thus, if clock genes can modulate the number of inhibitory synapses or the receptor number on the post synaptic neuron then it is plausible that level and location of circadian gene expression might influence neural oscillator frequency.

In brain clock circuits of *Drosophila*, rhythmic secretion of neuropeptide F causes diurnal synaptic remodeling (Fernandez et al., 2008; Fernandez, Berni, and Ceriani, 2008). Further, Berni, Beckwith, Fernandez, and Ceriani (2008) show the *Drosophila* gene *roundabout* (involved in axon guidance) also regulates activity cycles through a diurnal remodeling of neuronal branching in clock circuit cells that is down stream of and determined by the circadian molecular clock. Specifically, the clock genes *per* and *tim* regulate the size of *Drosophila* synaptic motor terminal boutons in a circadian manner and alter the number of neuronal projections in motor neurons in a developmental manner. Flies with lower functional *tim* expression show hyper-branching of neuronal projections while *per* mutants show hypo-branching. Thus, *tim* suppresses branching while *per* promotes branching in these *Drosophila* motor neurons. The double mutant (*per–/tim–*), however, has a normal phenotype with regards to branching (Mehnert et al., 2007; Mehnert and Cantera, 2008).

Thus, in *Drosophila*, there is a *per/tim* dependent developmental effect on a neuronal branching phenotype. It is, therefore, conceivable that differential clock gene expression in the different cell types of the brain could developmentally regulate the degree of inter-arborization of adjacent neurons of different cell types. If these were the cellular elements of a neuronal oscillator circuit it is possible this could affect the overall frequency of the oscillator. Thus, in principle there is precedent for clock genes and particularly the period gene, playing both a developmental and diurnal role in defining neuronal architecture. This author, therefore, proposes the hypothesis that clock genes acting as developmental and diurnal morphogenetic agents of synaptic structure and plasticity tune certain neuronal oscillator circuits. The experiments of Menhert and Cantera (2007, 2008) mentioned above, show how *per*⁻ flies and *tim*⁻ flies show contrasting branching phenotypes but the double mutant fly does not. This intriguing observation demonstrates how alterations in the *balance* of expression of these clock genes and in specific tissues might be the critical factor in determining *per* associated short duration timing anomalies and indicates how defining phenotypes specifically linked to a given gene knock-out can be a complex process. These effects on neuronal architecture are produced by mutation of specific clock genes and this may help to explain why an experiment with *Clock* mutant mice did not show interval timing deficits (Cordes and Gallistel, 2008). Masking by entrainment in the interval timing experiments of Cordes and Gallistel (2008) and the regular feeding regime in the experiments by Papachristos, Jacobs, and Elgersma (2011) may possibly have obscured interval timing effects of the *Clock* and $Cry1^-$/ *Cry2*$^-$ mutants, respectively. Nevertheless, this masking argument notwithstanding, these mutations may not be considered to alter the ratio of CLOCK to *PER*.

In addition to the clock gene effects on neuronal architecture described above, the implication of the period gene in dopamine signaling pathways suggests another avenue of investigation into mechanisms whereby clock genes might regulate the frequency of neural oscillators. Dopamine supports neural network oscillations via a calcium dependent mechanism and studies of the pyloric oscillator circuit in the spiny lobster shows dopamine specifically excites or inhibits different sets of neurons in the circuit (Kloppenburg, Levini, and Harris-Warrick, 1999). This situation is analogous to the findings of Lagier et al. (2007), mentioned above, where alterations of receptor number affect excitability of specific cells with concomitant effects on oscillator frequency. Thus, altered dopamine levels can affect the frequency of neural oscillators, however, as this effect is cell type specific, the results of experiments to investigate the role of dopamine in phenotypes that involve neuronal oscillators should take account of this selectivity.

In Vivo, normal levels of dopamine in the globus pallidus and subthalamic nucleus produce gamma oscillations, ~70Hz and reduced dopamine levels in PD in these brain regions alters the frequency of these oscillations with pathogenic consequences (Brown et al., 2001; Schnitzler and Gross, 2005). The implication of PER1 in dopamine signaling and altered patterns of clock gene promoter methylation and clock gene expression seen in PD may contribute to the dopamine related pathology of the disorder (Anantharam et al., 2007; Cai et al., 2010; Ding et al., 2011; Lin et al., 2012).

The concept that circadian clock proteins modulate short period timing cannot, however, be generalized for all short duration timers in an organism. Gene mutations that produce specific short duration effects appear to have specific tissues/cell types as targets. This is presumably due to the differential tissue expression-patterns of the timing genes in question. Tissue specific patterns of gene expression are maintained primarily by the interplay between DNA methylation and chromatin remodeling. The methylation of certain nucleotides in critical regions of the DNA code may be construed as an imprint upon the DNA that signals that a gene be turned "on" (un-methylated) or "off" (methylated). These methylation patterns may be specific for a particular tissue type, developmental stage, or show sex specificity. Some of these epigenetic modifications are influenced by environmental factors and some epigenetic marks can be inherited. This suggests that epigenetic regulation, that determines the expression patterns of *PER1, CLOCK, NPAS2* etc., may thus be as critical to oscillator function, as genetic mutations in the DNA code. In this context it is noteworthy that the primary circadian gene regulatory element, the E-box is subject to methylation at the CG dinucleotide at the centre of this feature (CACGTG). This is posited in The Social Timing Hypothesis (Wimpory, Nicholas, and Nash, 2002). A methylation screen of monozygotic twins with discordant autism diagnoses (and their non autistic siblings) highlighted differential methylation of the clock regulated gene *BCL2* and the clock gene *RORA* in the autistic twin compared with the co-twin and unaffected sib. Further immunohistochemical analysis of autism brain tissue showed altered expression of BCL2 and RORA compared to controls and significant genome-wide epigenetic parent-of-origin effects for autism at the *CLOCK* locus tentatively implicate clock gene methylation in the disorder (Fradin et al., 2010; Nguyen, Rauch, Pfeifer, and Hu, 2010). A study of clock gene promoter methylation in mouse, found major changes in clock gene E-Box region methylation around the perinatal period of development, suggesting a role for developmentally programmed methylation of clock gene regulatory elements that could contribute to tissue specificity of clock gene expression patterns (Ji, Qin, Shu, and Li, 2010). This is in keeping with a role for clock gene methylation in later developmental programmes indicated by altered methylation of clock genes in obesity, cancer, and PD (Chen et al., 2005; Lin et al., 2012; Milagro et al., 2012).

7 Abnormal Circadian Signaling could Impact Neuronal Homeostasis

Altered sleep patterns and circadian rhythm anomalies accompany many neuropsychiatric, neurodevelopmental, and neurodegenerative disorders and this disruption contributes to the severity and/or development of these disorders. The process by which circadian disregulation contributes to these disorders is however not extensively explored. The neuronal architecture of brain oscillators and neural networks is maintained within strict limits, under homeostatic control and some of the genes that regulate this homeostatic process are clock-controlled genes. It is, therefore, plausible that this interaction between the homeostatic and circadian process may lead to homeostatic failure if the circadian signal is attenuated or boosted beyond a certain limit.

Ramocki and Zoghbi (2008) drawing on findings that certain gene copy number variants are associated with autism, schizophrenia, and Rett syndrome, stress that duplication *or* deletion of certain critical genes can give rise to remarkably similar neuropsychiatric phenotypes. Further, analysis of altered neuronal architecture that is a consequence of some of these deletions and duplications show hyper- and hypo-branching morphologies. From this, Ramocki and Zoghbi conclude that healthy neurons must maintain within limits, the number and structure of synapses and certain genes are critical in this respect, *MECP2* for example. They propose that failure of neuronal homeostasis is the basis of autism as a feature of several neurodevelopmental disorders. Neuronal homeostatic processes ensure normal synaptic function by maintaining neuronal architecture and excitability within certain critical and neurotypical limits. Genes that need to be tightly regulated to maintain neuronal homeostasis may also be clock controlled genes. If so, then mutations that alter the amplitude or location of the circadian signal could boost (or fail to boost) neuropsychiatric relevant genes beyond their normal working limits with pathological consequences.

The main circadian regulatory element, the gene promoter E-Box, is primarily an enhancer element. E-box binding by the cardinal positive element of the circadian molecular clock (CLOCK- or NPAS2-/ARNTL heterodimer), boosts transcription over the baseline constitutive level. A number of genes associated with neuropsychiatric disorders are known to contain at least one conserved E-Box in their promoter regions and to show circadian patterns of expression in brain tissues (Yang et al., 2007).

In this context, it is plausible that the influence of gene/epigenetic variants that alter constitutive levels of clock controlled gene expression, would be additive to variants of clock genes specifically, that affect the amplitude etc. of circadian cycles. Further, the dopamine receptor DRD2 potentiates CLOCK/

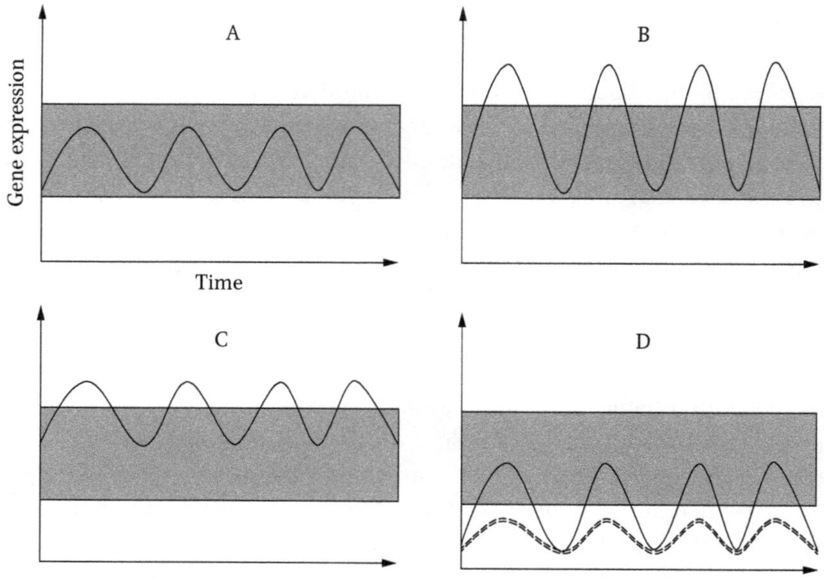

FIGURE 13.2 *A schematic representation of the interaction between Neuronal Homeostatic Limit (grey box) of expression of neuronal genes and circadian regulation of one of these neuronal genes (sinusoid); graphs A, B, and C. A, neurotypical; B, variation causing a circadian signal with increased amplitude; C, variation causing increased base line (constitutive) expression of clock controlled gene. D represents another clock-controlled gene where neurotypical circadian boosting (solid line) fails due to mutation that produces a week circadian signal (dashes).*

BMAL1 activation of E-boxes (Yujnovsky, Hirayama, Doi, Borrelli, and Sassone-Corsi, 2006), suggesting variation in the dopamine pathway could increase/decrease the amplitude of cycles of clock controlled gene expression.

It is clear that clock genes and the circadian clock regulate the expression levels of certain genes relevant to development and maintenance of neuronal architecture and excitability. Figure 13.2 describes three scenarios where anomalous expression of clock genes could cause the expression levels of clock-controlled genes to fall outside the normal limits of neuronal homeostasis. This hypothesis, of circadian effects on homeostatic limits, shows how genetic variation at a multitude of different points in the system could lead to the same effect, failure of homeostasis that is driven by the action of the circadian clock. In Figure 13.1, four scenarios are presented: A-neurotypical, B-high amplitude clock signal, C-normal clock signal plus increased constitutive expression level of clock controlled genes, and D-decreased amplitude of clock signal. Scenario D could pertain to situations such as circadian effects on memory consolidation where a circadian boost is required at a specific time point, during sleep for example, and where inadequate boosting would lead to inadequate expression to support consolidation. Under such scenarios, pinning down genetic effects in neuropsychiatric disorders such as autism, and at the level of populations, could be exceedingly complex, as appears to be the case. Examining DNA sequence variation alone would be inadequate, and risk could be spread thinly, over a large number of variants of small effect. Epigenetic mis-regulation of neuronal genes and clock genes (known to occur in cancer; Gery et al., 2007) and which dictates tissue expression patterns, would need to be taken into consideration. Thus, epigenetic and genetic effects, within clock genes or in clock-controlled genes could be pathologically additive. On the other hand, the involvement of epigenetic effects and common association of circadian disruption in neuropsychiatric disorders suggests possible avenues of treatment. Epigenetic effects are currently potentially more amenable to pharmacological reprogramming compared to DNA variations and circadian problems are already being addressed with drugs such as lithium and melatonin. As more personalised genomics become possible, there is growing potential for chrono-directed drug administration and chronotherapies that takes account of individual genetic differences in the circadian clock and the systems it regulates.

8 Summary

Circadian rhythms in physiology and behaviour are endogenous biological cycles of ~24h reliant on the cyclic expression of genes, the so-called clock genes, that encode transcription factors. Circadian rhythm disruption is apparent in disorders such as autism, schizophrenia, and bipolar disorder and genetic association studies indicate that variations in clock genes and/or clock related genes may contribute to the circadian disturbances in these disorders.

Experimental evidence supports the notion that clock genes may influence short period timing by an effect on neural oscillator networks mediated by dopamine signaling pathways or alterations in neuronal architecture or excitability. Association of short duration timing anomalies and sleep and circadian rhythm disturbance with schizophrenia and autism suggest these disturbances may possibly both derive from or be contributed to, by circadian molecular clock anomalies.

Psychologists have pondered the developmental effects of infants growing up with timing difficulties, especially a short duration timing deficit that affects processing speed during social interaction (see Chapter 3 of this book), and the cognitive impact of sleep disturbance is well known. Clearly disturbed sleep and circadian rhythms cannot of themselves cause autism in children or schizophrenia in adults, but the possibility that a timing deficit contributes to autism and schizophrenia is plausible. Clinically, the amenability of the circadian clock to treatments with low risk pharmaceuticals and the availability of behavioural interventions that structure and support timing in social interaction encourage further exploration.

References

Aalto, S. et al. 2005. "Frontal and Temporal Dopamine Release during Working Memory and Attention Tasks in Healthy Humans: A Positron Emission Tomography Study Using the High-Affinity Dopamine D-2 Receptor Ligand C-11 FLB 457" *Journal of Neuroscience* 25(10): 2471–7.

Akashi, M., and T. Takumi. 2005. "The Orphan Nuclear Receptor ROR Alpha Regulates Circadian Transcription of the Mammalian Core-Clock Bmal1" *Nature Structural and Molecular Biology* 12(5): 441–8.

Akhtar, R.A. et al. 2002. "Circadian Cycling of the Mouse Liver Transcriptome, as Revealed by cDNA Microarray, is Driven by the Suprachiasmatic Nucleus" *Current Biology* 12(7): 540–50.

Allman, M.J. 2011. "Deficits in Temporal Processing Associated with Autistic Disorder" *Frontiers in Integrative Neuroscience* 5.

Allman, M.J., I.G. DeLeon, and J.H. Wearden. 2011. "Psychophysical Assessment of Timing in Individuals with Autism" *American Journal on Intellectual and Developmental Disabilities* 116(2): 165–78.

Alt, S. et al. 1998. "The Period Gene Controls Courtship Song Cycles in Drosophila Melanogaster" *Animal Behaviour* 56: 87–97.

Anantharam, V. et al. 2007. "Microarray Analysis of Oxidative Stress Regulated Genes in Mesencephalic Dopaminergic Neuronal Cells: Relevance to Oxidation Damage in Parkinson's Disease" *Neurochemistry International* 50(6): 834–47.

Aschoff, J. 1965. "Circadian Rhythms in Man – A Self-Sustained Oscillator with an Inherent Frequency Underlies Human 24-Hour Periodicity" *Science* 148(3676): 1427–32.

Aschoff, J. et al. 1971. "Human Circadian Rhythms in Continuous Darkness – Entrainment by Social Cues" *Science* 171(3967): 213–5.

Asher, G. et al. 2008. "SIRT1 Regulates Circadian Clock Gene Expression through PER2 Deacetylation." *Cell* 134(2): 317–28.

Balsalobre, A. et al. 2000. "Resetting of Circadian Time Peripheral Tissues by Glucocorticoid Signaling" *Science* 289(5488): 2344–7.

Bangash, Park, Tatiana Melnikova et al. 2011. "Enhanced Polyubiquitination of Shank3 and NMDA Receptor in a Mouse Model of Autism" *Cell* 145(5): 758–72.

Barnard A.R., and P.M. Nolan. 2008. "When Clocks Go Bad: Neurobehavioural Consequences of Disrupted Circadian Timing" *PLoS Genetics*. 4(5)e1000040.

Beaver, L.M., and J.M. Giebultowicz. 2004. "Regulation of Copulation Duration by Period and Timeless in Drosophila Melanogaster" *Current Biology* 14(16): 1492–7.

Benca R. et al. 2009. "Biological Rhythms, Higher Brain Function, and Behavior: Gaps, Opportunities, and Challenges" *Brain Research Reviews* 62(1): 57–70.

Berni, J. et al. 2008. "The Axon-Guidance Roundabout Gene Alters the Pace of the Drosophila Circadian Clock" *European Journal of Neuroscience* 27(2): 396–407.

Berson, D.M., F.A. Dunn, and M. Takao. 2002. "Phototransduction by Retinal Ganglion Cells that Set the Circadian Clock" *Science* 295(5557): 1070–3.

Boucher, J. 2001. "'Lost in a Sea of Time': Time Parsing and Autism" In C. Hoerl, and T. Cormack (Eds.), *Time and Memory*. Oxford: Clarendon.

Brown, P. et al. 2001. "Dopamine Dependency of Oscillations between Subthalamic Nucleus and Pallidum in Parkinson's Disease" *Journal of Neuroscience* 21(3): 1033–8.

Brown, S.A. et al. 2002. "Rhythms of Mammalian Body Temperature Can Sustain Peripheral Circadian Clocks" *Current Biology* 12(18): 1574–83.

Buhusi, C.V., and W.H. Meck. 2005. "What Makes Us Tick? Functional and Neural Mechanisms of Interval Timing" *Nature Reviews Neuroscience* 6(10): 755–65.

Bunger, M.K. et al. 2000. "Mop3 is an Essential Component of the Master Circadian Pacemaker in Mammals" *Cell* 103(7): 1009–17.

Cai, Y. et al. 2010. "Expression of Clock Genes Per1 and Bmal1 in Total Leukocytes in Health and Parkinson's Disease" *European Journal of Neurology* 4: 550–4.

Chen, S.T. et al. 2005. "Deregulated Expression of the PER1, PER2 and PER3 Genes in Breast Cancers" *Carcinogenesis* 26(7): 1241–6.

Chen, Z., and S.L. McKnight. 2007. "A Conserved DNA Damage Response Pathway Responsible for Coupling the Cell Division Cycle to the Circadian and Metabolic Cycles" *Cell Cycle* 6(23): 2906–12.

Cheng, H.Y.M. et al. 2007. "microRNA Modulation of Circadian-Clock Period and Entrainment" *Neuron* 54(5): 813–29.

Cheng, R.-K., C.J. MacDonald, and W.H. Meck. 2006. "Differential Effects of Cocaine and Ketamine on Time Estimation: Implications for Neurobiological Models of Interval Timing" *Pharmacology Biochemistry and Behavior* 85(1): 114–22.

Cools, R. et al. 2008. "Working Memory Capacity Predicts Dopamine Synthesis Capacity in the Human Striatum" *Journal of Neuroscience* 28(5): 1208–12.

Cordes, S., and C.R. Gallistel. 2008. "Intact Interval Timing in Circadian CLOCK mutants" *Brain Research*, 1227.

Coull, J.T., R.-K. Cheng, and W.H. Meck. 2011. "Neuroanatomical and Neurochemical Substrates of Timing" *Neuropsychopharmacology* 36(1): 3–25.

Crumbley, C., and T.P. Burris. 2011. "Direct Regulation of CLOCK Expression by REV-ERB" *Plos One* 6(3): e17290.

Crumbley, C. et al. 2010. "Characterization of the Core Mammalian Clock Component, NPAS2, as a REV-ERB alpha/ROR Alpha Target Gene" *Journal of Biological Chemistry* 285(46): 35386–92.

Ding, H. et al. 2011. "Decreased Expression of Bmal2 in Patients with Parkinson's Disease" *Neuroscience Letters* 499(3): 186–8.

Dioum, E.M. et al. 2002. "NPAS2: A Gas-Responsive Transcription Factor" *Science* 298(5602): 2385–7.

Doi, M., J. Hirayama, and P. Sassone-Corsi. 2006. "Circadian Regulator CLOCK is a Histone Acetyltransferase" *Cell* 125(3): 497–508.

Dreher, J.C., and J. Grafman. 2002. "The Roles of the Cerebellum and Basal Ganglia in Timing and Error Prediction" *European Journal of Neuroscience* 16(8): 1609–19.

Duong, H.A. et al. 2011. "A Molecular Mechanism for Circadian Clock Negative Feedback" *Science* 332(6036): 1436–9.

Edgar, R.S. et al. 2012. "Peroxiredoxins are Conserved Markers of Circadian Rhythms" *Nature* 485(7399): 459–64.

Etchegaray, J.P. et al. 2003. "Rhythmic Histone Acetylation Underlies Transcription in the Mammalian Circadian Clock" *Nature* 421(6919): 177–82.

Etchegaray, J.P. et al. 2006. "The Polycomb Group Protein EZH2 is Required for Mammalian Circadian Clock Function" *Journal of Biological Chemistry* 281(30): 21209–15.

Feillet, C.A. et al. 2008. "Forebrain Oscillators Ticking with Different Clock Hands" *Molecular and Cellular Neuroscience* 37(2): 209–21.

Fernandez, M.P., J. Berni, and M.F. Ceriani. 2008. "Circadian Remodeling of Neuronal Circuits Involved in Rhythmic Behavior" *Plos Biology* 6(3): 518–24.

Fradin, D. et al. 2010. "Parent-of-Origin Effects in Autism Identified through Genome-Wide Linkage Analysis of 16,000 SNPs" *Plos One* 5(9): e12513.

Gekakis, N. et al. 1998. "Role of the CLOCK Protein in the Mammalian Circadian Mechanism" *Science* 280(5369): 1564–9.

Gerlach, J., and D.E. Casey. 1990. "Remoxipride, a New Sellective D2 Antagonist, and Haloperidol in Cebus Monkeys" *Progress in Neuro-Psychopharmacology and Biological Psychiatry* 14(1): 103–12.

Gery, S. et al. 2007. "Epigenetic Silencing of the Candidate Tumor Suppressor Gene Per1 in Non-Small Cell Lung Cancer" *Clinical Cancer Research* 13(5): 1399–404.

Goff, D.C., and J.T. Coyle. 2001 "The Emerging Role of Glutamate in the Pathophysiology and Treatment of Schizophrenia" *Ammerican Journal of Psychiatry* 158(9): 1367–77.

Hastings, M.H., A.B. Reddy, and E.S. Maywood. 2003. "A Clockwork Web: Circadian Timing in Brain and Periphery, in Health and Disease" *Nature Reviews Neuroscience* 4(8): 649–61.

Hata, T. 2011. "Glutamate – A Forgotten Target for Interval Timing" *Frontiers in Integrative Neuroscience* 5: 27.

Hattar, S. et al. 2002. "Melanopsin-Containing Retinal Ganglion Cells: Architecture, Projections, and Intrinsic Photosensitivity" *Science* 295(5557): 1065–70.

Hayes, K.R., J.E. Baggs, and J.B. Hogenesch. 2005. "Circadian Clocks are Seeing the Systems Biology Light" *Genome Biology* 6(5): 219.

Helfrich-Förster, C. 2002. "The Circadian System of Drosophila Melanogaster and Its Light Input Pathways" *Zoology* 105(4): 297–312.

Hirayama, J. et al. 2007. "CLOCK-Mediated Acetylation of BMAL1 Controls Circadian Function" *Nature* 450: 1086–90.

Hogenesch, J.B. et al. 1998. "The Basic-Helix-Loop-Helix-PAS Orphan MOP3 Forms Transcriptionally Active Complexes with Circadian and Hypoxia Factors" *Proceedings of the National Academy of Sciences of the United States of America* 95(10): 5474–9.

Honma, S. et al. 2002. "Dec1 and Dec2 are Regulators of the Mammalian Molecular Clock" *Nature* 419(6909): 841–4.

Hu, L.J. et al. 2006. "RACK1, A Novel hPER1-Interacting Protein" *Journal of Molecular Neuroscience* 29(1): 55–63.

Impey, S. et al. 2010. "An Activity-Induced microRNA Controls Dendritic Spine Formation by Regulating Rac1-PAK Signaling" *Molecular and Cellular Neuroscience* 43(1): 146–56.

Ji, Y. et al. 2010. "Methylation Analyses on Promoters of mPer1, mPer2, and mCry1 During Perinatal Development" *Biochemical and Biophysical Research Communications* 391(4): 1742–7.

Johansson C. et al. 2003. "Circadian Clock-Related Polymorphisms in Seasonal Affective Disorder and Their Relevance to Diurnal Preference" *Neuropsychopharmacology* 28(4): 734–9.

Kimberg, D.Y., M. D'Esposito, and M.J. Farah. 1997. "Cognitive Functions in the Prefrontal Cortex – Working Memory and Executive Control" *Current Directions in Psychological Science* 6(6): 185–92.

Kissling C. et al. 2008. "A Polymorphism at the 3'-Untranslated Region of the CLOCK Gene is Associated with Adult Attention-Deficit Hyperactivity Disorder" *American Journal of Medical Genetics B Neuropsychiatric Genetics* 147(3): 333–8.

Kloppenburg, P., R.M. Levini, and R.M. Harris-Warrick. 1999. "Dopamine Modulates Two Potassium Currents and Inhibits the Intrinsic Firing Properties of an Identified Motor Neuron in a Central Pattern Generator Network" *Journal of Neurophysiology* 81(1).

Konopka, R.J., and S. Benzer. 1971. "Clock Mutants of Drosophila-Melanogaster" *Proceedings of the National Academy of Sciences of the United States of America* 68(9): 2112–6.

Konopka, R.J., C.P. Kyriacou, and J.C. Hall. 1996. "Mosaic Analysis in the Drosophila CNS of Circadian and Courtship-Song Rhythms Affected by a Period Clock Mutation" *Journal of Neurogenetics* 11(1–2): 117–39.

Kozlov, S.V. et al. 2007. "The Imprinted Gene Magel2 Regulates Normal Circadian Output" *Nature Genetics* 39(10): 1266–72.

Kyriacou, C.P., and J.C. Hall. 1980. "Circadian-Rhythm Mutations in Drosophila-Melanogaster Affect Short-Term Fluctuations in the Males Courtship Song" *Proceedings of the National Academy of Sciences of the United States of America-Biological Sciences* 77(11): 6729–33.

Lagier, S. et al. 2007. "GABAergic Inhibition at Dendrodendritic Synapses Tunes Gamma Oscillations in the Olfactory Bulb" *Proceedings of the National Academy of Sciences of the United States of America* 104(17): 7259–64.

Lee, K.H. et al. 2004. "Syntaxin 1A and Receptor for Activated C Kinase Interact with the N-Terminal Region of Human Dopamine Transporter" *Neurochemical Research* 29(7): 1405–9.

Levine, J.D. et al. 2002. "Resetting the Circadian Clock by Social Experience in Drosophila Melanogaster" *Science* 298(5600): 2010–2.

Lidow, M.S., and P.S. Goldmanrakic. 1994. "A Common Action of Clozapine, Haloperidol, and Remoxipride on D-1-Dopaminergic and D-2-Dopaminergic Receptors in the Primate Cerebral-Cortex" *Proceedings of the National Academy of Sciences of the United States of America* 91(10): 4353–6.

Lin, Q. et al. 2012. "Promoter Methylation Analysis of Seven Clock Genes in Parkinson's Disease" *Neuroscience Letters* 507(2): 147–50.

Looby, P., and A.S.I. Loudon. 2005. "Gene Duplication and Complex Circadian Clocks in Mammals" *Trends in Genetics* 21(1): 46–53.

Loudon, A.S.I. et al. 1994. "Ultradian Endocrine Rhythms are Altered by a Circadian Mutation in the Syrian-Hamster" *Endocrinology* 135(2): 712–8.

Mansour, H.A. et al. 2006. "Association Study of Eight Circadian Genes with Bipolar I Disorder, Schizoaffective Disorder and Schizophrenia" *Genes Brain and Behavior* 5(2): 150–7.

Matsuki, T. et al. 2001. "A Novel Protein Interacts with a Clock-Related Protein, rPer1" *Brain Research* 916(1–2): 1–10.

Matsuo, T. et al. 2003. "Control Mechanism of the Circadian Clock for Timing of Cell Division in Vivo" *Science* 302(5643): 255–9.

Matsuoka, K. 2011. "Analysis of a Neural Oscillator" *Biological Cybernetics* 104(4–5): 297–304.

McClung, C.A. et al. 2005. "Regulation of Dopaminergic Transmission and Cocaine Reward by the Clock Gene" *Proceedings of the National Academy of Sciences of the United States of America* 102(26): 9377–81.

Meck, W.H. 1996. "Neuropharmacology of Timing and Time Perception" *Brain Research. Cognitive Brain Research* 3(3–4): 227–42.

Mehnert, K.I. et al. 2007. "Circadian Changes in Drosophila Motor Terminals" *Developmental Neurobiology* 67(4): 415–21.

Mehnert, K.I., and R. Cantera. 2008. "A Peripheral Pacemaker Drives the Circadian Rhythm of Synaptic Boutons in Drosophila Independently of Synaptic Activity" *Cell and Tissue Research* 334(1): 103–9.

Mercer, R.E. et al. 2009. "Regionally Reduced Brain Volume, Altered Serotonin Neurochemistry, and Abnormal Behavior in Mice Null for the Circadian Rhythm Output Gene Magel2" *American Journal of Medical Genetics Part B-Neuropsychiatric Genetics* 150B(8): 1085–99.

Middleton, F.A., and P.L. Strick. 2000. "Basal Ganglia Output and Cognition: Evidence from Anatomical, Behavioral, and Clinical Studies" *Brain and Cognition* 42(2): 183–200.

Milagro, F.I. et al. 2012. "CLOCK, PER2 and BMAL1 DNA Methylation: Association with Obesity and Metabolic Syndrome Characteristics and Monounsaturated Fat Intake" *Chronobiology International* 29(9): 1180–94.

Mistlberger, R.E., and D.J. Skene. 2004. "Social Influences on Mammalian Circadian Rhythms: Animal and Human Studies" *Biological Reviews* 79(3): 533–56.

Miyatake, T., and K. Kanmiya. 2004. "Male Courtship Song in Circadian Rhythm Mutants of Bactrocera Cucurbitae (Tephritidae:Diptera)" *Journal of Insect Physiology* 50(1): 85–91.

Monsalve, G.C., C. Van Buskirk, and A.R. Frand. 2011. "LIN-42/PERIOD Controls Cyclical and Developmental Progression of C-elegans Molts" *Current Biology* 21(24): 2033–45.

Moriya, T. et al. 2000. "Correlative Association between N-methyl-D-aspartate Receptor-Mediated Expression of Period Genes in the Suprachiasmatic Nucleus and Phase Shifts in Behavior with Photic Entrainment of Clock in Hamsters" *Molecular Pharmacology* 58(6): 1554–62.

Moser, E. et al. 2010. "G Protein-Coupled Receptor-Associated Sorting Protein 1 Regulates the Postendocytic Sorting of Seven-Transmembrane-Spanning G Protein-Coupled Receptors" *Pharmacology* 86(1): 22–9.

Nakahata, Y. et al. 2008. "The NAD(+)-dependent Deacetylase SIRT1 Modulates CLOCK-Mediated Chromatin Remodeling and Circadian Control" *Cell* 134(2): 329–40.

Nguyen, A. et al. 2010. "Global Methylation Profiling of Lymphoblastoid Cell Lines Reveals Epigenetic Contributions to Autism Spectrum Disorders and a Novel Autism Candidate Gene, RORA, Whose Protein Product is Reduced in Autistic Brain" *Faseb Journal* 24(8): 3036–51.

Nicholas, B. et al. 2007. "Association of Per1 and Npas2 with Autistic Disorder: Support for the Clock Genes/Social Timing Hypothesis" *Molecular Psychiatry* 12(6): 581–92.

Nitabach, M.N., and P.H. Taghert. 2008. "Organization of the Drosophila Circadian Control Circuit" *Current Biology* 18(2): R84–93.

O'Neill, J.S., and A.B. Reddy. 2011. "Circadian Clocks in Human Red Blood Cells" *Nature* 469(7331): 498–503.

Pallier, P.N. et al. 2007. "Pharmacological Imposition of Sleep Slows Cognitive Decline and Reverses Dysregulation of Circadian Gene Expression in a Transgenic Mouse Model of Huntington's Disease" *Journal of Neuroscience* 27(29): 7869–78.

Panda, S. et al. 2002. "Melanopsin (Opn4) Requirement for Normal Light-Induced Circadian Phase Shifting" *Science* 298(5601): 2213–6.

Papachristos, E.B., E.H. Jacobs, and Y. Elgersma. 2011. "Interval Timing Is Intact in Arrhythmic Cry1/Cry2-Deficient Mice" *Journal of Biological Rhythms* 26(4): 305–13.

Pierce, R.C., and V. Kumaresan. 2006. "The Mesolimbic Dopamine System: The Final Common Pathway for the Reinforcing Effect of Drugs of Abuse?" *Neuroscience and Biobehavioral Reviews* 30(2): 215–38.

Pittendrigh, C.S. 1966. "The Circadian Oscillation in Drosophila Pseudoobscura Pupae: A Model for the Photoperiodic Clock" *Z Pflanzenphysiol Z Bot* 54(4): 275–307.

Preitner, N. et al. 2002. "The Orphan Nuclear Receptor REV-ERB Alpha Controls Circadian Transcription within the Positive Limb of the Mammalian Circadian Oscillator" *Cell* 110(2): 251–60.

Qiu, J., and P.E. Hardin. 1996. "Developmental State and the Circadian Clock Interact to Influence the Timing of Eclosion in Drosophila Melanogaster" *Journal of Biological Rhythms* 11(1): 75–86.

Rammsayer, T.H. 1993. "On Dopaminergic Modulation of Temporal Information-Processing" *Biological Psychology* 36(3): 209–22.

Rammsayer, T., and W. Classen. 1997. "Impaired Temporal Discrimination in Parkinson's Disease: Temporal Processing of Brief Durations as an Indicator of Degeneration of Dopaminergic Neurons in the Basal Ganglia" *International Journal of Neuroscience* 91(1–2): 45–55.

Ramocki, M.B., and H.Y. Zoghbi. 2008. "Failure of Neuronal Homeostasis Results in Common Neuropsychiatric Phenotypes" *Nature* 455(7215): 912–8.

Ritchie, M.G., E.J. Halsey, and J.M. Gleason. 1999. "Drosophila Song as a Species-Specific Mating Signal and the Behavioural Importance of Kyriacou and Hall Cycles in D-Melanogaster Song" *Animal Behaviour* 58: 649–57.

Robles, M.S. et al. 2010. "Identification of RACK1 and Protein Kinase C Alpha as Integral Components of the Mammalian Circadian Clock" *Science* 327(5964): 463–6.

Sancar, A. et al. 2004. "Molecular Mechanisms of Mammalian DNA Repair and the DNA Damage Checkpoints" *Annual Review of Biochemistry* 73: 39–85.

Sato, T.K. et al. 2004. "A Functional Genomics Strategy Reveals Rora as a Component of the Mammalian Circadian Clock" *Neuron* 43(4): 527–37.

Schnitzler, A., and J. Gross. 2005. "Normal and Pathological Oscillatory Communication in the Brain" *Nature Reviews Neuroscience* 6(4): 285–96.

Schurov, I.L. et al. 1999. "Glutamatergic Induction of CREB Phosphorylation and Fos Expression in Primary Cultures of the Suprachiasmatic Hypothalamus in Vitro is Mediated by Co-ordinate Activity of NMDA and Non-NMDA Receptors" *Journal of Neuroendocrinology* 11(1): 43–51.

Shearman, L.P. et al. 2000. "Targeted Disruption of the mPer3 Gene: Subtle Effects on Circadian Clock Function" *Molecular and Cellular Biology* 20(17): 6269–75.

Stokkan, K.A. et al. 2001. "Entrainment of the Circadian Clock in the Liver by Feeding" *Science* 291(5503): 490–3.

Thome, J. et al. 2011. "Clock Genes and Circadian Rhythmicity in Alzheimer Disease" *Aging Research* 2011: 383091.

Toh, K.L. et al. 2001. "An hPer2 Phosphorylation Site Mutation in Familiar Advanced Sleep Phase Syndrome" Science 291(5506): 1040–3.

Uchida, Y., J. Hirayama, and H. Nishina. 2010. "A Common Origin: Signaling Similarities in the Regulation of the Circadian Clock and DNA Damage Responses" *Biological and Pharmaceutical Bulletin* 33(4): 535–44.

Utge, S.J. et al. 2010. "Systematic Analysis of Circadian Genes in a Population-Based Sample Reveals Association of TIMELESS with Depression and Sleep Disturbance" *Plos One* 5(2): e9259.

Ueda, H.R. et al. 2002. "A Transcription Factor Response Element for Gene Expression During Circadian Night" *Nature* 418(6897): 534–9.

Ueda, H.R. et al. 2005. "System-Level Identification of Transcriptional Circuits Underlying Mammalian Circadian Clocks" *Nature Genetics* 37(2): 187–92.

Ward, R.D. et al. 2012. "Timing as a Window on Cognition in Schizophrenia" *Neuropharmacology* 62(3): 1175–81.

Wearden, J.H., and I.S. Pentonvoak. 1995. "Feeling the Heat – Body-Temperature and the Rate of Subjective Time, Revisited" *Quarterly Journal of Experimental Psychology Section B-Comparative and Physiological Psychology* 48(2): 129–41.

Wiener, M., F.W. Lohoff, and H.B. Coslett. 2011. "Double Dissociation of Dopamine Genes and Timing in Humans" *Journal of Cognitive Neuroscience* 23(10): 2811–21.

Williamson, L.L. et al. 2008. "'Speed' Warps Time: Methamphetamine's Interactive Roles in Drug Abuse, Habit Formation, and the Biological Clocks of Circadian and Interval Timing" *Current Drug Abuse Reviews* 1(2): 203–12.

Wimpory, D., B. Nicholas, and S. Nash. 2002. "Social Timing, Clock Genes and Autism: A New Hypothesis" *Journal of Intellectual Disability Research* 46: 352–8.

Yang, S.Z. et al. 2007. "Genome-Wide Expression Profiling and Bioinformatics Analysis of Diurnally Regulated Genes in the Mouse Prefrontal Cortex" *Genome Biology* 8(11): R247.

Yin, L., and M.A. Lazar. 2005. "The Orphan Nuclear Receptor Rev-erb Alpha Recruits the N-CoR/histone Deacetylase 3 Corepressor to Regulate the Circadian Bmalı Gene" *Molecular Endocrinology* 19(6): 1452–9.

Yin, L. et al. 2006. "Nuclear Receptor Rev-erba is a Critical Lithium-Sensitive Component of the Circadian Clock" *Science* 311(5763): 1002–5.

Yujnovsky, I. et al. 2006. "Signaling Mediated by the Dopamine D2 Receptor Potentiates Circadian Regulation by CLOCK:BMAL1" *Proceedings of the National Academy of Sciences of the United States of America* 103(16): 6386–91.

Zeidner, L.C., J.L. Buescher, and C.J. Phiel. 2011. "A Novel Interaction between Glycogen Synthase Kinase-3alpha (GSK-3alpha) and the Scaffold Protein Receptor for Activated C-Kinase 1 (RACK1) Regulates the Circadian Clock" *International Journal of Biochemistry and Molecular Biology* 2(4): 318–27.

Index

www.ingramcontent.com/pod-product-compliance
Lightning Source LLC
Chambersburg PA
CBHW071252180926
51613CB00001B/20
9789004230644